READING CHINESE SCRIPT

A Cognitive Analysis

READING CHINESE SCRIPT

A Cognitive Analysis

Edited by

Jian Wang
Zhejiang University

Albrecht W. Inhoff
State University of New York at Binghamton

Hsuan-Chih Chen
Chinese University of Hong Kong

LEA LAWRENCE ERLBAUM ASSOCIATES, PUBLISHERS
1999 Mahwah, New Jersey London

Lawrence Erlbaum Associates, Inc., Publishers
10 Industrial Avenue
Mahwah, New Jersey 07430

Cover design by Kathryn Houghtaling Lacey

Library of Congress Cataloging-in-Publication Data

Reading Chinese Script : a cognitive analysis / edited by Jian Wang, Albrecht Inhoff, Hsuan-Chih Chen.
p. cm.
Includes bibliographical references and index.
ISBN 0-8058-2478-2 (alk. paper)
1. Chinese language--Psychological aspects. 2. Psycholinguistics--China. I. Wang, Jian. II. Inhoff, Albrecht. III. Chen, Hsuan-Chih.
PL1071.R43 1998
495.1'01'9--dc21 98-6065
CIP

Books published by Lawrence Erlbaum Associates are printed on acid-free paper, and their bindings are chosen for strength and durability.

Printed in the United States of America
10 9 8 7 6 5 4 3 2 1

Contents

Preface

Recent years have seen a dramatic increase in the study of Chinese language. To some extent, this increase is related to economic developments in mainland China, which has seen double-digit economic growth during each of the past 10 years. Increased trade, increased communication, and increased investment in mainland China have all contributed to a boom in Chinese language courses and in the development of Chinese language software.

Chinese writing, much more so than spoken Chinese, provides access to its culture (and economics), as it has provided a major means of unifying a diverse country. Spoken dialects differ substantially within the mainland, to the extent that a speaker of Mandarin is unable to understand a speaker of Cantonese or other southern dialects. Yet, irrespective of dialect, the same written characters are used to express meaning. Language researchers and psycholinguists also have shown increased interest in the Chinese language, reflected in the growing number of publications in which Chinese characters were used as stimuli. Chinese writing differs in critical aspects from alphabetic writing and thus provides an elegant means of testing hypotheses that are exceedingly difficult to examine in alphabetic writing systems.

This book examines the reading of Chinese text from a cognitive perspective; that is, it attempts to determine the nature of mental processes and structures that are used to assign meaning to Chinese character symbols during reading. A similar perspective was assumed in several other recently edited books, although we feel that this book is unique in at least two aspects. First, it focuses on the use and role of morphological knowledge. This differs from earlier volumes that focused on the role of phonology (although several of the current chapters either directly refer to or have implications for the representation of phonological knowledge). Second, the book includes a relatively large proportion of chapters with contributors from mainland China who have traditionally emphasized the study of meaning-related processes. Earlier volumes primarily featured contributors from outside the mainland.

The first chapter, by Shu and Anderson, examines issues related to reading development. It shows that learning to read includes the acquisition of metalinguistic skills that are used to guide assignment of character meaning. Chapters 2 to 5 (Feldman and Siok, chap. 2; Zhou and Marslen-White, chap. 3; Peng, Liu, and Wang, chap. 4; and Taft, Liu, and Shou, chap. 5) all examine the morphological parsing of complex characters into effective constituent units. The results from these studies converge in showing, first, that character decomposition is obligatory in Chinese and, second, that decomposition occurs in a principled manner.

Chapters 6 and 7, by Perfetti and Tan (chap. 6) and Zhou (chap. 7), examine the usage of phonological knowledge in character recognition, and Liu, in chapter 8, reviews work on the informational value of different spatial character locations. Several of the character-recognition chapters, in particular chapters 3, 5, 6, 7, and 8, also delineate models of Chinese character recognition.

Chapters 9 to 14 are concerned with the reading of multicharacter expressions. In chapters 9, 10, and 11 (Sun and Feng, chap. 9; Yang and McConkie, chap. 10; Inhoff, Liu, and Tang, chap. 11), the authors define basic properties of Chinese readers' eye movements during sentence reading. Sun and Feng also compare eye movements of readers of Chinese and alphabetic (English) text and attempt to define the functional role of eye movements during reading. Yang and McConkie's chapter is primarily concerned with oculomotor control during the reading of Chinese text, and Inhoff et al. delineate character and word-level analyses as they are performed on-line, that is, while these characters and words are fixated during sentence reading.

In chapter 12, Jenn-Yeu Chen examines the recognition of two-character words. The core result of the reported experiments show that multicharacter units are discerned automatically, irrespective of sentence context. Chapter 13, by Hsuan-Chih Chen, uses a moving window method to appraise the time course of lexical, syntactic, and pragmatic processes during sentence reading. The volume concludes with a chapter by Miao (chap. 14) that examines the assignment of sentence meaning. Miao concludes that sentence meaning is primarily determined by the use of semantic and pragmatic constraints, rather than by syntactic cues.

Each of the chapters is designed to be self-contained so that virtually any chapter can be read without having read other chapters. This design also incurs some costs. There are some experimenter-specific preferences for the reporting of data that may be confusing. For instance, eye movements (saccades) during reading are measured in 8-pixel units in Yang and McConkie's study, but saccades in Sun and Feng's chapter and in Inhoff, Liu, and Tang's chapters are measured in character units. As each chapter defines its key terms, there is some repetition in the description of script properties. Moreover, the same term (e.g., morpheme, radical) may not refer to the identical concept across chapters. This heterogeneity in the usage of linguistic terminology reflects the current lack of standards, and we felt that it would have been premature–and perhaps impossible–to prescribe a particular usage of linguistic terms.

As noted before, written Chinese, more so than spoken language, provided a major means for unifying Chinese culture. What is often neglected, however, is that there are currently two, rather than one, Chinese scripts: simplified Chinese script, introduced after the Revolution in mainland China, and traditional script, used in Taiwan, Hong Kong, Singapore, and Chinese minorities in other countries (e.g., Malaysia, Indonesia, United States, Canada).

Structurally, simplified and traditional Chinese script are similar, but simplified script is less ornamental (visually less complex), as it was aimed at making reading and writing available to the "masses." Successive characters of text are written from left-to-right in simplified script and successive lines of text are ordered from top to bottom. There are fewer and less stringent spatial conventions for writing traditional script, where characters can either be ordered left-to-right or right-to-left, or top-to-bottom. When written from top-to-bottom, successive columns successive columns can be ordered from left-to-right or right-to-left. Articles on the front page of a Hong Kong newspaper may, for instance, be written following different spatial layouts. None of the chapters in this book examined effects of script type, but differences between script types need to be kept in mind when results of studies with mainland Chinese readers are compared to the results of studies with Taiwanese students.

This book provides a state-of-the art synopsis of Chinese reading research and should be of interest to cognitive psychologists and psycholinguists. Because the volume addresses a wide range of topics, it may also be used for graduate and advanced undergraduate courses on Chinese language processing, Chinese reading, or reading in general.

All chapters went through a journal-type review process prior to being included in the book. We would like to extend our special thanks to the following reviewers who did not contribute chapters to the volume: Professors Keith Rayner, Charles Clifton, Susan Lima, Ralph Radach, Celia Klin, and Tom Urbach. We also acknowledge support from NIMH (Grant 503870) to Albrecht Inhoff and from the Research Grants Council (Earmarked Grants) of Hong Kong to H.-C. Chen.

Jian Wang
Albrecht W. Inhoff
Hsuan-Chih Chen

READING CHINESE SCRIPT

A Cognitive Analysis

CHAPTER

1

Learning to Read Chinese: The Development of Metalinguistic Awareness

Hua Shu
Beijing Normal University

Richard C. Anderson
University of Illinois

This chapter examines the role of metalinguistic awareness in learning to read Chinese. Metalinguistic awareness is the ability to identify and reflect on the structural features of language. Our working hypothesis is that learning to read is "fundamentally metalinguistic" (Nagy & Anderson, 1998). Learning to read requires becoming aware of the basic units of spoken language, the basic units of the writing system, and the mapping between the two.

Research over the past two decades has documented a strong relationship between metalinguistic awareness and reading development. It is well established that to learn to read alphabetic languages such as English children need to gain the metalinguistic insight that *phonemes*, the basic units of the spoken language, map onto *letters*, the basic units of the writing system (Cossu, Shankweiler, Liberman, Katz, & Tola, 1986; Goswami & Bryant, 1992; Nagy & Anderson, 1998). The metalinguistic insights needed to learn Chinese are less well established. This chapter describes some of the important structural features of Chinese, summarizes a program of research that has sought to determine when Chinese children become aware of these features, and reports data on how aspects of metalinguistic awareness relate to reading development.

Chinese is called a logographic writing system, but this is not entirely accurate. A logographic language is one in which each word in the language is represented by a distinct symbol in the writing system. The basic symbols of written Chinese are characters; however, most Chinese words are represented by two or more characters. Thus, it is more accurate to say that

Chinese characters represent morphemes than to say they represent words. The vocabulary of Chinese is represented with a vast number of visually complex characters, as many as 50,000 of them (Taylor & Taylor, 1995). Children are expected to learn more than 2,500 characters by the time they leave the sixth grade, and thousands more are learned by high school and college students. Because of the large number of characters, because of their visual complexity, and because Chinese does not have the grapheme–phoneme correspondence rules of alphabetic languages, it is commonly assumed that Chinese children acquire characters by memorization, learning each one as an unanalyzed whole. We make a different assumption: We assume that, as children achieve insight into the structure of characters, they begin to systematically use the fairly rich information in character components. This enables accelerated progress in learning to read and means that learning need not be wholly, or even largely, a matter of memorization.

Over the past half dozen years, we have completed a series of studies in the Peoples' Republic of China designed to answer several questions: Which aspects of metalinguistic awareness are important for learning to read Chinese? Is awareness of the internal structure of characters associated with the reading development of Chinese children? What are the stages in Chinese children's development of different aspects of metalinguistic awareness?

This chapter has five parts. First, we describe features of the Chinese writing system. Second, we report an analysis of the characters taught in the national Chinese language curriculum in the first through the sixth grades. Third, we present findings from a study of children's *awareness of the structure of the standard compound character*. Fourth, we summarize research on *radical awareness* and its role in reading development. Fifth, we describe research on *phonetic awareness* and reading development.

THE CHINESE WRITING SYSTEM

In the foreigner's view, the most important feature of Chinese is that the characters directly represent meanings. Ancient Chinese did contain pictographs that were easily seen as picturing objects. For example,  was the character for moon. However, pictographs represent only a small percentage of present-day characters, and over centuries of use, pictographs have become more stylized so that, now, few of them clearly represent the objects they denote, anymore. For instance, in present-day Chinese, 月 is the character for moon.

Another type of character considered to be directly meaningful is the ideograph. Examples of ideographs include the characters for *up* 上 and *down* 下, which may be thought of as expressing the ideas of up and down from the directions in which they point. Modern Chinese also contains quite a few compound ideographs in which the components denote the elements in an event or complex idea. Consider, for instance, the characters for *tree* 木, *woods* 林, and *forest* 森. Another interesting example is 尖. The component on the top 小 means *small* while the component on the bottom 大 means *large*; together they mean *sharp*, as of a knife blade. The types of more or less directly meaningful characters that we have described so far, along with a few other special types, comprise about 18% of the total number of characters.

The remaining 82% of modern Chinese characters are what we will call *standard compound characters*, and this percentage has increased over the centuries. The standard compound character consists of two parts, a component called a radical, which carries information about meaning, and a component called a phonetic, which provides information about pronunciation (Hoosain, 1991; Li, 1977; Taylor & Taylor, 1995).

In most cases, the radical indicates the conceptual category of the character. In other words, there is usually a member–set relationship between a character and its radical. For example, *sister* 姐 and *aunt* 姨 are among the many characters that contain the *female* radical 女. In other cases, the relationship of the radical to the meaning of the character is indirect. For example the character 唱, which means *sing*, has the *mouth* 口 radical; it is easy to see that singing has a connection with mouth, but this is not a member–set relationship. There are, in addition, irregular characters in which the radical provides no discernible information about meaning. Chinese has many homophones, since the language contains only about 400 syllables (excluding tones). Radicals provide the principal basis for distinguishing among homophone characters.

The phonetic component of a standard compound character gives a clue to the pronunciation of the character, although not a completely reliable one. According to Yin's (1991) analysis, of all the phonetics in Chinese, 36% give clear information about a character's pronunciation, 48% give partial information, and 16% give no useful information.

Most radicals and phonetics are *free morphemes*; that is, in addition to being components of compound characters, they are themselves simple characters. The remaining radicals and phonetics are *bound forms*; they never appear alone, but only appear as components of compound characters. The contribution of bound forms to the meaning or pronunciation of compound characters must be abstracted, or resolved by analogy with other compound characters containing the same component, which we assume is difficult and slow to develop among children.

Written Chinese contains about 190 radicals and 1,100 phonetics. Thousands of compound characters are formed from combinations and recombinations of these components. Phonetics may be further subdivided into about 650 subcomponents (Fu, 1989). A phonetic may consist of two or more of these subcomponents. The subcomponents are not systematically related to pronunciation, but it is reasonable to suppose that knowledge of these stable configurations facilitates visual processing. The stroke is the smallest unit of the Chinese writing system. There are eight types of strokes.

The positioning of elements within characters is critical. Different patterns of the same strokes, such as 田 and 旧, must be discriminated. Likewise, different configurations of the same components must be attended to because each configuration forms a distinct character, such as 杏 and 呆 which mean *apricot* and *dull*, respectively. Most radicals and phonetics have a fixed position within compound characters. For example, the radical signifying *water* 氵 always appears on the left of characters. In fact, the radical usually occupies the position on the left side of a character, although it may appear in other positions or even surround the phonetic.

In summary, although the Chinese writing system is complex, it does have a clear and useful logic. It stands to reason that children who understand this logic have an easier time learning to read.

CHARACTERS TAUGHT IN ELEMENTARY SCHOOL

We have completed an analysis of all the characters in the national language textbooks used in elementary schools throughout China. Altogether, 2,557 characters were coded for three features, as follows:

1. *Frequency of usage*. Number of occurrences per million was determined from a large character frequency study (Project Group on Statistical Analysis of Chinese, 1988). In this study, books and other written materials were sampled from various fields, such as industry, agriculture, military, science and technology, politics, economy, literature and art, education, medicine, astronomy, geography, nature, chemistry, archaeology, and from 192 newspapers and magazines. The frequency of 6,374 characters (types) was tabulated from 21,629,372 running characters of text (total tokens), making this study considerably larger than the largest study of English word frequency (Carroll, Davies, & Richman, 1971).

2. *Semester taught*. The semester in which the character is introduced in the national reading textbooks was coded 1–12 for the 12 semesters in the 6 grades of elementary school.

3. *Regularity of pronunciation*. The pronunciation of characters was coded in eight categories: 1—Pronunciation of the character is the same as the pronunciation of its phonetic, including the tone; 2—Pronunciation of the character is the same as the pronunciation of its phonetic, except for the tone; 3—The vowel of the character is the same as the vowel of the phonetic; 4—The initial consonant of the character is the same as the initial consonant of the phonetic; 5—Pronunciation of the character is totally different from the pronunciation of the phonetic; 6—Nonstandard compound character;

TABLE 1.1
Proportion of Standard Compound Characters With Regular Pronunciation

Grade	*1st*	*2nd*	*3rd*	*4th*	*5th*	*6th*
Standard Compounds	.48	.71	.79	.86	88	.87
Regular Pronunciation	.29	.40	.42	.45	.43	.43

7—The character or the phonetic has more than one pronunciation; 8—The character has lost its phonological cue because it was simplified during reform of the Chinese writing system. In our database, standard compound characters include characters coded as 1–5.

Table 1.1 presents an analysis of the new characters introduced in each grade. The table shows that the percentage of standard compound characters rises over the first few grades and then levels off. The trend is similar for percentage of standard compounds that have a regular pronunciation, where regular pronunciation is defined as including Categories 1 and 2 above. Notice, especially, that a relatively small percentage of the characters taught in the first grade are standard compounds with regular pronunciations.

Figure 1.1 presents the data from the analysis of the characters in Chinese language textbooks according to frequency of use. For this figure characters were divided into four frequency bands: above 100 per million, 10–99 per million, 1–9 per million, and below 1 per million. As you can see, regularity decreases as frequency increases. This result agrees with the finding of Perfetti, Zhang, and Berant (1992), who asked raters to judge the pronunciation regularity and semantic transparency of a sample of 300 characters. They found that both measures decrease as the frequency of the characters increases.

In summary, an analysis of 2,557 characters Chinese children are expected to learn during the 6 years of elementary school shows that the majority of characters introduced after the first grade are standard compound characters. The analysis also shows that the percentage of standard compounds with a regular pronunciation increases through the fourth grade and then levels off. A relatively small percentage of the characters introduced in the first grade are compounds with a predictable structure and a regular pronunciation. This implies that first graders have limited opportunities to acquire insight into the internal structure of Chinese characters.

AWARENESS OF THE STRUCTURE OF COMPOUND CHARACTERS

The first study of metalinguistic awareness that we describe investigated when children become aware of the basic structure of standard compound characters. Previous research has demonstrated that skilled readers of Chinese are sensitive to the position of components within characters

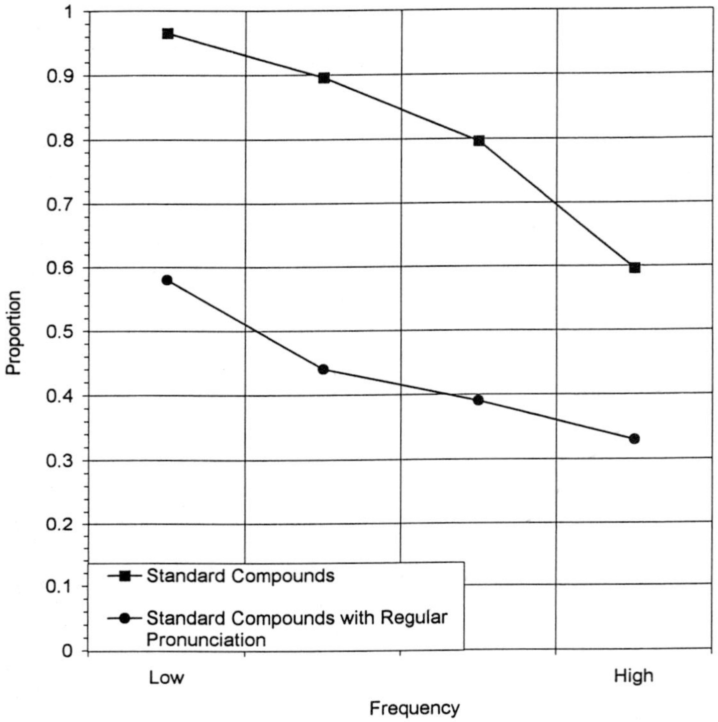

FIG. 1.1. Proportion of standard compounds with regular pronunciation as a function of frequency.

(Cheng, 1982; Peng, Li, & Huang 1995). Children have been reported to gradually acquire knowledge of the positions of radicals within characters over the elementary school years (Cheng & Huang, 1995; Li & Peng, 1995).

The questions our experiment asked were: As children learn more characters, what insights do they acquire about the structure of characters? When do children become aware of the different aspects of the structure of standard compound characters? The subjects were 143 children in the first, second, fourth, and sixth grades from a working-class school in Beijing. They were asked to decide which of a series of items could be Chinese characters and which could not be. Half of the items were compound characters of high, medium, and low frequency. The remaining half were compound pseudocharacters of one of three types. One third had a well-formed structure, one third had an ill-formed structure, and one third had ill-formed components, defined as follows:

Well-formed structure. These items contain real components of Chinese, the students are familiar with the components, and the components are in the correct position; but the items are not actual characters. For example, the pseudocharacter 秷 has a real radical 禾 (standing grain), which is in the left position where it is supposed to be, and a real phonetic 至 /zhi/, but there is no such character in Chinese. A high rate of false alarms, declining somewhat over the elementary school years, was expected because there is nothing wrong with the structure of these items. Rejecting well-formed pseudocharacters requires a lack-of-knowledge inference (Collins, Brown, & Larkin, 1980) of the form: (a) If this symbol were a character, I would know it. (b) I do not know it. (c) Therefore, it is not a character. Since young children are aware that there are many, many characters they do not know, they do not have a good basis for rejecting these items. In other words, the base of knowledge to reject these items is hypothesized to be late developing.

Ill-formed structure. Items of this type contain familiar, real components of Chinese, but the radicals are in the wrong position. For example, the radical 糹 (silk) is always on the left of a character. However, it is on the right in the pseudocharacter 工糹. A low rate of false alarms was expected because these items violate a fundamental structural constraint of compound Chinese characters. It is hypothesized that this insight develops early. False alarms were expected only from the youngest and least able students.

Ill-formed components. These items look like standard compound characters, but the components are not real components in the Chinese writing system. For example, the pseudocharacter 衤乃 has a real radical 衤 (cloth) on the left side. The component 乃 on the right side looks like the phonetic 乃 /nai/, but is not a real phonetic. The rate of false alarms was expected to decline over the elementary school years. This is another type of item that requires a lack-of-knowledge inference: (a) If this part were a proper component of Chinese characters, I would know it. (b) I do not know it. (c) Therefore, the symbol containing this part is not a character. By the time they reach the upper elementary school grades, it is hypothesized that most students know—and are aware that they know—most of the proper components of common characters, so they have the knowledge to reject pseudocharacters with ill-formed components. Insight into the detailed internal structure of characters is expected to develop gradually over the elementary grades.

The most interesting aspect of the data is the rate of false alarms on the pseudocharacters, where a false alarm means indicating that a pseudocharacter might be a real character. Figure 1.2 shows the false alarm rates on the three types of pseudocharacters. As anticipated, the false alarm rate was very low on items with an ill-formed structure, even for first and second graders. This shows early development of insight into the basic structure

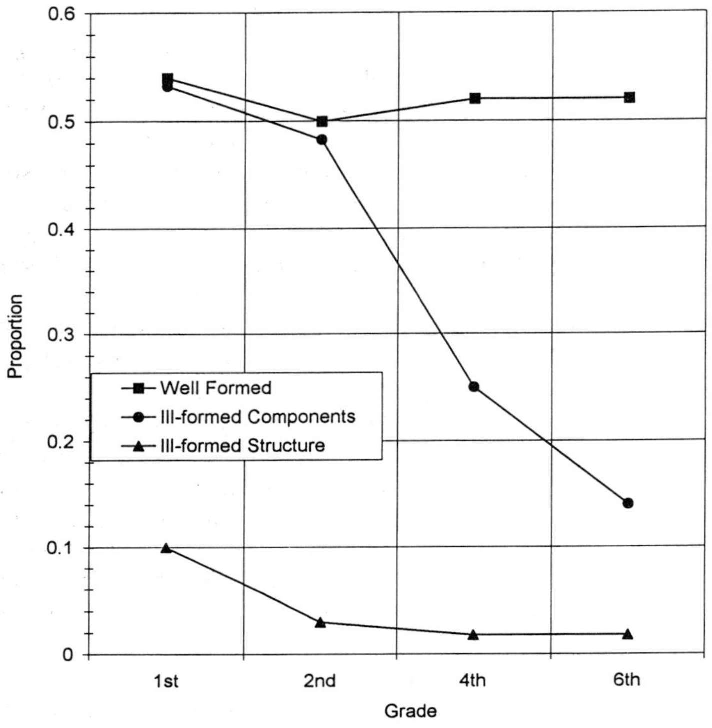

FIG. 1.2. Proportion false alarms on pseudocharacters.

of characters. Also as anticipated, there was a steady decline in the false alarm rate for items with ill-formed components. This shows gradual development of insight into the detailed internal structure of characters. The rate of false alarms was high on well-formed items, as expected. However, we thought the rate would decline by sixth grade. This did not happen. Perhaps high school students or college students would be able to reject these items.

RADICAL AWARENESS AND ITS ROLE IN READING DEVELOPMENT

By analogy with the role of phonemic awareness in alphabetic literacy, in Chinese, in which characters map onto morphemes, it might be expected that morphological awareness would be the aspect of metalinguistic awareness that is most closely related to literacy. Morphological relationships among words influence the way words are represented in memory and the process by which skilled readers recognize complex words and derive their meanings (Anshen & Aronoff, 1988; Nagy, Anderson, Schommer, Scott, &

Stallman, 1989; Taft, 1985). This conclusion has been reached in studies involving several languages (Feldman & Fowler, 1987; Grainger, Colé, & Segui, 1991; Schriefers, Friederici, & Graetz, 1992). There is also evidence that morphology may influence children's vocabulary acquisition (Carlisle & Nomanbhoy, 1993; Derwing & Baker, 1979; Tunmer & Hoover, 1992; Tyler & Nagy, 1989; White, Power, & White, 1989; Wysocki & Jenkins, 1987), but this research has been done only with English-speaking children. What our studies seek to answer is: Does the morphological structure of characters influence children's character and word acquisition? How are age and ability related to knowledge of morphology?

We completed two experiments to examine the *radical awareness* of children (see Shu & Anderson, 1997). By radical awareness, we mean the insight that radicals convey information about the meanings of characters. In the Chinese writing system, large groups of characters, sometimes numbering as many as 100, share the same radical and are semantically related. The child who is aware of the function of radicals has a powerful tool for literacy learning.

A total of 220 first, third, and fifth graders from a working-class school in Beijing participated in the first experiment. The children were divided into groups of high, average, and low verbal ability, based on their classroom teachers' ratings.

For each grade, 90 characters were selected from the national reading textbooks. Thirty of the characters were classified as familiar because they had been introduced in the textbook two grades earlier, and the children had encountered them numerous times. The familiar characters for first graders were those introduced one semester before. At each grade, 30 characters had been recently learned. These came from the textbook students were using in the present semester. Finally, 30 characters were classified as unfamiliar to students because they would not appear in the textbook until two grade levels above the students' current grade.

Within each degree of familiarity, there were three types of characters:

1. Morphologically transparent characters contained radicals helpful for understanding the meaning of the whole character. For example, the character 烛 (*candle*) contains the radical 火 (*fire*).
2. Morphologically opaque characters contained radicals that were unrelated to the meaning of the whole character. An example is the character 错 (*mistake*) that contains the radical 钅 (*metal*).
3. Unanalyzable characters, such as 欠 (*owe*), could not be analyzed into parts.

The opaque and unanalyzable characters did not contain helpful morphology, so the difference between these two types and the transparent characters represents the value students were able to derive from using morphology.

The children were presented with two-syllable words familiar from oral language. One syllable was written as a character; the other was written in *Pinyin*. Pinyin is an alphabetic script that provides the pronunciation of characters. In mainland China, children learn Pinyin during the first 10 weeks of first grade.

The children's task was to circle a character to replace the Pinyin in each two-syllable word. The correct character contained a radical consistent with the meaning of the two-syllable word. There were three distractor characters with either the same pronunciation or the same phonetic as the correct character, but a different radical.

Figure 1.3 presents the main results of the study. As you can see, morphology was not very important for familiar characters, but had a large effect on performance with recently learned characters, $F(1, 208) = 281.72$, $P < .01$, and an even larger effect on performance with unfamiliar characters, $F(1, 208) = 402.86$, $P < .01$. Transparent morphology mainly helped older students. Third and fifth graders assuredly were using radicals to derive unfamiliar characters, as well as to learn and remember characters recently

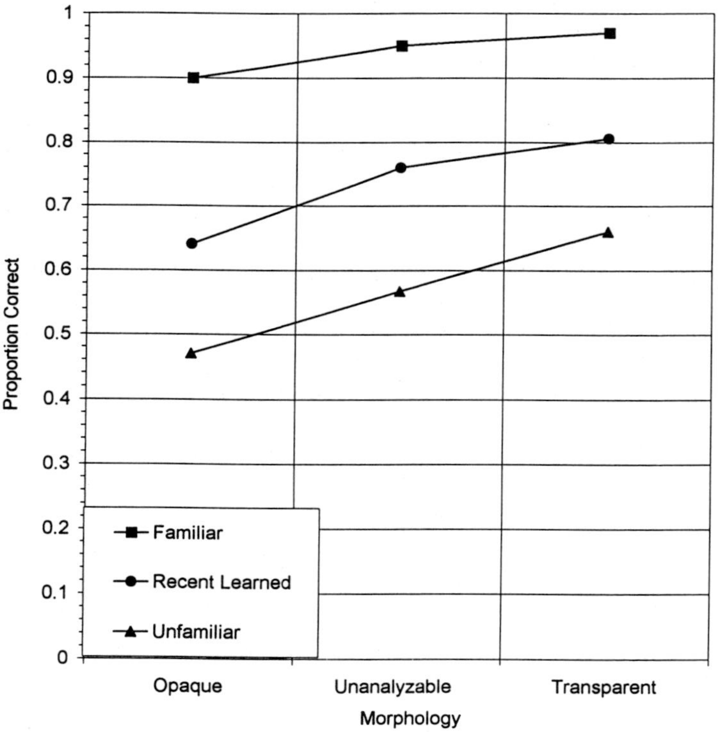

FIG. 1.3. Proportion correct as a function of character familiarity and morphology.

introduced in school. There was little indication of productive use of morphology by first graders.

Figure 1.4 indicates that children with different levels of ability did equally well on the familiar characters, but as familiarity decreased, the performance of higher and lower ability children diverged. Higher ability children got significantly better scores than lower ability children on both recently learned characters, $F(2, 208) = 35.97$, $P < .01$, and unfamiliar characters, $F(2, 208) = 25.44$, $P < .01$. Two possible explanations are: High ability children may have more specific character knowledge than low ability children, including knowledge of some characters not yet introduced in school. Alternatively, high ability children may have more facility than low ability children in decomposing complex characters and using morphological knowledge to assimilate new characters. The first experiment did not permit a choice between these two explanations.

In the second experiment, we varied the conceptual difficulty of the words and used only unfamiliar characters and only third and fifth graders. Conceptually easy words were ones judged by teachers to be in children's

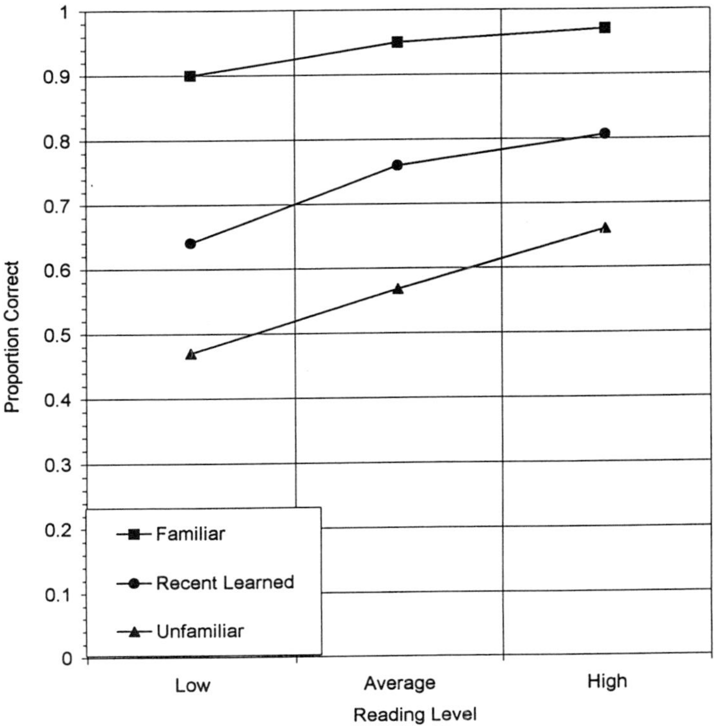

FIG. 1.4. Proportion correct as a function of character familiarity and reading level.

oral vocabulary or, if not, to involve familiar concepts. Three types of characters were investigated: (a) Morphologically transparent characters with radicals familiar to children; (b) Morphologically transparent characters with radicals unfamiliar to children; and (c) Morphologically opaque characters. As in the first experiment, the task was to select the character to replace the Pinyin in two-syllable words familiar from oral language.

Figure 1.5 portrays the chief results of the second study. Neither the opaque characters nor the transparent characters with unfamiliar radicals contain information accessible to children. Thus, the difference between these types and the transparent characters with familiar radicals represents the increment attributable to the use of morphology. The results show that morphology was of greatest value when the words were conceptually easy. This finding implies that character and word acquisition are enhanced when students are able to triangulate different sources of information.

Figure 1.6 shows that higher ability children did better than lower ability children only with morphologically transparent characters with familiar radicals, $F(2, 64) = 9.65$, $P < .01$. The advantage of higher ability children

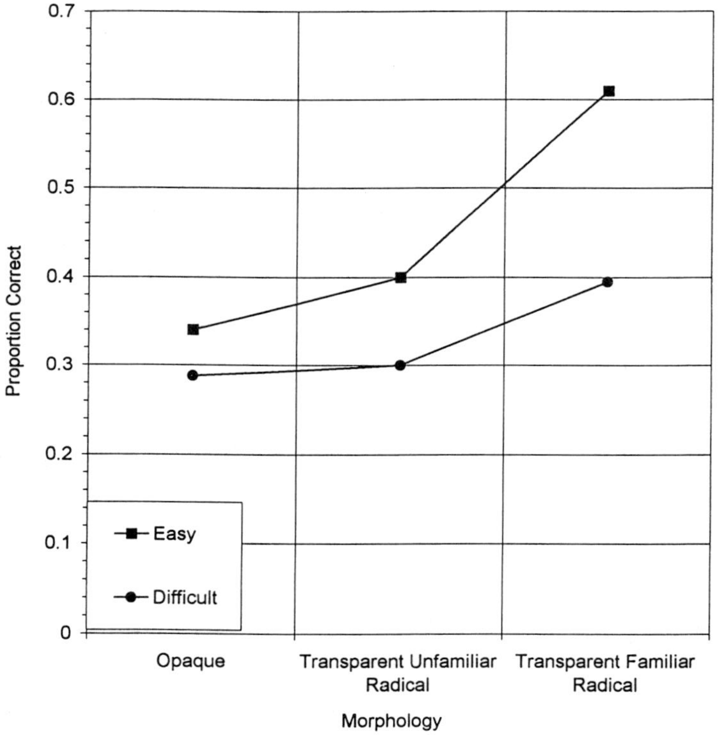

FIG. 1.5. Proportion correct as a function of morphology and conceptual difficulty.

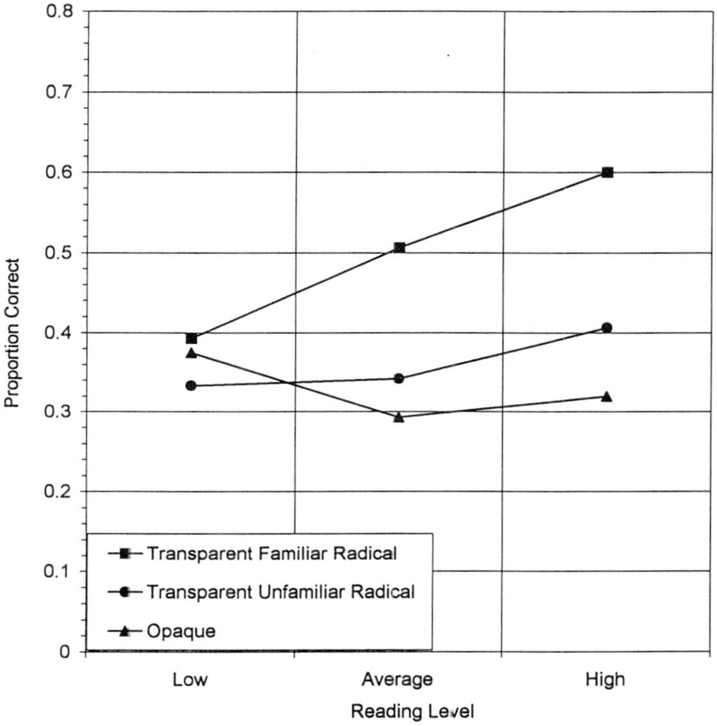

FIG. 1.6. Proportion correct as a function of morphology and reading level.

could not have been due to a general superiority in character knowledge or they would have done better on all three types of characters. The result suggests, instead, that higher ability children could make use of morphology, but lower ability children could not.

The major finding of these two experiments is that many Chinese children have a functional awareness of the relationship between the radical in a character and the character's meaning. Radical awareness was found to develop with grade and to be related to reading ability. Young readers and poor readers either have not discovered the basic morphological features of Chinese, or they do not spontaneously use what they know about morphology to process familiar characters or derive the meanings of new characters.

PHONETIC AWARENESS AND READING DEVELOPMENT

A classic issue in reading research concerns the functions of phonological information in skilled reading and in learning to read (Gough, Juel, & Griffith, 1992; Perfetti & Zhang, 1995). Previous research has reported that skilled adult Chinese readers use the phonological information in the phonetic component for pronunciation of characters (Fang, Horng, & Tzeng, 1986; Hue, 1992; Peng, Yang, & Chen, 1994; Seidenberg, 1985). The next experiment we describe extends this research to children. The questions we asked were: Are children aware of and able to use the phonological information in phonetic components? What is the variation in phonetic awareness among children of different grades and abilities?

A total of 113 second-, fourth-, and sixth-grade children from a school in Beijing, all of whom were native speakers of Mandarin, participated in the experiment. The children were divided into groups of high, average, and low reading ability, based on ratings by their classroom teachers.

At each grade, children were tested on 60 characters, 30 familiar and 30 unfamiliar. Familiarity was defined with reference to the national Chinese language textbooks. Familiar characters had appeared in the language textbook one grade below the grade in which the children were currently enrolled. Unfamiliar characters would appear in the textbook one grade above the children's current grade.

At each grade and degree of familiarity, there were three types of compound characters:

1. Regular characters, in which the pronunciation of the whole is the same as the pronunciation of the phonetic alone. For example, the pronunciation of the character 绘 is /hui/ and the pronunciation of its phonetic 会 is also /hui/.

2. Irregular characters, in which the pronunciation of the character is different from the pronunciation of its phonetic. For example, the pronunciation of the character 略 is /lue/, but the pronunciation of its phonetic 各 is /ge/.

3. Characters with bound phonetics. In these characters, the phonetic is not a freestanding character and, therefore, the pronunciation must be resolved by analogy with other characters containing the phonetic. For example, the pronunciation of the character 忱 is /chen/ but its phonetic by itself is not a character and does not have a determinate pronunciation.

The children's task was to write down the pronunciation of the 60 characters in Pinyin; or, if they were not sure of the Pinyin, they could write a familiar character with the same pronunciation, instead.

Familiarity and regularity strongly affected accuracy of pronunciation, and as expected, there was a robust familiarity by regularity interaction. When characters were familiar, children were almost as accurate at representing the pronunciation of irregular characters as they were regular ones. However, when characters were unfamiliar, regularity made a larger differ-

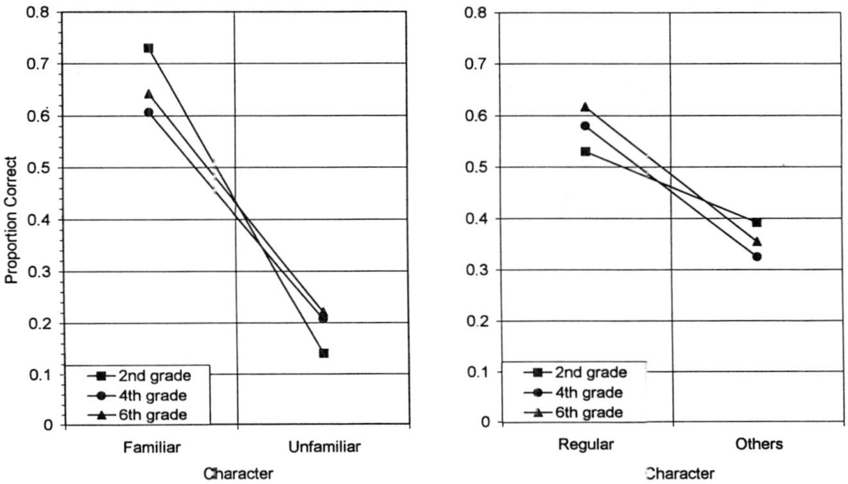

FIG. 1.7. Proportion correct by grade as a function of character familiarity and regularity.

ence in performance, especially for high ability children who were able to make use of the information in regular characters.

Both the familiarity by grade, $F(2, 107) = 23.35$, $P < .01$, and regularity by grade, $F(4, 214) = 13.48$, $P < .01$, interactions were significant. The left panel in Fig. 1.7 shows the influence of character familiarity on pronunciation; the right panel shows the influence of regularity. The figure reveals that the effects of familiarity are much stronger than the effects of regularity. Figure 1.7 also shows that second graders are more influenced by familiarity, whereas sixth graders are more influenced by regularity. This implies that, as they gain more experience with written Chinese, children acquire the ability to use the information in the phonetics in regular compound characters.

The same conclusion is suggested by an analysis of pronunciation errors. Many errors reflected attempts to use phonetic information. A frequent mistake was to give an irregular character the same pronunciation as its phonetic, for example, pronouncing 琼 as /jing/ instead of /qiong/. Another common error was mispronouncing a character with a bound phonetic by analogy with another character with a different pronunciation that contains the same phonetic. Figure 1.8 shows variation in phonetic-based errors as a function of grade, $F(2, 107) = 73.41$, $P < .01$, and ability, $F(2, 107) = 22.80$, $P < .01$. The error analysis suggests that even high ability second graders make little use of phonetic information. In contrast, most sixth graders, even those of low ability, are making systematic use of phonetic cues. Fourth graders are in transition. High and average fourth-grade students are making substantial use of phonetic information, but low ability students lag behind.

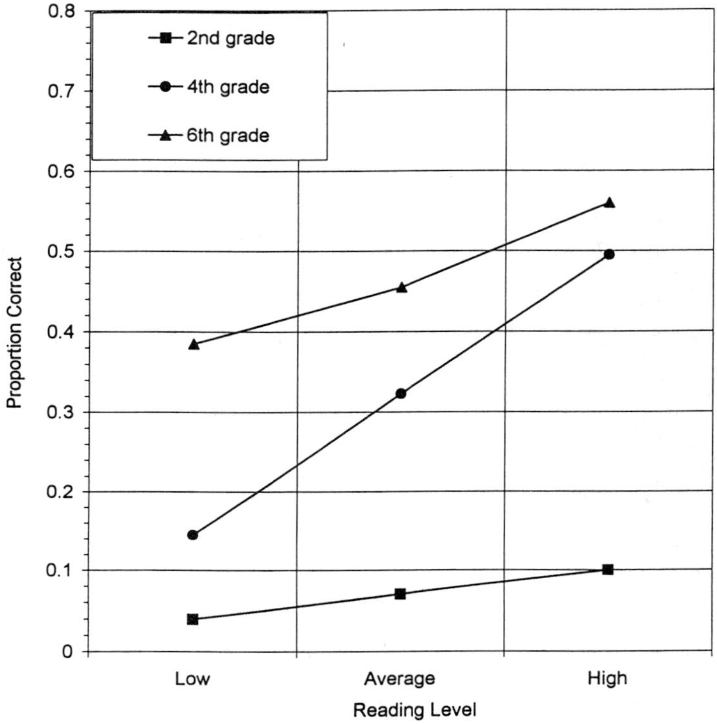

FIG. 1.8. Proportion of errors showing use of phonetic cues by ability and grade.

SUMMARY AND CONCLUSIONS

The most frequent Chinese characters are the least regular. The characters children first learn in school are least likely to be standard compound characters with regular pronunciations. The characters introduced over the elementary grades become increasingly regular.

The basic configuration of the standard compound character is understood by most Chinese first graders and almost all Chinese second graders. However, knowledge of the detailed structure of the components and subcomponents of characters is acquired gradually over the elementary grades.

By the time they reach the third grade, most Chinese children rated as average or high ability are functionally aware that the radicals in compound characters contain information about meaning. They use this information to learn and remember characters introduced in school and to derive the meanings of unfamiliar characters. First graders, and older children rated low in ability, either are not aware of the function of radicals or they do not systematically use the information.

Awareness of the function of phonetics in the pronunciation of compound characters develops gradually over the elementary grades. Phonetic awareness is not clearly present in second graders. It is displayed by fourth graders of high ability, but in lesser degrees by fourth graders of lower ability. Most sixth graders, including those rated low in ability, show basic awareness of the function of phonetics. However, high ability sixth graders continue to show greater mastery of the details of phonetics and their mappings with speech syllables than do children lower in ability.

REFERENCES

Anshen, F., & Aronoff, M. (1988). Producing morphologically complex words. *Linguistics, 26*, 641–655.

Carlisle, J., & Nomanbhoy, D. (1993). Phonological and morphological awareness in first graders. *Applied Psycholinguistics, 14*, 177–195.

Carroll, J. B., Davies, P., & Richman, B. (1971). *Word frequency book*. Boston: Houghton Mifflin.

Cheng, C. M. (1982). The processing of Chinese character recognition. In H. S. R. Kao & C. M. Cheng (Eds.), *Psychological research on Chinese language*. Taipei, Taiwan: WenHe Publish Limited Co.

Cheng, C. M., & Huang, H. M. (1995). *The acquisition of general lexical knowledge of Chinese characters in school children*. Paper presented at the Seventh International Conference on the Cognitive Processing of Chinese and Other Asian Languages, Hong Kong.

Collins, A., Brown, J. S., & Larkin, K. M. (1980). Inference in text understanding. In R. Spiro, B. Bruce, & W. Brewer (Eds.), *Theoretical issues in reading comprehension*. Hillsdale, NJ: Lawrence Erlbaum Associates.

Cossu, G., Shankweiler, D., Liberman, I., Katz, L., & Tola, G. (1986). Awareness of phonological segments and reading ability in Italian children. *Applied Psycholinguistics, 9*, 1–16.

Derwing, B. L., & Baker, W. J. (1979). Recent research on the acquisition of English morphology. In P. Fletcher & M. Garman (Eds.), *Language acquisition*. Cambridge, England: Cambridge University Press.

Fang, S.-P., Horng, R.-Y., & Tzeng, O. J. L. (1986). Consistency effect and pseudo-character naming task. In H. S. R. Kao & R. Hoosain (Eds.), *Linguistics, psychology and the Chinese language*. Hong Kong: Chinese University of Hong Kong Centre of Asian Studies.

Feldman, L. B., & Fowler, C. A. (1987). The inflected noun system in Serbo-Croatian: Lexical representation of morphological structure. *Memory & Cognition, 15*, 1–12.

Fu, Y. H. (1989). A statistical analysis of Chinese character components In Y. Chen (Ed.), *Quantitative analysis of modern Chinese*. Shanghai: Shanghai Educational Press.

Gough, P. B., Juel, C., & Griffith, P. L. (1992). Reading, spelling, and orthographic cipher. In P. B. Gough, L. C. Ehri, & R. Treiman (Eds.), *Reading acquisition*. Hillsdale, NJ: Lawrence Erlbaum Associates.

Goswami, U., & Bryant, P. (1992). Rhyme, analogy, and children's reading. In P. B. Gough, L. C. Ehri, & R. Treiman (Eds.), *Reading acquisition*. Hillsdale, NJ: Lawrence Erlbaum Associates.

Grainger, J., Colé, P., & Segui, J. (1991). Masked morphological priming in visual word recognition. *Journal of Memory and Language, 30*, 370–384.

Hue, C.-W. (1992). Recognition processing in character naming. In H.-C. Chen & O. J. L. Tzeng (Eds.), *Language processing in Chinese*. Amsterdam: North-Holland.

Hoosain, R. (1991). *Psycholinguistic implications for linguistic relativity: A case study of Chinese*. Hillsdale, NJ: Lawrence Erlbaum Associates.

Li, H. T. (1977). *The history of Chinese characters*. Taipei, Taiwan: Lian-Jian.

Nagy, W. E., & Anderson, R. C. (1998). Metalinguistic awareness and literacy acquisition in different languages. In D. Wagner, R. Venezky, & B. Street (Eds.), *Literacy: An international handbook*. New York: Garland.

Nagy, W. E., Anderson, R. C., Schommer, M., Scott, J. A., & Stallman, A. C. (1989). Morphological families and word recognition. *Reading Research Quarterly, 24*, 262–282.

Peng, D.-L., Li, Y.-P., & Huang, F. M. (1995). *Orthographic information in identification of Chinese characters*. Paper presented at the Seventh International Conference on the Cognitive Processing of Chinese and Other Asian Languages, Hong Kong.

Peng, D.-L., Yang, H., & Chen, Y. (1994). Consistency and phonetic independency effects in naming Chinese phonograms. In Q. C. Jing & H. C. Zhang (Eds.), *Information processing of Chinese language*. Beijing: Beijing Normal University Publishing Co.

Perfetti, C. A., & Zhang, S. (1995). The universal word identification reflex. In G. Bower (Ed.), *Psychology of learning and motivation* (Vol. 33, pp. 159–189).

Perfetti, C. A., Zhang, S., & Berant, I. (1992). Reading in English and Chinese: Evidence for a "universal" phonological principal. In R. Frost & J. Katz (Eds.), *Orthography, phonology, morphology, and meaning*. Amsterdam: Elsevier.

Project Group on Statistical Analysis of Chinese. (1988). *Dictionary of Chinese character information*. Beijing: Science Publishers.

Schriefers, H., Friederici, A., & Graetz, P. (1992). Inflectional and derivational morphology in the mental lexicon: Symmetries and asymmetries in repetition priming. *The Quarterly Journal of Experimental Psychology, 44A*, 373–390.

Seidenberg, M. S. (1985). The time course of phonological code activation in two writing systems. *Cognition, 19*, 1–30.

Shu, H., & Anderson, R. C. (1997). Role of radical awareness in the character and word acquisition of Chinese children. *Reading Research Quarterly, 32*, 78–89.

Taft, M. (1985). The decoding of words in lexical access: A review of the morphographic approach. In D. Besner, T. G. Waller, & G. E. Mackinnon (Eds.), *Reading research: Advances in theory and practice* (Vol. V). New York: Academic Press.

Taylor, I., & Taylor, M. M. (1995). *Writing and literacy in Chinese, Korean, and Japanese*. Philadelphia: John Benjamins.

Tunmer, W. E., & Hoover, W. A. (1992). Cognitive and linguistic factors in learning to read. In P. B. Gough, L. C. Ehri, & R. Treiman (Eds.), *Reading acquisition*. Hillsdale, NJ: Lawrence Erlbaum Associates.

Tyler, A., & Nagy, W. E. (1989). The acquisition of English derivational morphology. *Journal of Memory and Language, 28*, 649–667.

White, T. G., Power, M., & White, S. (1989). Morphological analysis: Implications for teaching and understanding vocabulary growth. *Reading Research Quarterly, 24*, 283–304.

Wysocki, K., & Jenkins, J. R. (1987). Deriving word meanings through morphological generalization. *Reading Research Quarterly, 22*, 66–81.

Yin, W. (1991). *On reading Chinese characters—An experimental and neuropsychological study*. Unpublished dissertation, University of London.

CHAPTER

2

Semantic Radicals in Phonetic Compounds: Implications for Visual Character Recognition in Chinese

Laurie B. Feldman
Witina W. T. Siok
University at Albany and Haskins Laboratories

The Chinese writing system is generally considered to be logographic. Accordingly, the graphemes of written Chinese do not map onto individual phonemic units of the spoken language. Instead, each logogram corresponds to one morpheme. Moreover, the morphemes of the language are typically monosyllabic. Therefore, Chinese writing is sometimes described as a morphosyllabic or morphophonological system (DeFrancis, 1989; Perfetti & Zhang, 1995). The interweaving of component strokes within a character, together with spatial separation between characters, makes each Chinese character a salient and an integrated visual unit (Hoosain, 1991). Nevertheless, there is structure internal to the logographic character. One structural element of Chinese writing, the *semantic radical*, is evident only in the written language. Semantic radicals specify aspects of meaning, and they are not generally realized phonologically. A second structural element, the *phonetic component*, specifies aspects of a character's pronunciation, but does not reflect the meaning of the character. Because the semantic and phonetic components of Chinese compounds differ in function, the experimental study of reading in Chinese provides an approach to the study of reading that is inaccessible from the study of alphabetic writing systems. In an alphabetic orthography, letters combine to represent syllabic as well as morphological units. In contrast, in the logographic orthography of Chinese, components differ in function such that some components correspond to aspects of whole character phonology and others correspond to aspects of character meaning. In this chapter, we consider the functional differences between components and examine some of the implications for visual character recognition. We begin by describing this distinctive attribute of Chinese, that is, the differing contributions of the semantic and phonetic components to the phonological and semantic integrity of the phonetic compound.

SEMANTIC AND PHONETIC COMPONENTS CONTRASTED

Phonetic compounds constitute 90% of 4,574 characters in a 1,800,000-character corpus, according to the *Xiandai Hanyu Pinlu Cidian* [*Modern Chinese Frequency Dictionary*] (1986). This type of compound always includes both a phonetic component and a semantic radical. There are approximately 200 semantic radicals in modern Chinese, most of which cannot appear as characters and have no associated pronunciation. Therefore, on average, about 20 phonetic compounds are formed from each semantic radical. The *combinability* of a semantic radical reflects its tendency to enter into compounds. Although the combinability of semantic radicals in Chinese tends to be quite high, it is important to point out that there is substantial variability. Some semantic radicals appear in many compounds; others appear in only a very few. For example, the semantic radical 扌 (i.e., *hand*, or *the actions of hand*) appears in 328 phonetic compounds, whereas the semantic radical 身 (i.e., *body*) appears in only 6 phonetic compounds.

As when any language evolves, in Chinese there is a tendency for the semantic contribution of meaningful components to the meaning of the whole form to become progressively more opaque (Hoosain, 1991). The general consensus is that most Chinese semantic radicals still provide a useful cue to the meaning of whole characters. Jin (1985) observed that more than 90% of the phonetic compounds with the semantic radical 扌 have a meaning related to its semantic radical (i.e., *hand*, or *the actions of hand*). For example, 打 meaning *to hit* or *to beat*, 拉 meaning *to pull*, and 抱 meaning *to embrace*, all include the *hand* radical and have a meaning associated with the hand. In these characters, the semantic radical is semantically transparent.[1] Fan (1986) found that about 65% of phonetic compounds containing the semantic radical 纟 (meaning *silk* or *things related to silk fabric*) have at least aspects of their meaning that are consistent with the concept of silk. These studies indicate that the semantic contribution of

[1]Note that the definition of semantic radical transparency or opacity depends on the characters in which it appears. A radical can be semantically transparent in some characters and opaque in others. For example, the semantic radical 扌 that appears in 打 (meaning "to hit" or "to beat"), 拉 (meaning "to pull"), and 抱 (meaning "to embrace") is defined as semantically transparent in these characters. However, it is defined as semantically opaque when it appears in 捌 (meaning "eight").

radicals is systematic with respect to the meaning of the compound. Although it is important to point out that the meaning of a compound itself may vary in its precision (Tan, Hoosain, & Peng, 1995), the meaning of a phonetic compound does tend to be at least partially transparent with respect to the meaning of its semantic radical. Nevertheless, there are exceptions. For example, 捌 meaning *eight* is also formed from the semantic radical 扌 (*hand*, or *the actions of hand*). In this example, the semantic radical is semantically opaque. In summary, somewhat like the morphemes of an alphabetic orthography, in written Chinese characters there exist meaningful orthographic components that tend to be related in a nonarbitrary manner to the meaning of the whole character.

The phonetic component, usually located in the right part of phonetic compounds, typically provides clues as to the pronunciation of the whole character. By one estimate, there are approximately 800 phonetic components (Taylor & Taylor, 1983). Relative to semantic radicals, the reliability with which phonetic components of characters specify the pronunciation of compounds is low. For example, Fan, Gao, and Ao (1984) found that 26.3% of 5,990 phonetic compounds are regular in that they have pronunciations identical to that of their phonetic component. This assessment is apparently based on a strict criterion, requiring both identical pronunciation and identical tone. Considering consistency and regularity of phonetics,[2] Zhou (1978) estimated that 38% of phonetic components have pronunciations that are consistent with that of their whole characters (see also Perfetti, Zhang, & Berent, 1992). Although assessments of the validity of phonetic components depend on researcher-defined criteria, the general consensus is that the phonetic cueing value of phonetic components is low (but see Mattingly & Ni, 1998), at least relative to the cueing value of semantic radicals. Extending the terminology of Perfetti and Tan (1998), one might claim that the mapping between written phonetic components and whole character phonology for compounds is less systematic than the mapping between written semantic radicals and whole character meaning.[3]

---

[2]A *consistent* phonetic has a pronunciation that is identical to the pronunciation of all characters that contain this phonetic. For an *inconsistent* phonetic, some whole characters containing this phonetic have pronunciation identical to that of its phonetic, but others do not. For example, all characters consisting of 家 (pronounced as /jia/, 稼 嫁 镓) are pronounced as /jia/, without exception. Therefore, 家 is a consistent phonetic. By contrast, some whole characters with the phonetic 加 (/jia/) are pronounced as /jia/ (e.g., 枷 驾 跏); some are pronounced as /qie/ (e.g., 茄 伽). As a result, 加 is an inconsistent phonetic. *Regularity* refers to whether a particular compound character has a pronunciation identical to that of its phonetic. If they have identical pronunciation, the character is considered *regular*. If they do not, it is *irregular*.

[3]This claim is based on the proportion of components that are related to the whole character. It assumes that similarity can be treated in an all-or-nothing fashion.

Much of the recent research in Chinese psycholinguistics focused on the issue of whether phonology contributes to character identification. Within the phonological domain, effects of sublexical phonology (i.e., the phonological correlate of the phonetic component) must be differentiated from the effects of whole character phonology. As for the role of phonetic components, the consensus is that there is no difference between regular and irregular phonetic components or between consistent and inconsistent phonetics in identifying frequently used compound characters in such tasks as naming and character decision. However, regular and consistent phonetic components do facilitate responses to low-frequency characters relative to irregular and inconsistent characters (Fang, Horng, & Tzeng, 1986; Hue, 1992; Seidenberg, 1985). These observations in Chinese parallel the word recognition research on phonological regularity and consistency in English (e.g., Baron & Strawson, 1976). It is also the case that with controls for surface frequency, character decision latencies to phonetic compounds with phonetics that are high in phonetic combinability (and therefore appear in many compounds) are faster than to compounds whose phonetics are low in combinability (Feldman & Siok, 1997). On the other hand, the time course of whole character phonological activation in Chinese character recognition remains less clear. One position is that phonological processing does not occur prior to lexical access for Chinese characters presented in isolation at relatively long exposure durations (e.g., Baron & Strawson, 1976; Chen, Yung, & Ng, 1988; Hoosain & Osgood, 1983; Leck, Weekes, & Chen, 1995). In contrast, studies with data-limited presentation conditions (e.g., backward masking) demonstrate an unequivocal role of whole character phonological information early in the course of character identification (Perfetti & Zhang, 1991, 1995; Tan et al., 1995; Tan, Hoosain, & Siok, 1996).

Research on the role of semantic radicals in character recognition is sparse relative to the work on the phonetic. Nevertheless, the results of two studies that examined the time course of orthographic and phonological information can be interpreted as suggesting an influence of semantic radicals in character recognition. One study (Leck et al., 1995) used a semantic categorization procedure that required subjects to decide whether or not a particular target character (e.g., 狐 meaning *fox* whose radical is 犭 meaning *beast* or *wild animal*) belongs to a particular semantic category (e.g., animal). For phonetic compounds, the false positive response rate to foils that shared the same semantic radical as the target (e.g., 猜 meaning *guess* whose semantic radical is 犭) was higher than that to control foils. However, the false positive response rate to foils that shared identical phonetic components with the target (e.g., 呱 /gua/ meaning *noisy* whose phonetic is 瓜 /gua/) was not different from that of controls. Because repetition of the semantic radical and of the phonetic component both introduce partial graphic similarity, it is unlikely that the effect is orthographic in origin.

In a backward-masking perceptual paradigm (Ju & Jackson, 1995), a target and a linguistic mask were presented in succession, and each target–linguistic mask pair was preceded and followed immediately by a pattern mask. The masks were presented for 30 ms. The exposure duration of the target was adjusted individually to maintain a 40% to 60% target identification rate. In the orthographic mask condition, more than 90% of the targets had a semantic radical identical to that of the mask, and the mask facilitated target recognition relative to controls. Ju and Jackson explained their finding as revealing the role of orthographic information in target recognition. Alternatively, this outcome, in conjunction with that of Leck et al. (1995), could be interpreted as evidence that the semantic characteristics of semantic radicals influence target recognition. That semantic radicals play a part in character categorization is compatible with the finding from compound Japanese kanji (Flores d'Arcais, Saito, & Kawakami, 1995), whereby the meaning of the component radicals was retrieved in a naming task, even when the radicals were not semantically related to the meaning of the whole character.

In summary, Chinese phonetic compounds are composed of two types of components that differ with respect to their function. We argue that an appreciation of the different nature of the components is essential to any investigation of how the visual units of Chinese are processed (Feldman & Siok, 1997; cf. Taft & Zhu, 1997). In the remainder of the chapter, we first describe descriptive and experimental variables that may influence the identification of components of compound characters. Then we present some evidence of our own for the processing of the semantic radical by skilled readers of Mandarin (Putonghua).

CHINESE CHARACTER RECOGNITION: DESCRIPTIVE AND EXPERIMENTAL VARIABLES

In written Chinese, there is little consensus about the psychologically relevant units for character recognition. Because characters are distinguished by fixed spaces in text and occupy a square area, irrespective of the number of strokes they possess, some researchers have assumed that characters are the basic units of visual recognition and mental representation (e.g., Hoosain, 1991; Morton, Sasanuma, Patterson, & Sakuma, 1992). Similarly, because the number of strokes in a character affects naming (Leong, Cheng, & Mulcahy, 1987) and character decision latencies (Tan & Peng, 1990), others have posited analysis of features such that the number of strokes in a character is an important factor in printed character identification (Flores d'Arcais, 1992; Huang, 1986; Leong et al., 1987; Tan et al., 1995). Finally, it is suggested that combinations of strokes constrained with respect to position (i.e., stroke patterns) are the relevant units of analysis (Chen, Allport, & Marshall, 1996)

Strokes combine to form components, and some researchers (e.g., Taft & Zhu, 1997) claimed that the components[4] of compound characters are decomposed in the course of recognizing a character. In a character decision task (Taft & Zhu, 1997), subjects made judgments about whether or not compound characters that consisted of two components were legal Chinese characters. Components differed with respect to type frequency. Components that appeared in more than 280 characters were classified as high in frequency, whereas components that appeared in fewer than 69 were low in frequency (Taft & Zhu, 1997, Experiment 1). They examined differences in character decision latencies for targets composed of high- and low-frequency components (i.e., a differential frequency effect). Because compound characters contained two components, the differential frequency effects were examined for the right position as well as for the left. Results indicated that the type frequency (the number of characters containing a particular component) of the right component affected decision times, whereas the type frequency of the left component did not. Taft and Zhu explained this outcome as suggesting that all components of a compound are activated sequentially in the process of character recognition. Their view is reminiscent of the left-to-right processing account for visually presented words in alphabetic orthographies (Taft & Forster, 1976) in that components of a character are activated in series. As originally described, this account is neutral with respect to the role of lexical variables (e.g., character frequency) in processing and focuses instead on sublexical events.

According to Taft and Zhu's view, the sublexical components that comprise a character must be analyzed position by position in order to recognize a character. Within this serial framework, some positions exert a greater influence on character identification latencies than do others. Our view, also analytic in nature, focuses on the function of components. A simple survey conducted in our lab, based on a sample of 4,516 phonetic compounds from *Xiandai Hanyu Cidian* [*Modern Chinese Dictionary*] (1992), estimates that approximately 75% of Chinese phonetic compounds have their semantic radical on the left. Of the remaining compounds, approximately 5% have the semantic on the right, 15% have the semantic on top, 4% have the semantic on the bottom, and 1% have the semantic in the periphery (see also Hoosain, 1991, for another simple survey). In other words, semantic radicals tend to be on the left, and it is atypical for semantics to appear on the right. Serial accounts of how compounds are processed must consider the function (viz., semantic or phonetic) of the component.

In the Taft and Zhu study, frequency or combinability of components was based on the *Chinese Radical Position Frequency Dictionary* (1984). Listed in

[4]Taft and Zhu referred to all components as *radicals* irrespective of function.

that source are 541 *bujian*. Bujian are components of a character, unconstrained with respect to both function and position. In essence, the count is based predominantly on graphic components of characters and only coincidentally includes semantic radicals and phonetic components. Because they based their statistics on this source, the frequency counts in Taft and Zhu (1997) reflect the tendency of both phonetic components and semantic radicals (as well as other units) to enter into compound characters. It is possible, therefore, that the position-sensitive differential frequency effect they observed reflects the way in which frequency was computed.

EXPERIMENTAL EVIDENCE FOR THE PROCESSING OF RADICALS

As previously noted, there is little direct experimental evidence to date for the role of the semantic radical in recognition of Chinese characters, and the available evidence is generally tenuous (see also Zhou & Marslen-Wilson, chapter 3, this volume). Some experimental work of our own looks at the role of semantic radicals in the recognition of Chinese compound characters. Most of it is based on variants of the primed character decision task. Because radicals are written units that map onto meaning, the research entails comparing effects of semantic and graphic similarity among compound characters with the effects of a shared radical. The basic strategy in the second, third, and fourth experiments described later is to compare patterns of facilitation (and inhibition) for targets following primes that share various dimensions of similarity with the target, and to manipulate the temporal relationship between prime and target. The first experiment, however, looks at identification latencies for isolated compounds where properties of the radical are manipulated. In particular, the semantic radicals of characters differed with respect to the number of combinations they entered into to form characters (i.e., combinability). Effects significant by both subjects and by items are reported unless otherwise indicated.

Experiment 1: Radicals Influence Isolated Character Recognition: Effects of Combinability

One experiment conducted in our laboratory was motivated by the outcome of an experiment (Taft & Zhu, 1997) in which combinability of the right component influenced target decision time, whereas combinability of the left component did not. An effect of combinability for components on the right but not for components on the left is intriguing. While it is possible that the effect of combinability on target decision latencies is position sensitive, as Taft and Zhu claimed, it is also possible that the confound between position and function inherent in their frequency count contaminated the effect of combinability. Experiment 1 attempted to differentiate effects of position and function by using a combinability measure that is sensitive to function (see Feldman & Siok, 1997, for a more complete account).

We presented compound characters, all of which were composed of a radical and a phonetic, to subjects from the Beijing Business College in a character decision task. Semantic radical combinability and position were manipulated. For example, 诲 has a radical, 讠, that enters into many combinations (160) and appears on the left. By contrast, 躺 has a radical, 身, that enters into relatively few combinations (6) and appears on the left. 鸭 has a radical, 鸟, that enters into many combinations (98), but it appears on the right. Finally, 瓶 has a radical, 瓦, that enters into relatively few combinations (17) and appears on the right. All target characters had semantically transparent radicals with high or medium frequencies (no fewer than 6 occurrences per million), according to *Xiandai Hanyu Pinlu Cidian* [*Modern Chinese Frequency Dictionary*] (1986). Four sets of 10 characters, each set matched for surface frequency (171, 173, 161, 169, respectively) as well as number of strokes, combinability of the phonetic, and phonological regularity of the phonetic, were created. Combinability and position of the semantic radical were manipulated within subjects.

Character decision latencies revealed a main effect of combinability (33 ms) for radicals on the left, but not for radicals on the right (2 ms). High radical combinability benefited processing. However, the interaction of position by combinability was significant by participants (although it did not even approach significance by items), and therefore it is possible that position of the semantic also plays a role in character recognition.[5] Analogously, in a post hoc analysis, when combinability of the phonetic was manipulated, there was an effect of combinability but no interaction with position. In summary, when combinability of semantic radicals and phonetic components are counted separately, combinability effects in character recognition are robust. Position effects, if they exist, are function dependent.

Experiment 2: Radicals Influence Primed Character Decision Latencies: Effects of Semantic Transparency

A second experiment was directed at semantic processing of the radical. Participants were native speakers of Mandarin (Putonghua) attending South China (Guangzhou) Normal University. Sixty-four targets, half with high-combinability semantic radicals and half with low-combinability semantic radicals, were presented in a character decision task. Targets followed primes at an SOA (stimulus onset asynchrony) of 243 ms. Across experimental lists, each target (e.g., 论 meaning *to say* or *to talk* and formed from the

[5]Taft and Zhu (1997) did not report the F values for the interaction effect between position and combinability (frequency), nor the main effects of these two variables.

radical 讠 meaning *say* or *related to the action of saying*) was preceded by four types of primes: (a) R+S+ primes (e.g., 评 meaning *to comment*) shared an identical semantic radical and were semantically related to the target. In this condition, the meaning of semantic radical was consistent with the meaning of both the target and the prime. That is, the radicals were semantically transparent; (b) R+S– primes (e.g., 诸 meaning *some* or *various*) shared an identical semantic radical with the target but were semantically unrelated to the target. In this condition, the semantic radical of the prime did not provide any clue as to the prime's meaning, thus it was semantically opaque; (c) R–S+ primes (e.g., 述 meaning *to speak*) were semantically related to the target but did not share a radical; and (d) R–S– control primes (e.g., 竿 meaning *stick*) were neither semantically related nor shared any component with the target (for further experimental details, consult Feldman & Siok, in press; Siok, 1996).

All targets were compounds composed of a phonetic component and a semantic radical. For high-combinability radicals, their occurrence in a corpus of about 6,000 characters contained in *Xiandai Hanyu Cidian* [*Modern Chinese Dictionary*] (1992) was no fewer than 56, with an average of 197 (*SD* = 215). For low-combinability radicals, their occurrence was no more than 41, with an average of 22 (*SD* = 9). All target characters had semantically transparent radicals with high or medium frequencies (no less than 10 occurrences per million), according to *Xiandai Hanyu Pinlu Cidian* [*Modern Chinese Frequency Dictionary*] (1986). The average surface frequency of the targets was 246 (*SD* = 294) for the high-combinability radical characters, and 254 (*SD* = 273) for the low-combinability radical characters. All primes had frequencies of no fewer than 4 occurrences per million. The average frequency among the four prime types did not differ significantly, and the average frequency of each prime type was lower than the average target frequency. Each radical appeared in only one prime-target pair.

For primes that shared a radical with the target, semantic transparencies ratings were obtained from 10 informants who were native Mandarin (Putonghua) speakers. They indicated their judgment by position on a 7-point scale where 1 indicated very low and 7 indicated very high. For primes in the R+S+ conditions, the average semantic cueing value was 5.7 (*SD* = .46). For primes in the R–S+ conditions, the average semantic cueing value was 1.8 (*SD* = .19). Finally, another 10 informants were asked to assess the semantic similarity of each prime-target pair in the R+S+ and R–S+ conditions, on a 7-point rating scale. The mean semantic relatedness value for the 64 targets and their R+S+ primes was 5.8 (*SD* = .51). For the same targets and their corresponding R–S+ primes, the mean semantic relatedness value was 5.9 (*SD* = .51). Based on semantic cueing ratings, semantically related primes with and without a shared radical were matched for semantic similarity to the target.

Decision latencies are summarized in Table 2.1. Results indicated that the effect of prime type was significant (see Table 2.1). At an SOA of 243 ms, semantically related primes with (R+S+) and without (R–S+) a shared radical produced equivalent (14 ms) and statistically significant facilitation relative to the unrelated control condition. By contrast, primes formed from the same radical where the meaning of the radical was semantically opaque (R+S–) produced significant inhibition (15 ms) relative to the unrelated control condition.

In addition to matching primes on dimensions of similarity to the target, combinability, a variable more directly reflective of properties of the radical, was explored. This variable is reminiscent of a measure on base morpheme productivity (in alphabetic studies of morphology) in that it reflects the size of the set of words in which a particular morpheme (more specifically, the set of characters in which a particular radical) can appear (Feldman, Frost, & Pnini, 1995).

The experimental manipulation of combinability entailed contrasting decision latencies for (different) isolated phonetic compounds formed from high- and low-combinability radicals. As previously discussed, a similar variable, radical frequency, was studied by Taft and Zhu (1997), although it is important to point out that the present definition differs from that of Taft and Zhu in that only semantic radicals entered into the count. In Experiment 1, there was evidence that radical combinability influenced character identification such that recognition latencies varied as a function of the number of compound character combinations a radical can enter into (see also Feldman & Siok, 1997). Analogous to a reciprocal activation perspective (see Andrews, 1989), characters with radicals that enter into many combinations would receive more reciprocal activation, or feedback, from their character "neighbors" and hence would be identified earlier. From an alternative perspective, density of representations for characters formed from the same

TABLE 2.1
Mean Response Latencies (ms) and Error Rates (%) for Targets as a Function of Radical Combinability and Prime Type in Siok (1996)

	Prime Type			
	R+S+	*R+S–*	*R–S+*	*Control*
High Combinability				
Mean	530	555	527	542
Facilitation	12*	–13*	15*	
Low Combinability				
Mean	537	569	540	552
Facilitation	15*	–17*	12*	

*$p < .05$.

semantic radical may be important such that characters that are members of large clusters benefit relative to those from clusters that are more sparse. In a priming task, processing of compound targets could be constrained by prior identification of the same radical in the prime, so that magnitude of facilitation varies with the combinability of the radical.

In Experiment 2, at an SOA of 243 ms, there was a main effect of combinability indicating that decision latencies for targets were sensitive to the number of combinations in which its radical could appear. With controls for character frequency, across all prime conditions, high-combinability targets were recognized significantly faster than targets in the low-combinability conditions. Importantly, combinability and prime type did not interact. That is, radicals influence how compounds are processed, but similar effects are observed across all prime types.

Finally, semantic transparency scores (from semantic cueing values) were correlated with the difference in response times between R+S+ and controls and between R+S– and controls. The correlation (based on item means) was significant ($r = .43$, $p < .001$). That is, semantic transparency of the radical is associated with the direction of priming, either facilitation or inhibition.

In summary, effects of the radical of the prime on target decision latencies were demonstrated. Varying dimensions of similarity were assessed relative to the same target, and only prime characteristics varied across conditions. Target latencies were significantly faster following semantically related primes (R–S+) relative to controls. This outcome demonstrates whole character semantic relatedness at an SOA of 243 ms. More relevant to our program of research on processing of the semantic radical, the effect on the target of repeating the semantic radical in prime and target varied as a function of semantic transparency. When the meaning of the radical was transparent in both the prime and the target, significant facilitation was observed. When the meaning of the radical was transparent in the target but opaque in the prime, inhibition was observed. This outcome is critical in that it reveals that it is the semantic dimension of the radical that is detected when primes and targets are offset by an SOA of 243 ms. In Experiment 2, it is important to note that we cannot exclude the possibility that the orthographic form of the radical also plays a role. In particular, facilitation due to whole character semantic relatedness and inhibition due to orthographic similarity based on a shared semantic can account for much of the data.[6] Experiment 3 was designed to differentiate more directly the

[6]Without facilitation following R+S+ primes (see Zhou & Marslen-Wilson, chapter 3, this volume), there would be no compelling evidence for semantic processing of the radical. A null effect following R+S+ primes in conjunction with R–S+ facilitation and R+S– inhibition can be interpreted as evidence of whole character semantic facilitation (S+) combined with orthographic inhibition due to repetition of the radical (R+). We suspect that multiple presentations of the same radical within the same session may have attenuated R+S+ facilitation in the Zhou and Marslen-Wilson study (chapter 3, this volume).

contribution of orthographic similarity from semantic transparency of the radical in the character recognition task.

Experiment 3: Radicals Influence Character Decision Latencies: Temporally Constrained Orthographic Facilitation

Further support for the claim that radicals are processed semantically comes from another study (Siok & Feldman, 1998) that included visually similar primes in addition to R+S+, R+S–, and R–S+ primes. Visually similar primes were generally formed from a radical that differed from the radical of the target by only one or two strokes. For example, the radical 礻 (meaning *god, ceremony*, or *sacred*) differs from the radical 衤 (meaning *clothes*) by the addition of only one stroke. Construction of experimental materials was based on judgments of visual similarity by 10 informants from Beijing Business College. Surface frequency and number of strokes were matched across conditions. Materials were presented to 145 subjects from Beijing Business College at SOAs of 243 ms and 43 ms. SOA was manipulated between subjects.

As summarized in Table 2.2, results indicated that at a 243 ms SOA, visually similar primes had no significant effect (5 ms), whereas, in replication of the previous experiment, R+S+ primes facilitated (22 ms) and R+S– primes inhibited (–22 ms). Because R+S+ and R+S– primes were equated for visual similarity to the target, and because visually similar primes had no effect on target decision latencies, it is unlikely that the inhibition in the R+S– condition can be attributed to graphic similarity. Nevertheless, when the same primes and targets were presented with an SOA of 43 ms, visually similar primes and R+S– primes both produced statistically equivalent facilitation. Evidently, orthographic similarity between characters can provide a source of facilitation at very short SOAs. Moreover, this outcome suggests that the visually similar primes were in fact comparable to R+S– primes in their similarity to targets. In summary, whereas R+S– primes produced facilitation at 43 ms, they produced inhibition at 243 ms. The comparison across SOAs suggests that the effect of semantic inconsistency of the radical is evident by 243 ms. Before that, only its orthographic similarity with the target has an effect.

A fuller interpretation of the effect of semantic inconsistency in R+S– compounds rests on further investigation. In particular, it is not yet clear whether inhibition reflects complexities of recognizing the prime itself or the effect of recognizing the prime on subsequent recognition of the target.

In summary, Experiment 3 provides further evidence for the semantic radical as a psychologically relevant unit of processing in the visual recognition of Chinese compounds. Primes that were visually similar to the target

TABLE 2.2
Mean Response Latencies (ms) and Error Rates (%) for Targets as a Function of Prime Type in Siok and Feldman (1998)

	Prime type				
	R+S+	*R+S–*	*Visually Similar*	*R–S+*	*Control*
SOA = 43 ms					
Mean	700	723	726	723	760
Facilitation	+60*	+37*	+34*	+37*	
SOA = 243 ms					
Mean	682	726	699	683	704
Facilitation	+22*	–22*	+5	+21*	

*$p < .05$.

had a facilitatory effect on target recognition latencies at 43 ms and had no effect at 243 ms, whereas R+S– primes had a facilitatory effect at 43 ms and an inhibitory effect at 243 ms. Facilitation following semantically related primes (R–S+) was evident even at an SOA of 43 ms.[7] Finally, target latencies following semantically related primes with (R+S+) and without (R–S+) a shared radical differed, but only at the 43 ms SOA. Collectively, these findings suggest that, although the orthographic information of semantic radicals is available in the 43 ms SOA condition, by 243 ms, the orthographic characteristics of the prime and target radical do not dominate processing. With the longer SOA, the semantic aspects of the radical play a critical role in character decision.

Experiment 4: Effects of Repeating a Radical at Long Lags

The temporal relation between prime and target in Experiment 3 was manipulated in order to track the time courses of the orthographic, semantic, and shared radical dimensions of similarity between prime and target. In contrast, in Experiment 4, the temporal relationship between prime and target was defined so as to minimize the independent contributions of orthographic and semantic similarity.

From the study of morphological processing in alphabetically transcribed languages, we know that when many (viz., 7 to 13) items intervene between prime and target, the prior presentation of a word that is similar in form but morphologically unrelated has no effect on target recognition (e.g.,

[7]Perfetti and Tan (1998) report that the effect of semantic relatedness in the naming task becomes evident after 85 ms. This value must be contingent on attributes of the experimental materials (e.g., frequency, semantic vagueness, homophony) as well as on the task. In no way does our finding conflict with their claim.

Feldman & Andjelković, 1992; Feldman & Moskovljević, 1987; Grainger, Colé, & Segui, 1991; Hanson & Wilkenfeld, 1985; Laudanna, Badecker, & Caramazza, 1992; Murrell & Morton, 1974; Napps & Fowler, 1987; Segui & Grainger, 1990; Stolz & Feldman, 1995). Similarly, effects of semantic relatedness between prime and target are also absent at intervals of 7 to 13 items (Bentin & Feldman, 1990). By analogy with the work on morphological processing, increasing the interval between presentation of prime and target should eliminate effects of orthographic similarity and whole form semantic relatedness.

In Experiment 4, participants responded to every character they saw, and the designations of "prime" and "target" simply reflected earlier or later presentations of a character formed from a particular radical within the experimental list. Related pairs were separated by an average of 10 intervening items. The materials and the task were the same as in Experiment 2.

Results of Experiment 4 revealed a main effect of prime type (R+S+, R+S–, R–S+, and unrelated), but no effect of radical combinability and no interaction of prime type by radical combinability when several items separated prime and target. Post hoc multiple comparisons indicated that R+S+ primes produced significant facilitation (16 ms). R+S– primes produced no significant effect (7 ms), nor did R–S+ primes (3 ms). That is, at long lags, only semantically transparent radicals in primes facilitated target decision latencies.

The outcome of Experiment 4 provided further evidence for the radical as a unit of processing in written Chinese compounds. Facilitation was evident under temporal conditions that do not reveal effects of orthographic similarity due to a shared radical or semantic relatedness between full forms that do not share a radical. Nevertheless, in the semantically transparent (R+S+) condition, orthographic similarity occurred in conjunction with semantic similarity. Under these conditions, long-term facilitation due to repetition of a radical with multiple intervening items was observed.

CONCLUDING REMARKS

Across four experiments, we demonstrated that the semantic radical is a unit of processing in the character recognition task in Chinese. An effect of radical combinability was demonstrated for compounds both in a character decision task for single characters and in a primed character decision task. The comparison of target latencies following various types of primes and at various SOAs provides insight into when particular dimensions of the semantic radical are psychologically salient.

When the prime is processed for only a very short period (43 ms) before onset of the target, the visual attributes of the radical seem to predominate.

When processing of the prime proceeds for 243 ms before presentation of the target, semantic transparency of the radical plays a salient role in target identification. Semantic transparency in both the prime and target (or the consistency of the semantic contribution of the radical to the meaning of the prime and the target) facilitates recognition, whereas semantic opacity (or the inconsistency of the semantic contribution of the radical to the prime relative to the target) typically hurts. Finally, after several intervening items, effects of a radical shared by prime and target are evident only when orthographic similarity and semantic relatedness co-occur. This condition arises with semantically transparent radicals.

In summary, we demonstrated that combinability, a function-specific attribute of components (e.g., semantic radicals), influences latencies in the character decision task, and that the semantic attributes of semantic radicals are activated in the course of character recognition in Chinese.

ACKNOWLEDGMENTS

This research was supported in part by funds from National Institute of Child Health and Development Grant HD-01994 to Haskins Laboratories. W. W. T. Siok presented Experiments 2 and 4 to the Department of Psychology at The University at Albany, State University of New York, in partial fulfillment of her first year (1995–1996) research requirement. Portions were also presented at The Ninth Annual CUNY Sentence Processing Conference (New York, March 1996).

REFERENCES

Andrews, S. (1989). Frequency and neighborhood effects on lexical access: Activation or search? *Journal of Experimental Psychology: Learning, Memory, and Cognition*, *15*, 802–814.

Baron, J., & Strawson, C. (1976). Use of orthographic and word specific knowledge in reading words aloud. *Journal of Experimental Psychology: Human Perception and Performance, 2*, 386–393.

Bentin, S., & Feldman, L. B. (1990). The contribution of morphological and semantic relatedness to repetition priming at short and long lags: Evidence from Hebrew. *The Quarterly Journal of Experimental Psychology*, *42A*, 693–711.

Chen, M. J., Yung, Y. F., & Ng, T. W. (1988). The effect of context on the perception of Chinese characters. In I. M. Liu, H.-C. Chen, & M. J. Chen (Eds.), *Cognitive aspects of the Chinese language* (pp. 27–39). Hong Kong: Asian Research Service.

Chen, Y. P., Allport, D. A., & Marshall, J. C. (1996). What are the functional orthographic units in Chinese word recognition: The stroke or the stroke pattern? *The Quarterly Journal of Experimental Psychology*, *49A*, 1024–1043.

Chinese radical position frequency dictionary. (1984). Changsha: Hunan Institute of Computer Science.

DeFrancis, J. (1989). *Visible speech: The diverse oneness of writing systems*. Honolulu: University of Hawaii.

Fan, K. Y. (1986, September). *Graphic symbol of the Chinese character.* Paper presented at the Symposium of Chinese Character Modernization, Beijing, China.

Fan, K. Y., Gao, J. Y., & Ao, X. P. (1984). Pronunciation principles of the Chinese character and alphabetic writing scripts. *Chinese character reform, 3* (pp. 19–22). Beijing: National Commission of Chinese Character Reform. (In Chinese)

Fang, S. P., Horng, R.-Y., & Tzeng, O. J. L. (1986). Consistency effects in the Chinese character and pseudo-character naming tests. In H. S. R. Kao & R. Hoosain (Eds.), *Linguistics, psychology and the Chinese language* (pp. 11–21). Hong Kong: University of Hong Kong Press.

Feldman, L. B., & Andjelković, D. (1992). Morphological analysis in word recognition. In L. Katz & R. Frost (Eds.), *Orthography, phonology, morphology, and meaning: An overview* (pp. 343–360). Amsterdam: Elsevier.

Feldman, L. B., & Moskovljević, J. (1987). Repetition priming is not purely episodic in origin. *Journal of Experimental Psychology: Learning, Memory, and Cognition, 13,* 573–581.

Feldman, L. B., & Siok, W. W. T. (1997). The role of component function in visual recognition of Chinese characters. *Journal of Experimental Psychology: Learning, Memory, and Cognition, 23,* 776–781.

Feldman, L. B., & Siok, W. W. T. (in press). Semantic radicals contribute to the visual identification of Chinese characters. *Journal of Memory and Language.*

Feldman, L. B., Frost, R., & Pnini, T. (1995). Decomposing words into their constituent morphemes: Evidence from English and Hebrew. *Journal of Experimental Psychology: Learning, Memory, and Cognition, 21,* 947–960.

Flores d'Arcais, G. B. (1992). Graphemic, phonological and semantic activation processes during the recognition of Chinese characters. In H.-C. Chen & O. J. L. Tzeng (Eds.), *Language processing in Chinese* (pp. 37–66). Amsterdam: Elsevier.

Flores d'Arcais, G. B., Saito, H., & Kawakami, M. (1995). Phonological and semantic activation in reading Kanji characters. *Journal of Experimental Psychology: Learning, Memory, and Cognition, 21,* 34–42.

Grainger, J., Colé, P., & Segui, J. (1991). Masked morphological priming in visual word recognition. *Journal of Memory and Language, 30,* 370–384.

Hanson, V. L., & Wilkenfeld, D. (1985). Morphophonology and lexical organization in deaf readers. *Language and Speech, 28,* 269–280.

Hoosain, R. (1991). *Psycholinguistic implications for linguistic relativity: A case study of Chinese.* Hillsdale, NJ: Lawrence Erlbaum Associates.

Hoosain, R., & Osgood, C. E. (1983). Information processing times for English and Chinese words. *Perception and Psychophysics, 34,* 573–577.

Huang, J.-T. (1986). Visual integration process in recognizing fragmented Chinese characters. In H. S. R. Kao & R. Hoosain (Eds.), *Linguistics, psychology, and the Chinese language* (pp. 45–54). Hong Kong: Centre of Asian Studies.

Hue, C.-W. (1992). Recognition processes in character naming. In H.-C. Chen & O. J. L. Tzeng (Eds.), *Language processing in Chinese* (pp. 93–107). Amsterdam: North-Holland.

Jin, J. H. (1985). Properties of Chinese characters. *Chinese character reform, 5* (pp. 13–15). Beijing: National Commission of Chinese Character Reform. (In Chinese)

Ju, D., & Jackson, N. E. (1995). Graphic and phonological processing in Chinese character identification. *Journal of Reading Behavior, 27,* 299–213.

Laudanna, A., Badecker, W., & Caramazza, A. (1992). Processing inflectional and derivational morphology. *Journal of Memory and Language, 31,* 333–348.

Leck, K. J., Weekes, B. S., & Chen, M. J. (1995). Visual and phonological pathways to the lexicon: Evidence from Chinese readers. *Memory & Cognition, 23,* 68–476.

Leong, C. K., Cheng, P. W., & Mulcahy, R. (1987). Automatic processing of morphemic orthography by mature readers. *Language and Speech, 30,* 181–196.

Mattingly, I. G., & Ni, W. (1998). *Zhou Youguang and phonological mediation in Chinese.* Manuscript submitted for publication.

Morton, J., Sasanuma, S., Patterson, K., & Sakuma, N. (1992). The organization of the lexicon in Japanese: Single and compound kanji. *British Journal of Psychology, 83*, 517–531.

Murrell, G. A., & Morton, J. (1974). Word recognition and morphemic structure. *Journal of Experimental Psychology, 102*, 963–968.

Napps, S., & Fowler, C. A. (1987). The effect of orthography on the organization of the mental lexicon. *Journal of Psycholinguistic Research, 16*, 257–272.

Perfetti, C. A., & Tan, L. H. (1998). The time course of graphic, phonological, and semantic activation in Chinese character identification. *Journal of Experimental Psychology: Learning, Memory, and Cognition, 24*, 101–118.

Perfetti, C. A., & Zhang, S. (1991). Phonetic processes in reading Chinese words. *Journal of Experimental Psychology: Learning, Memory, and Cognition, 17*, 633–643.

Perfetti, C. A., & Zhang S. (1995). The universal word identification reflex. In D. Medin (Ed.), *The psychology of learning and motivation* (Vol. 33, pp. 159–180). New York: Academic Press.

Perfetti, C. A., Zhang, S., & Berent, I. (1992). Reading in English and Chinese: Evidence for a "universal" phonological principle. In R. Frost & L. Katz (Eds.), *Orthography, phonology, morphology, and meaning* (pp. 227–248). Amsterdam: North-Holland.

Segui, J., & Grainger, J. (1990). Priming word recognition with orthographic neighbors: Effects of relative prime-target frequency. *Journal of Experimental Psychology: Human Perception and Performance, 163*, 65–76.

Seidenberg, M. S. (1985). The time course of phonological code activation in two writing systems. *Cognition, 19*, 1–30.

Siok, W. W. T. (1996, September). *The decomposed representation of Chinese characters: Evidence from semantic radical processing*. Manuscript submitted to SUNY, Albany, in partial fulfillment of first-year course requirements.

Siok, W. W. T., & Feldman, L. B. (1996, February). *How are semantic radicals processed in Chinese?* Ninth Annual CUNY Sentence Processing Conference.

Siok, W. W. T., & Feldman, L. B. (1998). *Activation of orthographic and semantic information from sub-lexical units in Chinese character recognition*. Manuscript in preparation.

Stolz, J. A., & Feldman, L. B. (1995). The role of orthographic and semantic transparency of the base morpheme in morphological processing. In L. B. Feldman (Ed.), *Morphological aspects of language processing*. Hillsdale, NJ: Lawrence Erlbaum Associates.

Taft, M., & Zhu, X. (1997). Submorphemic processing in Chinese. *Journal of Experimental Psychology: Learning, Memory, and Cognition, 23*, 761–775.

Tan, L. H., & Peng, D.-L. (1990). Yujing dui hanyu danzici tezheng fenxi de yingxiang [The effects of semantic context on the feature analyses of single Chinese characters]. *Xin li xue dong tai [Journal of Psychology], 4*, 5–10.

Tan, L. H., Hoosain, R., & Peng, D.-L. (1995). Role of early presemantic phonologic code in Chinese character identification. *Journal of Experimental Psychology: Learning, Memory, and Cognition, 21*, 43–54.

Tan, L. H., Hoosain, R., & Siok, W. W. T. (1996). Activation of phonological codes before access to character meaning in written Chinese. *Journal of Experimental Psychology: Learning, Memory, and Cognition, 22*, 865–882.

Taylor, I., & Taylor, M. M. (1983). *The psychology of reading*. New York: Academic Press.

Xiandai Hanyu Cidian [Modern Chinese dictionary]. (1992). (2nd ed.). Beijing, China: Shangwu Press.

Xiandai Hanyu Pinlu Cidian [Modern Chinese frequency dictionary] (1986). Beijing, China: Beijing Language Institute Press.

Zhou, Y. G. (1978). Xiandai hanzihong shengpangde biaoyin gongneng wenti [To what degree are the "phonetics" of present-day Chinese characters still phonetic?]. *Zhongguo Yuwen, 146*, 172–177.

CHAPTER

3

Sublexical Processing in Reading Chinese

Xiaolin Zhou
William Marslen-Wilson
Birkbeck College
University of London

Theories of visual word recognition differ considerably in their assumptions about the nature of sublexical processing. In traditional dual-route models of reading (e.g., Coltheart, 1978; Coltheart, Curtis, Atkins, & Haller, 1993), sublexical orthographic units like *own* in real words (e.g., *down*, *shown*) or nonwords (e.g., *pown*, *thown*) are used mainly as inputs to grapheme–phoneme conversion rules whose end products are phonemes, although they are also involved in cascaded activation from letters to word representations. In most connectionist models of word reading (e.g., Plaut, McClelland, Seidenberg, & Patterson, 1996; Seidenberg & McClelland, 1989), the sublexical orthographic units are both the input to distributed networks computing phonological outputs and the input to orthography-to-semantics networks computing semantic outputs. However, these semantic outputs are primarily for the *whole* letter strings, not for the sublexical orthographic units, even when these units happen to be words themselves (e.g., *own* in *down* or *boy* in *boycott*). There is no obvious difference between the processing of these units and the processing of other units that do not happen to be words (e.g., *int* in *hint* or *pint* or *mur* in *murder*). In models of visual word recognition that stress the predominant role of phonology in access to meaning (e.g., Frost, in press; Lukatela & Turvey, 1994; van Orden, Pennington, & Stone, 1990), sublexical processing of orthographic units is strictly a phonological process, which activates the phonological representation of the whole word before the meaning of the word is accessed. Therefore, despite their considerable differences concerning whether the sublexical processing in reading alphabetic words is rule-based and how computation from orthography to phonology is conducted, most localist and connectionist theories of visual word recognition share the assumption that sublexical processing in reading alphabetic scripts is a process that derives the phonology and meaning of whole words from sublexical orthographic units. It is either primarily or strictly a phonological event, with no significant semantic activation associated with these sublexical units.

The main purpose of this chapter is to evaluate whether this phonological view of sublexical processing can be applied without modification to sublexical or subcharacter processing in reading Chinese complex or compound characters, which are composed of semantic and phonetic radicals. We concentrate on the processing of phonetic radicals, although we report a preliminary study on the processing of semantic radicals. We argue that sublexical processing in reading Chinese, restricted here to the processing of phonetic and semantic radicals of complex characters, is both phonological and semantic in nature. Semantic radicals are decomposed and used to access their own semantic representations in reading complex characters. The processing of phonetic radicals, which may provide phonological information about the pronunciation of the complex character that they are part of, not only contributes to the mapping from orthography to phonology in lexical access, but also generates activation and competition in the semantic system. In other words, the sublexical processing of phonetic radicals in reading complex Chinese characters is not only a phonological and orthographic event, but also a semantic event. There are no fundamental differences between sublexical processing of phonetic radicals and lexical processing of simple and complex characters in reading Chinese.

We first provide a brief introduction to the structural properties of Chinese characters. Previous research on sublexical phonological processing in reading Chinese is summarized and evaluated. We then move to four phonological and semantic priming experiments that demonstrate the existence of an automatic decomposition process in reading complex characters. Their phonetic radicals are used to activate not only phonological information in the lexicon, but also the semantic properties corresponding to the radicals, which are characters in their own right. This sublexical semantic activation decreases over the processing time of the complex characters. The fifth experiment, which addressed the role of semantic radicals in the recognition of complex characters, is reported in a separate section. In the final section, we discuss the implications of our data for lexical processing of Chinese.

STRUCTURAL PROPERTIES OF CHINESE CHARACTERS AND SUBLEXICAL PHONOLOGICAL PROCESSING

The Chinese writing system is often described as logographic, where the basic orthographic units, the characters, correspond directly to morphemic meanings and syllables. The characters can be broadly differentiated into two categories: simple and complex (or compound), both of which are composed of strokes and arranged in squares of similar size. Simple characters (e.g., 义, yi(4), *righteousness*)[1] are about 5% of the total characters in modern Chinese (Li, 1993), and they are holistic visual patterns that cannot be divided meaningfully into sublexical units. No phonological information can be represented at the sublexical level for simple characters. Most complex or compound characters, however, do encode phonological information at the subcharacter level. These characters (e.g., 议 yi(4), *discuss*) can be divided meaningfully into components, and most of them are composed of a semantic radical on the left and a phonetic radical on the right, although some arrange their radicals in different ways (e.g., Li, 1993; Yin & Rohsenow, 1994). Semantic radicals have the function of indicating the semantic category of morphemes corresponding to the complex characters, whereas phonetic radicals have the function of pointing to the pronunciations of whole characters. However, both functions are not complete, with exceptions and irregularities littered across the writing system (see also the discussion in Zhou, Shu, Bi, & Shi, chapter 7, this volume).

Semantic radicals historically evolved from simple characters, but in most cases modern semantic radicals do not form characters on their own and do not have definite pronunciations, although they may have commonly accepted "nicknames." Although semantic radicals usually do not correspond directly with phonology, they do correspond to semantic properties. For example, the left part of 海 (hai(3), *sea*), usually called "three drops," indicates that the morphemic meaning of the whole character has something to do with water. Over time, however, the meanings of many complex characters changed, creating opaque relations between semantic radicals and the meanings of complex characters. The degree of accuracy with which a radical conveys the meaning of a character can be divided roughly into four levels (Yin & Rohsenow, 1994): representing a meaning category into which the character fits (e.g., 铜 tong(2), *copper*); having a direct relationship with the character's meaning (e.g., 饭 fan(4), *cooked rice*); having an indirect relationship with the character's meaning (e.g., 城 cheng(2), *city*); and having

[1]Through the chapter, the pronunciations of Chinese words are given in Pinyin, the Chinese alphabetic system. Numbers in parentheses represent the tones of syllables.

no relationship at all with the meaning of the character (e.g., 汉, han(4), *Chinese*; 独 du(2), *sole*). Complex characters in the first two types can be taken as "semantically transparent," whereas characters in the last two types can be treated as "semantically opaque."

The phonological relations between phonetic radicals and the complex characters in which they are embedded can roughly be differentiated into five major types (Table 3.1). According to our analyses of the complex characters covered in a corpus study of modern Chinese (Institute of Language Teaching and Research, 1986), less than 30% of the complex characters composed of semantic and phonetic radicals have exactly the same pronunciations as their phonetic radicals (see also Fan, Gao, & Ao, 1984; Li & Kang, 1993, for similar statistical analyses with similar results), and about one third of complex characters are totally irregular and their pronunciations have no relations with their phonetic radicals. About another third of complex characters have either the same segmental templates (but different lexical tones, which are used to differentiate lexical items in Chinese) or the same segmental rhyming parts or the same initial consonants as their phonetic radicals. Because of this, the method of reading complex characters according to their phonetic radicals, which often leads to wrong pronunciations and regularization errors, has become a standard source of jokes in Chinese. Note also that the same phonetic radicals could be involved in all or part of the five categories, and, crucially, most phonetic radicals themselves are usually independent characters, having their own meanings, which are not related to the meanings of the whole characters.

TABLE 3.1
The Phonological Relations Between Phonetic Radicals and Complex Characters

Type	*Phonetic Radical*	*Complex Character*
Regular	羊 yang(2) (*sheep*)	洋 yang(2) (*ocean*)
Semiregular	青 qing(1) (*blue*)	情 qing(2) (*affection*)
Rhyming	亡 wang(2) (*die*)	忙 mang(2) (*busy*)
Alliteration	某 mou(3) (*certain*)	煤 mei(2) (*coal*)
Irregular	白 bai(2) (*white*)	怕 pa(4) (*fear*)

There have been a number of studies on sublexical or subcharacter processing in reading Chinese complex characters. The orthographic units studied include stroke (e.g., Chen & Yung, 1989), component (e.g., Taft & Zhu, 1997), semantic radical (e.g., Feldman & Siok, chapter 2, this volume; Flores d'Arcais, Saito, & Kawakami, 1995), and phonetic radical (e.g., Fang, Horng, & Tzeng, 1986; Flores d'Arcais et al., 1995; Hue, 1992; Peng, Yang, & Chen, 1994; Seidenberg, 1985; Zhou, 1994). Almost without exception, studies on phonetic radicals concentrate on the orthographic and phonological aspects of sublexical processing. The question of whether semantic properties of phonetic radicals are automatically activated in reading complex characters has not been addressed (but see Zhou, 1994). This is presumably due to the influence of research on sublexical processing of orthographic units in monomorphemic alphabetic words (e.g., *boy* in *boycott*), which do not produce detectable effects of their own in the semantic system (e.g., Sandra, 1990; Zhou, unpublished data, see General Discussion at the end of this chapter).

The evidence about the phonological processing of phonetic radicals comes mainly from studies on regularity and consistency effects in reading complex characters. Seidenberg (1985) found that regular complex characters, or characters having the same pronunciations as their phonetic radicals, are named faster than frequency-matched simple characters. However, this effect was observed only for low-frequency characters, not for high-frequency characters. Fang et al. (1986) investigated regularity and consistency effects in naming characters, where consistency was defined according to whether *all* complex characters with a particular phonetic radical are pronounced in the same way as the radical. The authors observed a consistency effect but not a regularity effect. Regular–consistent characters (e.g., 材 cai(2), *materials*) were named faster than regular–inconsistent characters (e.g., 油 you(2), *oil*), although the latter were not named faster than irregular-inconsistent characters (e.g., 抽 chou(1), *lash*). Subsequent studies using similar designs, however, found both regularity and consistency effects for low frequency characters (Hue, 1992; Peng et al., 1994).

Yin and Butterworth (1992) studied brain-damaged patients with reading disorders and concluded that reading aloud Chinese complex characters, like reading alphabetic scripts, can be accomplished using two "distinct" routines: one associating a whole written word with its complete pronunciation and one utilizing parts of the written word. They observed that the "deep" dyslexics, that is, patients who made a lot of semantic errors in reading, did not regularize the pronunciations of (irregular) characters and could not pronounce pseudocharacters made of real semantic and phonetic radicals. The "surface" dyslexics, patients who made fewer semantic errors, on the other hand, made a lot of regularization errors and could "read" about 50% of pseudocharacters—presumably according to the pronunciations of phonetic radicals.

The typical account of these regularity and consistency effects in reading complex characters was in terms of Glushko's (1979) *activation-synthesis* model or Seidenberg's (1985) time course model of orthographic and phonological activation. According to these models, lexical access begins with the extraction of salient orthographic units from the visual input, which are then used to activate in parallel their corresponding orthographic and phonological representations in the lexicon. The phonological activation is synthesized according to interactive principles to manifest regularity or consistency effects in naming. The interaction between these effects and frequency is explained by the relative time course of orthographic and phonological activation, where the orthographic processing of high-frequency words is sufficiently fast to allow word recognition before phonological activation, but the orthographic processing of low-frequency words is not efficient and phonological activation has time to accrue to influence word recognition. Put slightly differently, the phonological output for high-frequency words is based mainly on their own phonological representations, which are activated efficiently through direct visual mapping and/or via semantics, whereas the phonological output for low-frequency words is influenced by phonological representations of other words activated in parallel from orthography. The connectionist models of reading account for regularity and consistency effects in reading Chinese and their interaction with frequency using essentially the same principles just outlined (see Chen & Peng, 1994; Seidenberg & McClelland, 1989).

The question for the present research is whether the sublexical processing of phonetic radicals is just a phonological event, involving just the computation from orthography to phonology, or whether it is also a semantic event, producing semantic activation of its own, and if so, what the time course is of this sublexical semantic activation. Although Zhou (1994) tried to address these questions by systematically manipulating the relative frequencies and the phonological relations between complex characters and their phonetic radicals in a primed naming study, the data were ambiguous, even though they led him to argue that sublexical processing of phonetic radicals in reading Chinese cannot be just a phonological event in the same way as the sublexical processing in reading alphabetic words. In this research, we use a primed naming technique to probe unequivocally into semantic activation of phonetic radicals in reading complex characters. Characters that are semantically related to the phonetic radicals of complex characters, but not to the complex characters themselves, are used as primes or targets. Any effects of this semantic priming on the processing of targets can be taken as evidence that, in reading complex characters, the phonetic radicals are decomposed and used to access their own semantic representations.

PHONOLOGICAL AND SEMANTIC PROCESSING OF PHONETIC RADICALS

We report four experiments that address the phonological and semantic processing of phonetic radicals in reading complex or compound characters. We used only complex character with left–right structure, because they are easier to decompose in perceptual analyses than characters with other structures (e.g., Yu, Feng, Cao, & Li, 1990) and hence the effects of their sublexical processing are easier to detect. Experiment 1 replicated previous findings about sublexical phonological processing, and the remaining experiments investigated the sublexical semantic processing of phonetic radicals in reading regular (Experiment 2), rhyming (Experiment 2), and irregular (Experiments 3 and 4) complex characters.

Experiment 1: Phonological Processing of Phonetic Radicals

Facilitatory priming effects between homophones have been found in previous research using the primed naming technique (e.g., Perfetti & Zhang, 1991; Zhou et al., chapter 7, this volume). The repeated activation of the phonological representations shared by primes and targets facilitates the processing of homophone targets. If reading Chinese involves decomposing phonetic radicals from complex characters and using them to access their own phonological representations in the lexicon, we should observe a phonological priming effect between phonetic radicals embedded in complex characters and characters homophonic to the radicals. Moreover, if the automatic decomposition of complex characters is modulated by the frequency of complex characters, as suggested in Seidenberg (1985) and other studies, we should find the homophone priming effect for low-frequency complex characters, but not for high-frequency ones.

Since regular and semiregular complex characters share phonological properties with their phonetic radicals, any priming effects between phonetic radicals and their homophone targets could be due, directly or indirectly, to the processing of whole characters. Thus only irregular complex characters were used in Experiment 1 to isolate any priming effects to the processing of phonetic radicals (Table 3.2). Target characters were preceded by complex characters whose phonetic radicals were homophonic to the targets. They were also preceded by the radicals on their own, which served as homophone primes. Because the independent measure of performance was based on subjects' responses to the targets, all the critical comparisons were within items, avoiding the difficulties of matching stimuli on different parameters that could influence naming performance.

TABLE 3.2
Experiment 1: Design and Sample Stimuli

Frequency	*Homophone*	*Complex*	*Control*	*Target*
Low	卒 zu(2) (*soldier*)	粹 cui(4) (*pure*)	恒 heng(2) (*permanent*)	族 zu(2) (*clan*)
High	欠 qian(4) (*owe*)	次 ci(4) (*sequence*)	比 bi(3) (*compare*)	倩 qian(4) (*pretty*)

Note. The frequency refers to that of the complex prime. The homophone prime is embedded in the complex prime as a phonetic radical.

The complex character primes were differentiated into types according to frequency. There were 51 low-frequency (with a mean of 27 per million) and 33 high-frequency (with a mean of 1,670 per million) complex characters. Control primes were chosen to match with complex primes on frequency, visual complexity (in terms of the number of strokes), and structure (all left–right composition). There were no orthographic similarities or semantic relations between targets and primes. There were also 100 pairs of orthographically and semantically unrelated fillers.

In each trial of testing, an eye fixation signal ("+") was first presented at the center of the screen for 300 ms, followed by a 300-ms blank interval. A prime was then presented for 100 ms and overwritten immediately by the corresponding target, which was presented for 400 ms. In other words, the SOA (stimulus onset asynchrony) between primes and targets was set at 100 ms. There was a 3-second interval between the disappearance of the last target and the appearance of the next eye fixation point.

Priming effects, as assessed against the baselines, are summarized in Fig. 3.1. The analyses of reaction times showed a significant main effect of prime type ($p < .01$). Post hoc Newman–Keuls tests indicated that the mean reaction time in homophone priming conditions was significantly faster than the mean reaction times in the complex priming condition and the control condition ($p < .01$ or $.05$). Separate analyses of reaction times for low- and high-frequency complex primes found different patterns for the complex priming conditions for the two types of complex characters. Although the mean naming latency for low-frequency complex primes was significantly faster than the control baseline ($p < .05$), the mean reaction time for high-frequency complex primes did not differ from the control primes ($p > .1$).

Four observations can be made from these analyses. First, there was a significant homophone priming effect between radical primes and targets, replicating Perfetti and Zhang (1991) and Zhou et al. (chapter 7, this volume). This indicates that the presentation of homophone primes activates

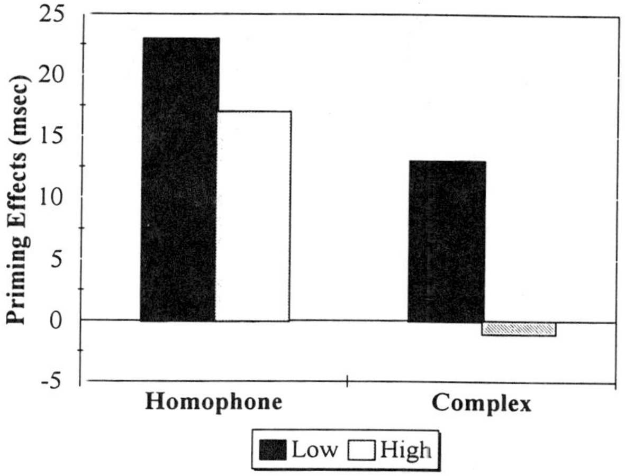

FIG. 3.1. Experiment 1: Phonological priming effects.

phonological representations they share with the targets. Although there is some room for doubt about the automaticity of phonological activation in reading Chinese (see Zhou et al., this volume, for discussion), more evidence suggests it is likely to be automatic. Second, the presentation of low-frequency complex characters facilitated the phonological processing of targets. Because complex characters themselves were neither phonologically, nor orthographically, nor semantically related to the targets, this facilitatory effect could only stem from the fact that phonetic radicals embedded in the complex characters were homophonic to the targets. It indicates that in processing complex characters, phonetic radicals are decomposed and used to access their own phonological representations. However, the ability to decompose phonetic radicals and gain access to their phonology is modulated by the frequency of complex characters, as demonstrated by the third finding that there were different patterns of priming effects for high- and low-frequency complex primes. Either reading high-frequency complex characters involves little decomposition, or the phonological activation of phonetic radicals embedded in complex characters was transient and was sufficiently suppressed by the phonological activation of complex characters themselves by the time the targets were presented. High-frequency complex characters are activated faster than low-frequency complex characters and hence have more chance to suppress competing phonological activation. Fourth, and related to this, the fact that phonological priming by phonetic radicals in complex characters was significantly smaller than priming by phonetic radicals presented alone suggested that in reading complex characters, there was indeed competition between phonological representations corresponding to complex characters and phonological representations corresponding to phonetic radicals. The phonological activation of radicals embedded in complex characters did not reach the same level as when these radicals were standing alone.

The findings of phonological activation of phonetic radicals embedded in complex characters and its interaction with the frequency of complex characters were in accordance with regularity and consistency effects in naming (e.g., Hue, 1992; Seidenberg, 1985). They were also consistent with the view that sublexical processing of phonetic radicals in reading complex characters is (at least) a phonological event. We believe that access to phonological representations of phonetic radicals is conducted in the same way as access to phonological representations of the whole complex characters, that is, mapping from orthographic input to phonological representations. This view leaves open whether there is also a mapping from orthography to semantics for phonetic radicals in reading complex characters. There are suggestions that, because the radical-sound correspondences were overlearnt, they could be used in the same way as the grapheme–phoneme correspondences in an alphabetic language (Flores d'Arcais et al., 1995; but see Zhou, 1994). If so, the sublexical processing of phonetic radicals in reading complex characters would be, as for sublexical processing in alphabetic languages, primarily a phonological event and should not be able to produce detectable effects of their own in the semantic system. Experiments 2, 3, and 4 provide evidence against this argument.

Experiment 2: Semantic Processing of Phonetic Radicals: Regular and Rhyming Characters

This and the following two experiments investigated whether sublexical processing in reading complex characters automatically activates the semantic properties of phonetic radicals, even though this semantic activation does not contribute to the processing of complex characters themselves. To index the activation of semantic properties of phonetic radicals in complex characters, we used the semantic associates of phonetic radicals as targets and complex characters as primes. If the phonetic radicals are automatically decomposed and used to access their own semantic representations in reading complex characters, the activation of these semantic representations should facilitate the processing of their semantic associates. To make sure that the associate targets *were* semantically related to the phonetic radicals and capable of detecting the semantic activation of phonetic radicals, the experiments also had a direct semantic priming condition where the same targets were preceded by the phonetic radicals presented alone, as illustrated in Table 3.3.

This experiment used regular and rhyming complex characters as primes and characters that were semantically related to their phonetic radicals, but

TABLE 3.3
Experiment 2: Design and Sample Stimuli

Type	*Semantic*	*Complex*	*Control*	*Target*
Regular	风 feng(1) (*wind*)	枫 feng(1) (*maple*)	柿 shi(4) (*persimmon*)	雨 yu(3) (*rain*)
Rhyming	少 shao(3) (*few*)	抄 chao(1) (*transcribe*)	扶 fu(2) (*prop up*)	多 duo(1) (*many*)

Note. The type refers to that of the complex prime. The semantic prime is contained in the complex prime as a phonetic radical.

not to themselves, as targets. Only low-frequency complex characters (with means of 18 per million for regular characters and 34 per million for rhyming characters) were used as primes. There were 42 regular characters and 30 rhyming characters. Their phonetic radicals (with mean frequencies of 840 and 858 per million, respectively) were also used as semantic primes for the target characters. Control primes were matched with complex primes in terms of frequency, visual complexity, and structure.

Three SOAs between primes and targets were used to track the time course of semantic activation: 57 ms, 100 ms, and 200 ms. We expected to find a facilitatory priming effect between complex primes and semantic targets. This effect should decrease as more time is allowed to process the complex primes. Figures 3.2 and 3.3 report respectively the priming effects for regular and rhyming complex characters, as assessed against the control priming conditions.

Statistical analyses of naming latencies were conducted separately for priming effects for the two types of complex characters. There was a significant main effect ($p < .01$) of prime type for regular characters (see Fig. 3.2). Post hoc tests showed that both the semantic priming effect (22 ms)

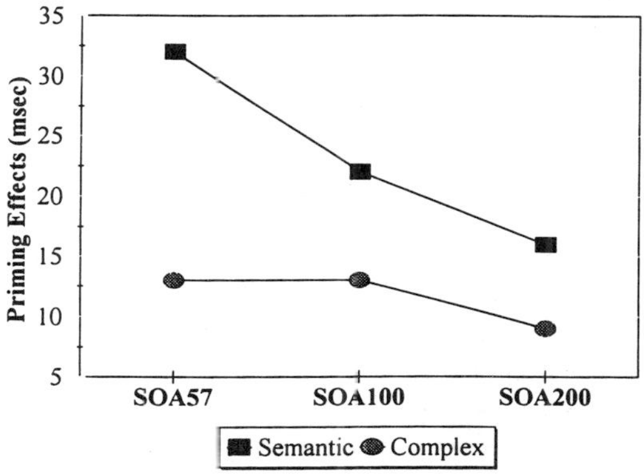

FIG. 3.2. Experiment 2: Semantic priming effects for regular characters.

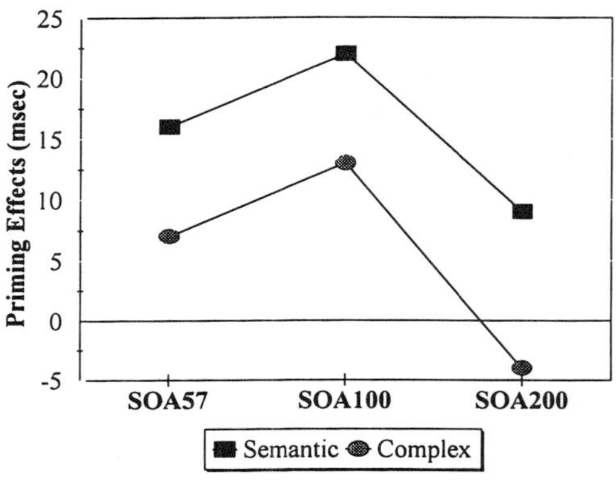

FIG. 3.3. Experiment 2: Semantic priming effects for rhyming characters.

and the complex priming effect (10 ms) against the control baseline was significant ($p < .01$). Moreover, the difference between semantic priming and complex priming (12 ms) was also significant ($p < .05$ or .01). The interaction between prime type and the SOA was only marginally significant ($p < .05$ in subject test and $p > .1$ in the item test). This is likely to be due to an overestimation of semantic priming effects at short SOAs. More importantly, regular complex primes could facilitate the processing of targets that were semantically related to the phonetic radicals of the complex primes, but not to the complex primes themselves. This facilitatory priming effect appeared at all three SOA conditions, although it dropped off slightly when the SOA was long.

The analyses of priming effects for rhyming complex characters produced different results (see Fig. 3.3). The main effect of prime type was significant ($p < .05$ or .01). Newman–Keuls tests showed that the 15 ms main effect of semantic priming against the control baseline was significant ($p <$ .05 or .01). However, the overall 5 ms main effect of complex priming was not significant ($p > .1$), nor was the interaction between prime type and SOA ($p > .1$). Nevertheless, there was a clear tendency for complex primes to speed up the processing of targets at the SOA of 57 ms or 100 ms, but this facilitation disappeared at the long SOA.

These facilitatory effects for complex primes indicate that, in reading complex characters, the phonetic radicals are decomposed and used to access their own semantic representations, whose activation spreads to other related words, including the semantic targets used in this experiment. This sublexical semantic priming effect was weaker than the direct semantic priming effect produced by phonetic radicals presented on their own. This indicates that the semantic activation of phonetic radicals in complex characters has to compete with the semantic activation of whole characters. Apparently, this competition becomes stronger and sublexical semantic activation becomes weaker over time.

However, there exists an alternative account for the priming effect between complex characters and semantic targets that does not need to assume sublexical semantic processing. This account is based on the assumption that access to semantics in reading Chinese, like reading alphabetic languages (e.g., Frost, in press; van Orden et al., 1990), has to go through phonology. The presence of a regular complex prime activates its phonological representation, which is shared with all other homophonic characters, including its phonetic radical on its own. This phonological activation spreads to all the corresponding semantic representations of homophonic morphemes. The semantic activation of the phonetic radical spreads further to its semantic associates, facilitating the processing of the target used in the experiment. Although Zhou et al. (chapter 7, this volume) provided evidence that access to semantics in reading Chinese is predominantly orthographically constrained and does not rely solely on phonological mediation, access to semantics through phonology is more plausible if phonological activation is very strong, for example, when both lexical and sublexical phonological processing produce consistent phonological activation (Zhou & Marslen-Wilson, 1998a). Therefore, it is possible that the relatively strong priming effect between regular complex primes and target characters in the present experiment was due to this phonological mediation, rather than to the decomposition of phonetic radicals and access to their semantics from orthography.

In order to rule out this alternative account, we used irregular complex characters as primes in Experiment 3. Since phonetic radicals of irregular character do not share the phonological representations with the complex characters, the activation of the phonological representations of the whole characters should not affect directly the processing of phonetic radicals. If decomposed phonetic radicals are used to access only phonology, we should not observe a facilitatory priming effect between irregular complex characters and target characters that are semantically, but not phonologically or orthographically, related to their phonetic radicals.

Experiment 3: Semantic Processing of Phonetic Radicals: Irregular Characters

The experimental design and sample stimuli are presented in Table 3.4. Forty-eight irregular complex characters (e.g., ■ cai(1), *guess*), whose phonetic radicals (e.g., 青 qing(1), *blue*) have clear semantic associates (e.g., 紫 zi(3), *purple*), were selected. Because of restrictions on the selection of stimuli, only low- and medium-frequency complex characters (with a mean of 216 per million) were used, and no attempt was made to investigate the potential effect of frequency of complex characters on the semantic activation of phonetic radicals. Control primes (e.g., 波 bo(1), *wave*) were chosen to match

TABLE 3.4
Experiment 3: Design and Sample Stimuli

Semantic	*Complex*	*Control*	*Target*
青 qing(1) (*blue*)	猜 cai(1) (*guess*)	波 bo(1) (*wave*)	紫 zi(3) (*purple*)

Note. The semantic prime is contained in the complex prime as a phonetic radical.

with complex primes on frequency, visual complexity, and structure. Thus, the same target (e.g., 紫) was preceded either by a semantic prime (e.g., 青), or a complex character prime (e.g., 猜) that contained the semantic prime as its phonetic radical, or a control prime (e.g., 波).

Three SOAs (57 ms, 100 ms, and 200 ms) were again used to track the time course of sublexical semantic activation. Priming effects, as assessed against the baselines in different SOA conditions, are depicted in Fig. 3.4. It is clear from this that the semantic priming effect between radical primes and target characters remains strong across different SOAs between primes and targets. In contrast, the priming effect between complex primes and targets was relatively strong initially, but disappeared at the long SOA. This echoed the pattern for rhyming complex characters in Experiment 2 (see Fig. 3.3), but the effects here were more solid. There was a highly significant main effect of prime type across three SOAs ($p < .001$). The priming effects between the irregular complex primes and targets were significant at SOAs of 57 ms ($p < .01$) and 100 ms ($.1 < p < .05$), but not at the SOA of 200 ms ($p > .1$).

The strong effects of semantic primes replicated our previous findings and suggested that the semantic activation can efficiently flow to phonology, resulting in the facilitatory effects in naming. More importantly, the significant priming effects between complex characters and the semantic associ-

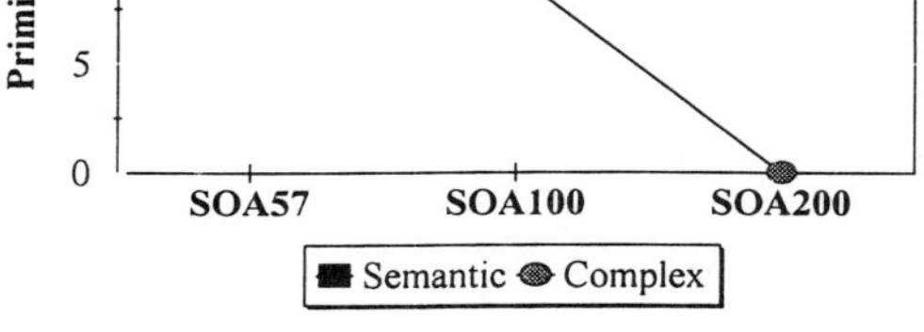

FIG. 3.4. Experiment 3: Semantic priming effects for irregular characters.

ates of their phonetic radicals at shorter SOA demonstrated that the phonetic radicals of complex characters are not only used to access their corresponding form representations in the lexicon, but also their semantic properties, which are not related to the complex characters. The smaller priming effect between complex primes and targets, compared with direct semantic priming, suggests that there is competition in the semantic system between representations corresponding to the whole complex characters and representations corresponding to the phonetic radicals contained within the characters. Within 200 ms, the semantic activation of phonetic radicals disappeared, leaving only the semantic activation of the whole complex characters in the system.

Experiment 4: Semantic Processing of Phonetic Radicals: Irregular Characters Revisited

Experiment 3 demonstrated that the priming effect between complex characters and semantic associates of their phonetic radicals cannot be attributed to the spread of activation from the phonological processing of the whole complex characters because the phonetic radicals did not have the same phonological forms as the whole characters. The main purpose of this experiment was to obtain converging evidence for the argument that the sublexical processing of phonetic radicals in complex characters is not only a phonological event, but also a semantic event.

The present experiment used irregular complex characters as targets, instead of as primes as in Experiment 3. If reading complex characters involves decomposing phonetic radicals and using them to access their own semantic representations, the presence of semantic primes, related to the phonetic radicals but not to the complex characters, should facilitate this decompositional process and increase the semantic activation level of phonetic radicals. That will in turn increase the competition between semantic representations corresponding to complex characters and semantic representations corresponding to their phonetic radicals, and increase the naming latencies of complex targets.

The design and sample stimuli are presented in Table 3.5. Forty-five complex characters (e.g., 猜 cai(1), *guess*) that were used as primes in Experiment 3 were taken as targets in this experiment, preceded by semantic associates (e.g., 紫 zi(3), *purple*) that were related to their phonetic radicals. The same complex characters were also primed by their phonetic radicals (e.g., 青 qing(1), *blue*) in another condition. The three SOAs between primes and targets used in Experiments 2 and 3 were also used in this experiment. Priming effects for radical and complex primes are reported in Fig. 3.5.

The naming latencies showed a significant main effect of prime type ($p < .001$) and a significant interaction between prime type and SOA ($p < .05$).

TABLE 3.5
Experiment 4: Design and Sample Stimuli

Radical	*Semantic*	*Control*	*Target*
青 qing(1) *(blue)*	紫 zi(3) *(purple)*	裂 lie(4) *(split)*	猜 cai(1) *(guess)*

Note. The radical prime is contained in the target as a phonetic radical.

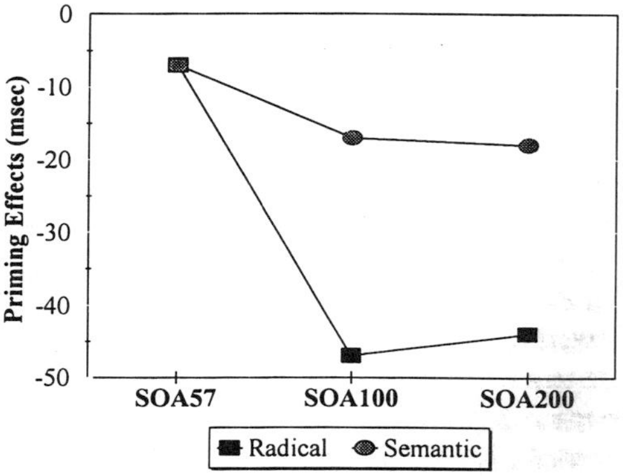

FIG. 3.5. Experiment 4: Irregular characters as primed targets.

There were no significant priming effects at the SOA of 57 ms, but highly significant effects at SOAs of 100 and 200 ms. Reaction times in the radical priming conditions were significantly slower than the mean reaction times in the other two conditions ($p < .01$). Crucially, the effects of semantic priming were marginally significant ($.05 < p < .1$) at the SOA of 100 ms and significant ($p < .05$) at the SOA of 200 ms.

The significant inhibitory effect of phonetic radicals on the processing of irregular complex characters at the longer SOA replicated Zhou (1994), who systematically manipulated the frequency of radicals and complex characters and the phonological relations between them. However, it is difficult from the present data to pinpoint the locus of this inhibitory effect, even though the effect was likely due to the increased phonetic decomposition in reading complex characters and the competition in computing phonology from orthography for both phonetic radicals and complex characters (Zhou, 1994). The presence of radicals serves to preactivate the phonological and/or semantic representations corresponding to radicals and to highlight the decompositional process in reading complex characters.

The more interesting finding in this experiment was that the presence of semantic primes related to phonetic radicals of complex characters could delay the activation of phonological representations of complex characters.

This finding is consistent with the data of Experiments 2 and 3 and with the view that there is an automatic decomposition process in reading complex characters, and both phonological and semantic properties of phonetic radicals are activated. The presence of semantic primes increases the semantic activation of radicals. The latter activation increases its competition with the semantic representations of complex characters and/or sends its activation to the corresponding phonological representations and intensifies their competition with the phonological representations of complex characters, delaying the vocalization of the complex targets.

PROCESSING SEMANTIC RADICALS IN READING CHINESE

If we accept that phonetic radicals are automatically decomposed from the whole complex characters and used to access their own phonological and semantic representations in the lexicon in initial lexical access, a natural extension of this argument is that the semantic radicals of the complex characters are also decomposed and used to access their representations. This section investigates the role of semantic radicals in reading complex characters.

In investigating the sublexical processing of semantic radicals in reading Chinese, we need to consider not only the functional properties of semantic radicals in conveying the semantic features of complex characters (see the second section of this chapter), but also how semantic radicals might be represented in the lexicon. If semantic radicals can form characters on their own, in the same way as phonetic radicals, they must be represented independently from the complex characters, having direct connections with their phonological and semantic properties. The crucial questions here are, first, whether semantic radicals are decomposed and used to access these representations, and second, how semantic radicals that do not form independent characters are represented in the lexicon. We assume that these semantic radicals have direct connections with the representations of their semantic properties, even though they do not have fixed phonological forms and may not have direct relations with phonological representations.

If semantic radicals are decomposed and used to access their own semantic representations in reading complex characters, the processing of semantic radicals may facilitate or inhibit the semantic activation of the complex characters containing them. This depends on whether the complex characters are semantically transparent—that is, whether their meanings are closely related to the meanings of their semantic radicals. If the complex characters are transparent, the semantic activation of semantic radicals facilitates the semantic processing of the whole characters. If the complex characters are semantically opaque, however, the semantic activation of semantic radicals can only compete with the semantic activation of the whole characters, as they have different sets of semantic features.

One way to examine the decomposition of semantic radicals and the activation of their semantic (and possibly phonological) representations in reading complex characters is to use experimental designs in analogy to Experiments 2, 3, and 4. A complex character (e.g., 狼 lang(2), *wolf*) is presented as a prime (or target), and a character semantically related to the semantic radical is presented as the target (or prime). However, although we are currently running such experiments, this strategy may only work unambiguously for semantically opaque complex characters and semantic radicals that can stand alone as characters. Consequently, in the following experiment, we used a direct priming technique in which complex characters sharing semantic radicals (e.g., 狼 lang(2), *wolf*; 狮 shi(1), *lion*) were used as primes and targets (see also Feldman & Siok, chapter 2, this volume). If the semantic radicals are decomposed in reading these primes and used to access their own semantic representations, this process should facilitate the same decomposition and access process in reading the targets.

However, we need to consider at least the following three points in assessing the direct priming effects between characters sharing common semantic radicals. First, the prime–target pairs (e.g., 狮 shi(1), *lion*–狼 lang(2), *wolf*) are, by definition, orthographically related. Orthographically similar characters tend to inhibit each other in certain experimental tasks (e.g., Zhou, 1994). This orthographically based inhibitory effect could obscure the facilitatory priming effects based on sharing common semantic radicals and on the relatedness between the meanings of the whole characters. Second, when complex characters are semantically opaque in relation to their semantic radicals, the semantic activation of the semantic radicals of opaque primes could be inhibited by the semantic activation of the whole character by the time the target characters containing these radicals are presented. Third, although a character decision task may be more direct than a naming task in tapping into the semantic activation of semantic radicals and complex characters (e.g., Feldman & Siok, chapter 2, this volume), the naming task is also efficient in revealing the differential level of activation in the semantic system. Consequently, we used the primed naming task in the following experiment, consistent with the previous four experiments.

Experiment 5: Processing Semantic Radicals

We employed a design where the same semantically transparent complex characters (e.g., 狼 lang(2), *wolf*) were preceded by four types of primes (see Table 3.6). The first type of primes (e.g., 狮 shi(1), *lion*), called *semantic*

TABLE 3.6
Experiment 5: Design and Sample Stimuli

Semantic + Radical	*Radical*	*Semantic*	*Control*	*Target*
狮 shi(1) (*lion*)	独 du(2) (*sole*)	虎 hu(3) (*tiger*)	洗 xi(3) (*wash*)	狼 lang(2) (*wolf*)

radical (SR) primes, shared semantic radicals with the targets. The semantic radicals in primes and targets refer to clear semantic categories, and the meanings of primes and targets are clearly related to these categories. The primes as wholes were also related to the targets at the semantic level. The second type of primes (e.g., 独 du(2), *sole, independent*) had the same semantic radicals as the targets. However, these primes, referred to as *radical* primes, were semantically opaque, with no obvious relations between the meanings of the semantic radicals and the meanings of the complex characters. The complex characters as wholes were not semantically related to the targets. The third type of primes were *semantic* (e.g., 虎 hu(3), *tiger*). These were semantically related to, but shared no semantic radicals or orthographic similarities with, the targets. Finally, there were *control* primes that were neither semantically nor orthographically related to the targets. The four types of primes were not phonologically related to targets.

Fifty-two targets and their primes were selected, with 25 different critical semantic radicals used. Each radical was used no more than four times. Overall, these radicals referred to commonly accepted semantic categories. According to a pretest, the semantic relatedness between the first three types of primes (SR, radical, and semantic) and targets were 6.8, 1.7, and 6.8 respectively in a 9-point scale (9 = most related, 1 = not related at all). There were also 100 unrelated prime–target pairs that were used as fillers. None of these characters used the 25 critical semantic radicals.

The SOA between primes and targets were set at 100 msec. There were few naming errors. The mean naming latencies are reported in Fig. 3.6. Analyses of naming latencies revealed a significant main effect of prime type ($p < .001$). Post hoc tests showed that although the naming of targets was significantly facilitated (17 ms) by the presence of semantic primes ($p < .05$), it was inhibited (–43 ms) by the presence of radical primes ($p < .01$). Moreover, the SR primes did not produce a significant priming effect against controls ($p > .1$), even though the former were semantically related to the targets. The difference between SR primes and radical primes (36 ms) was highly significant ($p < .01$).

The facilitatory semantic priming effect between, say 虎 (hu(3), *tiger*) and 狼 (lang(2), *wolf*), in the naming task demonstrated again that the activation at the semantic system can flow efficiently to phonological representations

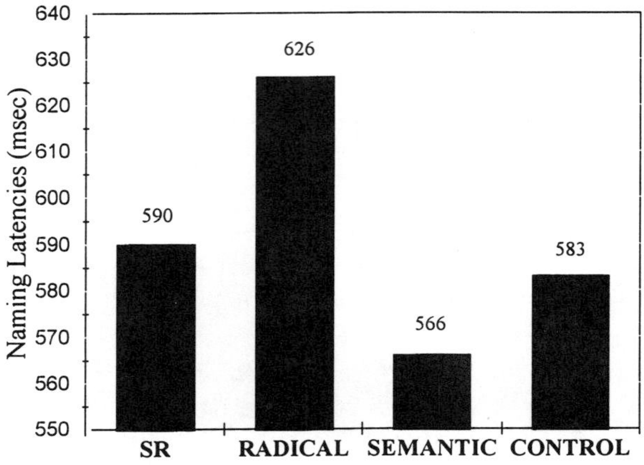

FIG. 3.6. Experiment 5: Mean naming latencies. SR = semantic plus radical.

in reading Chinese. The inhibitory effect between semantically opaque primes (e.g., 独 (du(2), *sole, independent*) and transparent targets (e.g., 狼 (lang(2), *wolf*) was presumably due to an orthographically based competition between primes and targets in computing phonology from orthography (e.g., Zhou, 1994; see also Experiment 4). The null effect between semantically related transparent characters sharing common semantic radicals (i.e., SR primes and targets) reflected the predominant orthography constraints on initial lexical access. Because these prime–target pairs (e.g., 狮 shi(1), *lion* and 狼 lang(2), *wolf*) were orthographically similar, they tended to inhibit each other. Because they were also semantically related, they tended to facilitate each other. The two conflicting effects may have canceled each other out, resulting in the overall null effect in this condition.

Does the pattern of priming effects here suggest decomposition of semantic radicals and access to their own representations in reading complex characters? The answer is "yes" when the priming effects for SR primes and for semantic primes were compared against appropriate baselines. Since the orthographic similarity between SR primes and targets produced an inhibitory priming effect in naming, the appropriate way to assess the semantic priming effects between them is to compare the mean naming latency for SR primes with the mean latency for radical primes. This produced a 36 msec facilitatory priming effect, which was significantly larger ($p < .01$) than the 17 ms effect for the semantic primes (against the control primes). Inasmuch as the SR primes and semantic primes were equally related to targets in whole word meanings, as indicated in the semantic relatedness judgment pretest, this extra priming effect for SR primes can only be attributed to the fact that SR primes, but not semantic primes, shared semantic radicals with the target characters. In reading SR primes, the semantic radicals were decomposed and used to access their own semantic representations, which shared many features with both the semantic representations of the SR primes and the targets. This process facilitated the decomposition process in reading targets, and the repeated access to the same semantic representations of the semantic radicals increased the semantic activation and hence the phonological activation of targets, resulting in the extra facilitatory priming effects for SR primes.

This suggestion of semantic radical decomposition and access to their own semantic representations is further supported by the data from Feldman and Siok (chapter 2, this volume). Using an experimental design very similar to this experiment, but with a character decision task, Feldmand and Siok found a facilitatory priming effect for semantic primes and an inhibitory effect for radical primes, as did the present experiment. Crucially, they found a larger facilitatory priming effect for SR primes, compared with the priming effect for semantic primes (but see Zhou & Marslen-Wilson, 1998b, who did not observe this extra effect in character decision with the same stimuli of the present experiment). This extra facilitatory effect in character decision can be interpreted, also, as due to decomposition and access to semantic representations of semantic radicals in reading complex characters.

GENERAL DISCUSSION

In this chapter, we obtained strong orthographically based inhibitory effects between characters sharing semantic or phonetic radicals (Experiments 4 and 5); we observed semantically mediated phonological activation (Experiments 2, 3, 4, and 5); we extended homophone priming to the sublexical level (Experiment 1); we found evidence of automatic decomposition of phonetic radicals in reading complex characters and access to both their phonological and semantic properties (Experiments 1, 2, 3, and 4); and we also found evidence for the automatic decomposition of semantic radicals and access to their own semantic representations. These findings are consistent with the claims of our lexical processing model in reading Chinese (e.g., Zhou et al., chapter 7, this volume), which stresses the predominant role of orthography in initial lexical access, the effective mapping from orthography to semantics, the automaticity of phonological activation, and the efficient flow of activation from semantics to phonology. They demonstrated that there are no fundamental differences between lexical processing and sublexical processing of phonetic and semantic radicals in reading Chinese. The sublexical processing of phonetic radicals is, like lexical level processing, not only a phonological event, but also a semantic event. The phonetic radicals are decomposed and mapped onto their semantic and phonological representations, in parallel to the mapping based on the whole complex characters. Semantic radicals in complex characters are also automatically decomposed and used to access their own semantic representations, although it is not clear yet whether they are used to access phonology as well (in cases when they are real characters on their own and have definite phonology).

Sublexical Phonological and Semantic Activation

The finding of automatic activation of phonological representations of phonetic radicals and its interaction with the frequencies of complex characters is consistent with what was found in single-character naming experiments (Fang et al., 1986; Hue, 1992; Peng et al., 1994; Seidenberg, 1985). This finding, together with the regularity and consistency effects in naming, can be easily explained in activation-synthesis or time course models (e.g., Glushko, 1979; Seidenberg, 1985) based on interactive-activation principles and in distributed representation models. More importantly, this research demonstrated that in initial lexical access, orthographic units of various sizes are mapped in parallel not only onto the corresponding phonological representations, but also onto their semantic representations—if they happen to be phonetic or semantic radicals. This sublexical semantic activation must be automatic because the semantic activation of phonetic radicals, unlike the semantic activation of semantic radicals (in transparent complex characters), does not aid, but competes with, the processing of the whole characters.

This evident sublexical semantic activation ın reading Chinese constrasts sharply with the absence of such activation in reading alphabetic languages. We ran a primed naming experiment to examine the potential sublexical semantic activation in English (Zhou, unpublished data). This experiment used a design similar to Experiment 2, with words (e.g., *boycott*) in which other words (i.e., *boy*) were embedded as primes and words that were semantically related to the embedded words as targets (e.g., *girl*). No priming effects were found for such targets, even when there were clear syllabic and orthographic (in terms of bigram frequency) boundaries between the embedded words and the remaining parts of primes. The absence of sublexical semantic activation in reading English (see also Sandra, 1990; Zwitserlood, 1994, for Dutch) is consistent with the assumptions of the lexical processing models (e.g., Coltheart et al., 1993; Frost, in press; Seidenberg & McClelland, 1989; van Orden et al., 1990) that do not allow the semantic system to have two separate patterns activated simultaneously.

This leads us to the question of why there should be a discrepancy of sublexical processing between Chinese and alphabetic languages. We believe that, although the underlying mechanisms for sublexical processing in logographic and alphabetic languages are essentially the same, the application or the performance of these mechanisms varies according to the structural properties of different writing systems. There are at least three contrasts that may lead the sublexical processing in Chinese to be both phonological and semantic and sublexical processing in English to be primarily phonological.

First, complex characters in Chinese can be thought of as analogous to compound words, with both semantic and phonetic radicals representing meanings. In contrast, sublexical orthographic units or letter clusters in English (monomorphemic) words do not usually represent meanings. Even when they do (e.g., *own* in *town*), after stripping these units, the remaining parts may not correspond to any words (except for a few pseudo-compound words like *shamrock* or *capsize*). The semicompoundhood of complex characters, together with other properties, may make the meanings of constituents more likely to be activated (see Zwitserlood, 1994, for discussion of semantic activation of constituents in Dutch compound words). Second, phonetic radicals are integrated orthographic units, whose constituents (e.g., strokes) have not systematic correspondences in phonology or semantics. There are usually clear visual separations between phonetic and semantic radicals, at least for complex characters with left–right structure. In contrast, sublexical orthographic units in English can be differentiated further into components (e.g., letters) that do have quasiregular correspondences to phonology, although not to semantics. Moreover, there are no clear visual cues for the separation between orthographic units, even taking into consideration "soft" constraints such as phonotactic rules or bigrame frequency. A letter string can be decomposed into different orthographic units of various size. Third, phonetic radicals are repeatedly used in different characters, and their frequencies are usually higher than the ones of the whole characters in which they are embedded. Moreover, the relations between phonetic and semantic radicals and the complex characters are either explicitly or implicitly taught to children when they learn characters. In English, however, the sublexical orthographic units that happen to correspond to real words are less likely to be used in different words, and it is rare, if not strange, for parents or teachers to point out to children that there is a word embedded in the new (monomorphemic) word.

These contrasts make phonetic radicals much more salient orthographic units in complex characters than letter clusters are in English words, even when these clusters happen to correspond to words. Phonetic radicals hence execute more power than letter clusters in computation from orthography to semantics. This efficient computation for phonetic radicals, especially for those in low-frequency complex characters, leads to both sublexical phonological activation and sublexical semantic activation, even though the semantic activation of phonetic radicals almost always interferes with the semantic activation of the whole characters.

The Time Course of Sublexical Semantic Activation

Although semantic representations corresponding to phonetic radicals in complex characters are initially activated, it is clear from the data on rhyming and irregular characters that this activation dies down quickly, probably within 200 ms of processing the complex characters (see Fig. 3.3 and Fig. 3.4). This is perhaps functionally required by the lexical processing system, given that the purpose of reading complex characters is to get the meanings of the whole characters, not the meanings of their phonetic radicals. Long-lasting semantic activation of phonetic radicals can only interfere with the normal reading processes.

The longer-lasting effect of semantic activation of phonetic radicals in Experiment 4 (Fig. 3.5) at the SOA of 200 ms was likely due to the experimental design and the priming paradigm. Because the irregular complex characters were used as targets, preceded by semantic cues to their phonetic radicals, the semantic primes increased the activation level of the embedded phonetic radicals and made the competition between phonetic radicals and whole characters even stronger. Moreover, subjects had more time and more attentional resources to process the phonetic radicals in naming the complex targets. The slightly larger and more consistent semantic activation of phonetic radicals in regular characters (see Fig. 3.2) than in irregular characters (Fig. 3.2 and Fig. 3.4) may reflect the contribution of phonological activation to semantic activation (Zhou et al., chapter 7, this volume; Zhou & Marslen-Wilson, in preparation). Although Zhou et al. (this volume) argued that phonology does not play a major role in access to meaning, phonology is nevertheless automatically activated in lexical access, and there are consistent interactions between different types of knowledge representations in the lexicon (Zhou & Marslen-Wilson, 1997). For regular characters, the sublexical phonological activation of phonetic radicals, supported by the phonological activation of the whole characters, may have oppotunities to spread its activation to the semantic representations of radicals (Zhou & Marslen-Wilson, 1998a).

Lexical Representation in the Lexicon

The finding of automatic activation of semantic representations of phonetic radicals in reading complex characters puts constraints on the structure of lexical representation and lexical processing models. A popular hierarchical model of visual word recognition in Chinese (e.g., Taft & Zhu, 1995; Taft, Liu, & Zhu, chapter 5, this volume; Zhu & Taft, 1994), which is along the lines of the interactive-activation framework, assumes that reading Chinese characters begins with the activation of strokes. This activation at the basic orthographic level passes to radicals, which connect to character representations.

The character representations connect to their corresponding syllable representations and to the whole word orthographic representations, which in turn connect to concept or semantic representations. Activation passes in step from lower levels to higher levels. Because phonetic radicals are themselves characters, this model has to represent phonetic radicals twice—once at the radical level, once at the character level (to connect to phonology).

Although the model may accommodate the finding of sublexical semantic activation of phonetic radicals by assuming that phonetic radical representations at the character level connect directly to semantics, it is not clear how semantic radicals should be represented in the model, as most semantic radicals are not characters themselves and yet they do have direct connections with semantics, as indicated in Experiment 5 and in Feldman & Siok (chapter 2, this volume).

Perfetti and Tan (1998) speculated that there are three interconnected "lexicons" in a multilexicon system: the character orthographic lexicon, the noncharacter component orthographic lexicon, and the phonological lexicon. Perfetti and Tan (1998) put phonetic and semantic radicals that are themselves characters into the character orthographic lexicon, which connects to the phonological lexicon, and the noncharacter components (a few phonetic radicals and many semantic radicals) into a separate lexicon. It is not clear in this speculation, however, what the function of a "noncharacter lexicon" is in lexical processing, and whether both phonological and semantic representations of phonetic radicals *embedded* in complex characters are automatically activated in reading complex characters.

Our thinking about the lexical representation of Chinese characters is more in line with the original time course model (e.g., Seidenberg, 1985) and with the connectionist theories of lexical processing (e.g., Plaut et al., 1996). We do not envisage different orthographic lexicons. Rather, we assume that all orthographic forms of Chinese morphemes, whether they are characters, phonetic or semantic radicals, and whether these radicals form real characters or not, are represented at the same level in the lexicon. These orthographic representations have direct links with representations in both the phonological and semantic systems. More importantly, we assume that in reading complex characters, the visual input is automatically decomposed into different orthographic units of *different sizes* (not just strokes as both Taft & Zhu, 1995, and Perfetti & Tan, 1998, assumed), and these units map *in parallel* to the orthographic representations in the lexicon. The activation of orthographic representations, whether they correspond to whole characters or to semantic or phonetic radicals, automatically leads to activation in both semantic and phonological systems. The idea of parallel mapping and activation is easier to translate into a distributed processing model than the idea of step-wise processing from low levels.

ACKNOWLEDGMENTS

The research reported here was supported in part by a grant from UK Economic and Social Research Council to William Marslen-Wilson and Xiaolin Zhou and in part by a grant from China National Science Foundation to Xiaolin Zhou. Part of the data were reported at the Advanced Workshop of Chinese Language Processing, Beijing, 1995 and the UK Experimental Psychological Society meeting, Bristol, 1996. We thank Yanchao Bi and Ningning Wu for helping us test subjects in Beijing and Marcus Taft and the editors of this book for their comments on an earlier version of this chapter.

REFERENCES

Chen, M. J., & Yung, Y. F. (1989). Reading Chinese: A holistic or piecemeal process? In A. F. Bennett & K. M. McConkey (Eds.), *Cognition in individual and social context* (pp. 91–100). Amsterdam: Elsevier.

Coltheart, M. (1978). Lexical access in simple reading tasks. In G. Underwood (Ed.), *Strategies of information processing* (pp. 151–216). San Diego, CA: Academic Press.

Coltheart, M., Curtis, B., Atkins, P., & Haller, M. (1993). Models of reading aloud: Dual-route and parallel-distributed-processing approaches. *Psychological Review, 100*, 589–608.

Fan, K, Y., Gao, J. Y., & Ao, X. P. (1984). Pronunciation principles of the Chinese character and alphabetic writing scripts. *Chinese Character Reform, 3*, 23–27.

Fang, S.-P., Horng, R.-Y., & Tzeng, O. (1986). Consistency effects in the Chinese character and pseudo-character naming tasks. In H. S. R. Kao & R. Hoosain (Eds.), *Linguistics, psychology, and the Chinese language* (pp. 11–21). Hong Kong: Centre of Asian Studies, University of Hong Kong.

Feldman, L. B., & Siok, W. W. T. (1997). The role of component function in visual recognition of Chinese characters. *Journal of Experimental Psychology: Learning, Memory, and Cognition, 23*, 776–781.

Flores d'Arcais, G. B., Saito, H., & Kawakami, M. (1995). Phonological and semantic activation in reading Kanji characters. *Journal of Experimental Psychology: Learning, Memory, and Cognition, 21*, 34–42.

Frost, R. (in press). Towards a strong phonological theory of visual word recognition: True issues and false trails. *Psychological Bulletin*.

Glushko, R. J. (1979). The organization and activation of orthographic knowledge in reading aloud. *Journal of Experimental Psychology: Human Perception and Performance, 5*, 674–691.

Hue, C.-W. (1992). Recognition processes in character naming. In H.-C. Chen & O. J. L. Tzeng (Eds.), *Language processing in Chinese* (pp. 93–107). Amsterdam: North-Holland.

Institute of Language Teaching and Research. (1986). *A frequency dictionary of modern Chinese*. Beijing: Beijing Language Institute Press.

Li, D. (1993). *A study of Chinese characters*. Beijing: Peking University Press.

Li, Y., & Kang, J. (1993). A statistic analysis of phonetic radicals in modern Chinese picto-phonetic characters. In *Information analyses of modern Chinese characters*. Shanghai: Shanghai Education Press.

Lukatela, G., & Turvey, M. T. (1994). Visual lexical access is initially phonological: 1. Evidence from associative priming by words, homophones, and pseudohomophones. *Journal of Experimental Psychology: General, 123*, 107–128.

Peng, D., Yang, H., & Chen, Y. (1994). Consistency and phonetic-independency effects in naming tasks of Chinese phonograms. In Q. Jing, H. Zhang, & D. Peng (Eds.), *Information processing of the Chinese language* (pp. 26–41). Beijing: Beijing Normal University Press.

Perfetti, C. A., & Tan, L. H. (1998). The time course of graphic, phonological, and semantic activation in Chinese character identification. *Journal of Experimental Psychology: Learning, Memory, and Cognition, 24*, 101–118.

Perfetti, C. A., & Zhang, S. (1991). Phonological processes in reading Chinese words. *Journal of Experimental Psychology: Learning, Memory, and Cognition, 17*, 633–643.

Plaut, D. C., McClelland, J. L., Seidenberg, M. S., & Patterson, K. E. (1996). Understanding normal and impaired word reading: Computational principles in quasi-regular domains. *Psychological Review, 103*, 56–115.

Sandra, D. (1990). On the representation and processing of compound words: Automatic access to constituent morphemes does not occur. *The Quarterly Journal of Experimental Psychology, 42A*, 529–567.

Seidenberg, M. S. (1985). The time course of phonological code activation in two writing systems. *Cognition, 19*, 1–30.

Seidenberg, M. S., & McClelland, J. L. (1989). A distributed, developmental model of visual word recognition and naming. *Psychological Review, 96*, 523–568.

Taft, M., & Zhu, X. (1995). The representation of bound morphemes in the lexicon: A Chinese study. In L. Feldman (Ed.), *Morphological aspects of language processing*. Hillsdale, NJ: Lawrence Erlbaum Associates.

Taft, M., & Zhu, X. (1997). Submorphemic processing in reading Chinese. *Journal of Experimental Psychology: Learning, Memory, and Cognition, 23*, 761–775.

van Orden, G. C., Pennington, B. F., & Stone, G. O. (1990). Word identification in reading and the promise of subsymbolic psycholinguistics. *Psychological Review, 97*, 488–522.

Yin, B., & Rohsenow, J. S. (1994). *Modern Chinese characters*. Beijing: Sinolingua.

Yin, W. G., & Butterworth, B. (1992). Deep and surface dyslexia in Chinese. In H.-C. Chen & O. J. L. Tzeng (Eds.), *Language processing in Chinese* (pp. 349–366). Amsterdam: North-Holland.

Yu, B., Feng, L., Cao, H., & Li, W. (1990). The perception of Chinese characters: Effects of tasks and properties of characters. *Acta Psychologica Sinica, 2*, 141–147.

Zhou, X. (1994). *Sub-lexical processing in reading Chinese*. Paper presented at the Inaugural Asian-Australian Workshop on Cognitive Processing of Asian Languages, Sydney.

Zhou, X., & Marslen-Wilson, W. (1997). Spread of activation in the mental lexicon. In M. G. Shafto & P. Langley (Eds.), *Proceedings of the 19th Annual Conference of the Cognitive Science Society* (pp. 838–843). Mahwah, NJ: Lawrence Erlbaum Associates.

Zhou, X., & Marslen-Wilson, W. (1998a). *Orthography, phonology, and lexical semantic activation in reading Chinese*. Manuscript submitted for publication.

Zhou, X., & Marslen-Wilson, W. (1998b). *Processing semantic radicals in reading Chinese characters*. Manuscript submitted for publication.

Zhu, X., & Taft, M. (1994). The influence of perceptual experience on Chinese character processing. In H.-W. Chang., J.-T. Huang, C.-W. Hue, & O. J. L. Tzeng (Eds.), *Advances in the study of Chinese language processing* (pp. 85–99). Taipei: Department of Psychology, National Taiwan University.

Zwitserlood, P. (1994). The role of semantic transparency in the processing and representation of Dutch compounds. *Language and Cognitive Processes, 9*, 341–368.

CHAPTER

4

How Is Access Representation Organized? The Relation of Polymorphemic Words and Their Morphemes in Chinese

Danling Peng
Ying Liu
Chunmao Wang
Beijing Normal University

ABSTRACT

A number of models of lexical access were developed in the past years. The major disagreement between models focuses on the issue of how lexical information is accessed. Does it use a strategy involving decomposing of elements of speech, or are they processed as a unit? No single model has yet been devised to settle such disagreement because both decomposed morphemes and whole words appear to exist in the access to word representation. The conflicting evidence in the literature suggests that hidden factors influence the organization of access representation. Surface frequency and transparency are two commonly mentioned factors (see a review by Sandra, 1994). Three experiments were designed to test for such factors in Chinese two-morpheme words. Results confirmed the presence of two factors. However, both interactive-activation models (Taft, 1994) and AAM (augmented addressed morpheme) models (Caramazza, Laudanna, & Romani, 1988) as well as other empirical research failed to account for our results. So, an IIC (Inter/Intra Connection) model was proposed and adapted to the problem of word processing.

A large number of psycholinguistic studies focused their analyses on morphemes, a construct borrowed from linguistics (Taft & Forster, 1975). Today most psycholinguists admit to the importance of morphemes or morphological structure in the lexicon or in the processing of words (but cf. Taft's BOSS, 1997a; Seidenberg's *bigram*, 1987, 1989). The morpheme effect is a robust phenomenon independent of form, phonology, and semantics (Bentins & Feldman, 1990; Feldman & Moskovljević, 1987; Feldman, 1992, 1994; Feldman & Andjelković, 1992; Grainger, Colé, & Segui, 1991; Napps, 1989; Napps & Fowler, 1987). The concern in research has been on the problem of structure and function of morphemes as reflected in the lexicon.

Most discussion in the literature is based on the hypothesis that there exist two levels of word processing. One level involves access representation. The approach and goal of attaining access utilizes either visual or auditory information inputs. The other level involves linguistic representation, including both semantic and syntactic information. The two levels are referred to as *access file* and *master file* by Taft and Forster (1975, 1976) and as access representation and lexical entry by Marslen-Wilson, Tyler, Waksler, and Older (1994). Our concern and major question deals with the issue of how access representation is organized.

EMPIRICAL RESEARCH

Three positions can be taken with respect to this issue. First is the full word position, which suggests that morphologically complex words are morphologically organized the same way as simple words (Osgood & Hoosain, 1974; Seidenberg, 1987, 1989; Seidenberg & McClelland, 1989). The second position considers the morpheme as the unit for lexical access (Taft, 1985, 1994; Taft & Forster, 1975, 1976). The third position combines some properties of the other two positions, although there are different versions of these hybrid models (Burani & Caramazza, 1987; Caramazza, Laudanna, & Romani, 1988; Sandra, 1990; Zhou, 1992; Zhou & Marslen-Wilson, 1992).

Much empirical data demonstrated the existence of morphological effects in word recognition processes across a variety of languages including English (Feldman, 1991; Taft, 1991), Chinese (Peng, Li, & Liu, 1994; Zhang & Peng, 1992), Dutch (Zwitserlood, 1994), and Hebrew (Feldman, Frost, & Pnini, 1995). It appears then that morphemic decomposition is an important step in the recognition of some polymorphemic words (Caramazza, Laudanna, & Romani, 1988; Taft, 1985; Taft & Forster, 1975). The nature of the content of a morpheme differs across studies. In linguistics, a *morpheme* is defined as the smallest linguistic unit, a unit that has a specific form and meaning (Sandra, 1994). Semantics is a critical factor in word processing. Feldman (1992), however, found that morphology is an independent factor in word processing because it cannot be inferred (or abstracted) from the relationship between phonological and semantic factors.

Three experimental paradigms are most frequently used in research on polymorphemic words.

1. Repetition and semantic priming. In repetition priming experiments, the target word is preceded by a prime that includes the same morpheme as the target word. In semantic priming experiments, there is a meaningful relationship between the prime and target or between components of the prime and target. Facilitation has been obtained with these two paradigms using English (Feldman, 1992, 1995), Dutch (Zwitserlood, 1994), and other languages.

2. Varying the whole word and morpheme frequency. A widely used method is to control whole word frequency and vary morpheme frequency, or the converse. With inflectional words, it was found that both word and stem frequency influence lexical decision time (Burani, Salmaso, & Caramazza, 1984; Burani & Caramazza, 1987; Taft, 1979b); however, Colé, Beauvillain, and Segui (1989) failed to find stem frequency effects with words that were prefixes.

Results for compound words are inconsistent. Taft and Forster (1976) and Andrews (1986) obtained cumulative morpheme frequency effects with compound words. Recognition time for high morpheme frequency words was faster than for low morpheme frequency words when whole word frequency was matched. However, experiments by van Jaarsfeld and Rattink (1988) and Frisson and Sandra (1992) obtained no morpheme frequency effects.

3. Research utilizing compound nonwords. Nonwords are frequently used as negative stimuli in lexical decision tasks. Taft and Forster (1976) evaluated recognition time in compound words and nonwords. Their logic suggested that if compounds are decomposed into morphemes in the lexical decision making process, there should be a similar decomposition for nonwords. Therefore, the lexical decision time for compound nonwords should be influenced by components that are words or nonwords. Compound nonwords containing a word as a component should be more difficult to reject as a word than compound nonwords that only contain nonwords as components.

Lima and Pollatsek (1983) found a significant difference in reaction time for words composed of two real words (RR) and words composed of a word and nonword (RN). Similar experiments by Monsell (1985) and van Jaarsfeld and Rattink (1988) demonstrated that the order of reaction times was RR > RN > NR > NN.

A problem for the compound nonword paradigm is that there is no representation of such "words" in the mental lexicon. Therefore, processing mechanisms for nonwords can be different from real words and cannot be used to explain differences in processing between real words.

What Happens in the Processing of Chinese Words?

Compared to English words, which include a large number of affixed words, more than 80% of Chinese words are compounds (*Dictionary of Usage Frequency of Modern Chinese Words*, 1987). The results from the inflected and derived words are not so reliable before being confirmed. For example, there is an affix stripping-off process in recognition of alphabetic words (Taft et al., 1975). However, this is not the case in Chinese.

Zhang and Peng (1992) conducted some of the earliest research on the morphological decomposition effect in Chinese by varying word and morpheme frequency. They demonstrated that both morpheme and word frequency influenced speed of lexical access as well as recognition accuracy. Coordinate compounds (such as 富强) and modifier compounds (such as 凉亭) were studied in their experiment. They found morpheme frequency influenced speed of word recognition. This data supported the existence of a morpheme level in Chinese language processing, they also found that whole word frequency significantly affects the processing speed of Chinese words. Their results demonstrated that both morphemes and words need to be considered as processing units. Their study did not examine whether morphemes and whole words are at the same level, or if one is subordinate to the other, and which one is accessed more rapidly.

Taft, Huang, and Zhu (1994) obtained similar component frequency effects. Their study utilized four types of two-character words: HH (both first and second component are high-frequency morphemes); HL (high–low), LH (low–high), and LL (low–low) words. Reaction times for HL and LH words were longer than for HH words. The results suggested that morpheme frequency affected the recognition of Chinese two-character words. An unexpected finding was that, under the LL condition, recognition time was as fast as for the HH condition. A possible explanation for these results was that most words in the LL condition were monomorphemic words (binding words) such as 蚯蚓. These words were monomorphemic and probably represented as one independent unit in the mental lexicon. It appears that there are morpheme (component) representation units in the mental lexicon. Morphemes influence the recognition processing of compounds.

Zhou and Marslen-Wilson (1994) failed to find a similar morpheme frequency effect with stimuli presented auditorily. The study suggested that morphemes did not contribute to the competition between homophonic morphemes.

Peng, Zhang, and Liu (1993) further investigated the lexical decomposition of Chinese two-character words using the semantic priming technique. In this experiment, subjects were required to specify the nature of both the prime and the target words. The study utilized a 2 × 2 design involving whole word related and unrelated and morpheme related/morpheme unrelated variables. The four groups were: (a) related whole words and related mor-

TABLE 4.1
The Mean Reaction Times (ms) and Correct Percentages (in parentheses) for Whole Word Related and Morpheme Related Conditions

	Whole Word Related	*Whole Word Unrelated*
Morpheme related	519 (94.0)	571 (89.2)
Morpheme unrelated	583 (92.3)	650 (92.6)

From Peng, Zhang, and Liu (1993).

phemes (e.g., 杰出-优异; 宽恕-原谅); (b) related whole words and unrelated morphemes (e.g., 恭维-奉承); (c) unrelated whole words and related morphemes (e.g., 敲打-拍卖; 制作-改造); and (d) unrelated whole words and unrelated morphemes (e.g., 耍弄-接替). Material in each group was well matched for semantic distance, frequency of usage of both words and characters, and syntactic function and lexical structure of the words. The average number of strokes per word was also balanced.

Results demonstrated both morpheme and whole word priming effects but no interaction effect (see Table 4.1). The authors suggested that decomposition is carried out at both the morphemic and whole word levels with separate storage units for each level. The representation of morphemes is suggested as occurring at an earlier stage of information processing than that for whole words.

Peng, Li, and Liu (1994) provided more evidence for the existence of processing at a morphemic level in an experiment that involved the use of both transparent and opaque two-character compound words. There were two conditions: (a) whole word repetition priming (轻柔 – 轻柔); and (b) one component repetition priming (轻 – 轻柔). Subjects were required to determine whether the target was or was not a word. SOA was 240 ms. The data suggested that compound words are decomposed in the word recognition process. No differences between transparent and opaque words were obtained.

The Present Experiments

Morpheme effects were obtained in Zhang and Peng's (1992) study. However, most of the compound words used were transparent compounds. Conclusions from that study cannot be generalized to opaque compounds.

Taft, Huang, and Zhu (1994) found, in a lexical decision task, that the response to a word containing two high-frequency components (HH) was faster than to words containing low-frequency components (HL and LH). It thus appears that access representation for Chinese words can be decomposed. Separate morphemes exist in access representation. Each component of a word makes a contribution to its processing. However, this hypothesis encounters difficulty in explaining one result that emerged in the same experiment, that is, the unusual fast reaction times of LL words.

These results suggest that the structure of access representation is not simple. At least, the access representation of these LL words cannot be regarded as decomposed, otherwise reaction times should have been longer to these words than to HH words. Therefore, it is possible that both decomposed and undecomposed representation occur in the mental lexicon. A possible solution is to assume that HH, HL, and LH words are represented in a decomposed manner. LL words, on the other hand, are represented as a unit. There is, however, no valid reason for such an assumption. Perhaps there were some uncontrolled variables in the Taft et al. experiment that affected response times to the LL words.

Sandra (1994) suggested two factors that contribute to the representation of polymorphemic words, namely, word frequency and semantic transparency. High-frequency and/or transparent words tend to be represented in a unitary manner, whereas low-frequency and opaque words are represented in a decomposed manner. In the Taft et al. experiment surface frequency of words was controlled, but transparency was not.

In Taft and Forster's earlier model (1975, 1976), whole words and morphemes are stored at different levels. Morpheme is a media for whole word access. AAM model (Caramazza et al., 1988) used a different processing framework. Morphemes are no longer necessary for word recognition, especially for those opaque compounds. In Taft and Zhu's new model (Taft & Zhu, 1995), there are connections between orthographic level and words level. Morphemes are no longer necessary for word access.

We suppose that both whole word units and morpheme units are stored in the access representation. There are connections between word units and morpheme units. The connections are developed during the recognition of different words that include a same morpheme. Surface frequency and semantic transparency of words may affect the development of these connections. As a result, the connection strength is different between different words and morphemes. It is what we wanted to delve into in this research.

In Experiment 1, we investigate the role of word units and morpheme units in the access of Chinese two-character words by varying surface frequency and cumulative frequency simultaneously. If there are only morpheme units in the access representation and no word units, then surface frequency cannot affect the recognition of words; that is, no *surface frequency effect* (SFE) emerges. On the other hand, if there are only word units and no morpheme units, then cumulative morpheme frequency cannot affect the recognition of words; that is, there is no *cumulative morpheme frequency effect* (CMFE).[1] Based on our hypothesis, we expect that both SFE and CMFE could be detected in a single experiment, and the relationship of

[1]However, even in this condition we have the possibility to attain a small CMFE. By any means, we can contribute the effect to visual input.

word units and morpheme units may be revealed. If the two effects are additive and not interactive, morpheme units and word units may exist in two different levels. Otherwise, we can conclude that the two types of units exist in the same level. Experiment 2 explored the function of semantic transparency by varying semantic transparency and cumulative frequency simultaneously. If transparency does not affect the structure of the access representation, it will not interfere with the CMFE; in other words, there is no interaction between transparency and cumulative frequency. On the contrary, transparency should interact with cumulative frequency.

In Experiment 3, another paradigm, repeated priming, was used to verify the result we obtained. A test to evaluate word transparency was conducted prior to these studies. Two hundred subjects were asked to evaluate the internal semantic consistency of 1,500 compound words on a 9-point scale: Each word was repeated 20 times to reduce variability (standard deviation) of the estimator. We ended up with two values, one representing the semantic relationship between the first morpheme and the whole word and the other representing that between the second morpheme and the whole word.

EXPERIMENT 1

The function of surface frequency effect (SFE) is a robust phenomenon confirmed by a great amount of research. The effect of cumulative morpheme frequency effect (CMFE) is also tested in much research. Zhang and Peng (1992) found CMFE in the lexical decision task of Chinese two-character words. Taft (1994) did, too. However, research seldom tested the two effects in a single experiment heretofore. It may be a relationship between surface frequency and cumulative morpheme frequency that could reveal the relationship of words and morphemes in the access representation. Hence, in the following experiment we tried to develop the relationship of morphemes and words by controlling surface frequency and cumulative frequency of Chinese two-character words. Different from previous research, we tried to balance semantic transparency of words across all groups of stimuli. In the experiments of Zhang and Peng (1992) and Taft (1994), semantic transparency was not regarded as a hint variable and was not controlled, so it is not easy to explain their discord result.

Method

Subjects. Twenty-one undergraduate students of Beijing Normal University were paid subjects in this study. All had normal or corrected-to-normal vision.

Materials and Design. Forty high- and 40 low-frequency (surface frequency) two-character compound words were selected (Ls and Hs). Each group of characters was divided into two subgroups dependent on the cumulative frequency of the first and second morpheme (Lm and Hm—see Appendix A for a list of words used). The four subgroups were balanced with respect to transparency, number of strokes, and radical numbers per character ($p > 0.10$; see Table 4.2). An equal number of nonwords were used as foils. Each nonword was made of two Chinese characters. To avoid the development of specific processing strategies, all nonwords were matched with words on the cumulative morpheme frequency of their two characters. Words and nonwords were arranged in random order, with the same order of presentation used with all subjects. No more than three words or nonwords were presented successively. Prior to the start of data collection, subjects were given demonstration trials with two words and two nonwords. The practice stimuli consisted of 18 words and nonwords, respectively.

Apparatus. Stimulus display and timing, as well as data collection, were controlled with an IBM compatible computer using Dmaster software. Stimuli were presented on a VGA monitor. The experimenter concurrently observed stimuli and responses on another monitor nearby.

Procedure. Subjects sat in front of a monitor and were instructed both in writing and verbally that on each trial, they would see two characters in the middle of the monitor; they should determine whether the characters were a two-character word or a two-character nonword and respond as rapidly as possible by pressing an appropriate response key. The key under the subject's dominant hand was used to signal word and that under the nondominant hand to signal nonword. Response latency was measured from

TABLE 4.2
The Material Data of Experiment 1

	SurFre	*1Fre*	*2Fre*	*1Tsp*	*2Tsp*	*1Strk*	*2Strk*	*1Rdc*	*2Rdc*
Ls & Lm	6.80	0.0624	0.0934	4.995	5.294	9.300	8.800	2.750	2.800
Hs & Lm	42.60	0.0947	0.0810	5.882	5.589	8.316	9.158	2.789	2.737
Ls & Hm	5.65	0.8367	0.9157	4.935	5.129	8.100	8.900	2.400	2.750
Hs & Hm	40.50	0.7557	0.9630	4.951	5.380	8.350	8.350	2.600	2.650

Note. SurFre: the means of surface frequency (token number in a corps of 1.31 million words, Beijing Language Institute Press, 1986); 1Fre, 2Fre: the means of first and second morpheme's cumulative frequency (thousandths), 1Tsp, 2Tsp: the means of first and second character's transparency (based on a 9-point scale); 1Strk, 2Strk: the means of first and second morpheme's stroke numbers; 1Rdc, 2Rdc: the means of first and second character's radical numbers. Ls: low surface frequency; Hs: high surface frequency; Lm: low cumulative morpheme frequency; Hm: high cumulative morpheme frequency.

stimulus onset to key press. Experimental trials were preceded by 36 practice trials (18 words and 18 nonwords).

The sequence of events was as follows: (a) a fixation sign appeared in the center of the screen; (b) following termination of the fixation stimulus there was a 300 ms period before; (c) presentation of a two-character stimulus for 500 ms; (d) the screen was then blanked for 3,000 ms; (e) subject's response occurred between (c) and (d).

Results

Only data from words were analyzed. Outliners were excluded, using the criterion that the response fell outside of a window ranging between 150 and 2,000 ms. This criterion excluded 0.24% of trials with correct responses. Thus most of the data was utilized in this analysis. The response latency, which was out of two standards from the means based on items or subjects, was replaced by the larger one. This criterion truncated 2.38% of all values. The mean response latencies for remaining correct trials were computed for each subject (averaging across items) as well as for each item (averaging across subjects). The error rate for each subject and item was computed as the percent of trials on which an error, including outliners as errors, occurred. The means of subject and item latencies and the means of subject and item error percentages are presented in Table 4.3.

TABLE 4.3
Mean Latencies (in ms) and Percentage of Error Rates in Experiment 1

	Mean Latency		*Error Rate (%)*	
	Lm	*Hm*	*Lm*	*Hm*
Ls				
Subjects				
M	568.79	566.40	5.48	5.00
SD	72.30	71.33	7.23	6.12
Items				
M	568.75	568.71	5.48	5.00
SD	45.42	44.34	5.84	5.46
Hs				
Subjects				
M	535.21	515.21	2.14	1.43
SD	60.99	54.55	2.99	2.80
Items				
M	536.12	516.09	2.14	1.43
SD	45.09	41.25	4.76	3.13

Note. Ls: low surface frequency; Hs: high surface frequency; Lm: low cumulative morpheme frequency; Hm: high cumulative morpheme frequency.

The latency and error data were submitted to two types of ANOVA, with cumulative frequency and surface frequency as the within-subject variables. One analysis was a repeated measures ANOVA with subjects as random variable. The second was an independent groups ANOVA with items as the random variable. The interaction between cumulative and surface frequency on mean latency by subject was significant ($MSE = 1628.88$, $F(1, 20) = 4.11$, $P = .056$), but not significant by items ($MSE = 1940.87$, $F(1, 76) = 1.03$, $P = .313$). The main effect of surface frequency was significant both by subjects ($MSE = 37722.76$, $F(1, 20) = 64.38$, $P = .000$) and by items ($MSE = 1940.87$, $F(1, 76) = 18.73$, $P = .000$). The simple effects analysis obtained no significant effects for the low surface frequency words evaluated either by items or subjects (Fs < 1), but there is a significant difference of 20 ms between low and high cumulative morpheme frequency for the high surface frequency words by subjects ($MSE = 4202.31$, $F(1, 20) = 23.15$, $P = .000$), though not significant by items ($MSE = 2387.67$, $F(1, 77) = 1.68$, $P = .199$).

The analysis of mean errors obtained no interactive effect between cumulative and surface frequency both by subjects and items (Fs < 1). The effect of surface frequency was significant both by subjects and by items ($MSE = 10.01$, $F(1, 20) = 8.62$, $P = .008$; $MSE = 1.06$, $F(1, 76) = 9.91$, $P = .002$). However, the effect of cumulative frequency failed to reach an acceptable level of significance (Fs < 1).

Discussion

From our study, both surface and cumulative frequency influenced response latencies for Chinese two-morpheme words. In fact, even though in the last years a large number of authors found CMFEs as well as robust SFEs (Burani, Salmaso, & Caramazza, 1984; Zhang & Peng, 1992), the relationship of surface frequency and cumulative frequency has seldom been taken into account. Based on these studies, we cannot arbitrarily conclude that only whole words *or* morphemes of words are represented in the access lexicon. A reasonable assumption is to admit that both words and morphemes are available to the access representation.

In Experiment 1, the average response latency of Hm words was 20 ms shorter than that for Lm words in the condition of high surface frequency groups, although they were not different from each other for the low surface frequency groups. One explanation is that high-frequency words tend to be represented as decomposed components, whereas low-frequency words tend to be represented in unitary fashion. This explanation is not satisfactory because we generally consider that low-frequency words are evaluated by decomposition. When presented frequently, words tend to be treated as a whole word unit (as stated by the AAM model, Burani, 1993; Burani & Laudanna, 1992; Caramazza et al., 1988; Frauenfelder & Schreuder, 1992; Laudanna, Burani, & Cermele, 1994).

A simple mixed access representation is inadequate to account for the obtained results. The access representation is not a pile of morpheme units and word units. There is a complicated connection between them. We generally assume that the CMFE of high surface frequency words comes from the strong connection between the word and its morphemes. For low-frequency words, on the other hand, the connection is weak and the influence from morpheme units to word units is small. As a result, no CMFE emerges in low-frequency words. It is easy to comprehend the development of the connection with the word usage. The more frequently a word presents itself, the stronger the connection between it and its morphemes.

Another question: Is the connection between morpheme and word units different level or same level? According to Interactive Activation Model (Taft, 1994) and Multi-level Cluster Model (Zhou et al., 1994, 1995), the access of compound words must transit morpheme level to word level sequentially. SFE and CMFE can be attributed to word level and morpheme level, respectively. The class of multilevel models, however, seems to have difficulty when they try to explain the interaction between surface frequency and morpheme frequency. If morphemes and words are at different levels, their effects should be independent, and with no interaction between surface frequency and cumulative frequency. CMFE should exist not only in high surface frequency words, but also in low surface frequency words. Unfortunately, it was not the case in this experiment.

Another opinion considers that morphemes and words are at the same level, such as AAM model (Caramazza et al., 1988). It can easily elucidate SFE and CMFE by postulating diverse thresholds of words and morphemes. Encountering the representation of high and low surface frequency words, however, it clogged. Caramazza (1988) deemed that high-frequency words tend to be represented as a whole, whereas low-frequency words tend to be represented morpheme-decomposedly.

Nevertheless, if we assume connections between words and morphemes, the obtained CMFE in high surface frequency words can be apprehended reasonably. Both morpheme units and word units in the access representation receive activation from a lower visual input level. Additionally, word units receive activation from morpheme units at the same level simultaneously. The connections between high-frequency words are stronger than those between low-frequency words. Consequently, the impact from morpheme units of high-frequency words is larger than that of low-frequency words. This is the reason that CMFE exists only in high-frequency words, but not in low-frequency words.

Besides, we can ponder the issue of a single level or two levels from another point. Given that morpheme level be the indispensable route in the access of a compound word, CMFE must be robust and should not be affected by other factors. That is, when surface frequency is balanced, the response latencies of high cumulative morpheme frequency words must be shorter than those of low cumulative morpheme frequency words. Contrarily, if morphemes and words belong to the same level, the connections between them may be influenced by the upper level, namely, the semantic level. This is the concerned question of semantic transparency. The meaning of a transparent word is similar to the meaning of its morphemes, whereas there is a weak relationship between the meaning of an opaque word and its morphemes. Does this difference impact the connection of words and morphemes in the access representation? That is the principal aim of the next experiment.

EXPERIMENT 2

The role of Experiment 2 was to explore how semantic transparency impacts the structure of access representation. If the access representation of transparent words is different from that of opaque words, this difference may reflect on the CMFE. Specifically, if morpheme units and word units belong to two sequent levels, semantic transparency may not influence CMFE. The response latency of high cumulative morpheme words must be shorter than that of low cumulative morpheme frequency words. On the contrary, if they are in the same level, semantic transparency may influence CMFE. For transparent and opaque words, CMFE may behave differently.

Method

Subjects. Seventeen subjects were paid undergraduate students from Beijing Normal University. All had normal or corrected-to-normal vision.

Materials and Design. Forty transparent and 40 opaque two-morpheme words were selected. Based on the cumulative frequencies of the first and second morpheme, each group of words was divided into two subgroups, namely Lm and Hm (see Appendix B). The four subgroups were completely balanced for surface frequency, number of strokes per character, and radical numbers ($P > .10$, see Table 4.4). the same number of nonwords were used as foils.

Apparatus and Procedure. They are similar to those used in Experiment 1.

Results

As in Experiment 1, only data from words were analyzed. Response latencies were adjusted using the same criteria used in Experiment 1. This criterion truncated 1.84% of all values. Average response latency and error percentages for subjects as well as items are displayed in Table 4.5.

TABLE 4.4
The Material Data of Experiment 2

	SurFre	*1Fre*	*2Fre*	*1Tsp*	*2Tsp*	*1Strk*	*2Strk*	*1Rdc*	*2Rdc*
OP & Lm	23.25	0.0752	0.1004	4.083	3.956	8.947	8.800	2.700	2.800
TR & Lm	20.70	0.0822	0.0734	6.839	7.005	8.600	9.900	2.850	2.900
OP & Hm	24.30	0.6712	1.0316	3.644	3.924	8.350	8.200	2.550	2.600
TR & Hm	23.60	0.8306	0.7744	6.219	6.748	8.050	8.850	2.450	2.650

Note. SurFre: the means of surface frequency (token number in a corps of 1.31 million words, Beijing Language Institute Press, 1986); 1Fre, 2Fre: the means of first and second character's cumulative frequency (thousandths), 1Tsp, 2Tsp: the means of first and second character's transparency (based on a 9-point scale); 1Strk, 2Strk: the means of first and second character's stroke numbers; 1Rdc, 2Rdc: the means of first and second character's radical numbers. OP: opaque; TR: transparent; Lm: low cumulative morpheme frequency; Hm: high cumulative morpheme frequency.

TABLE 4.5
Mean Latencies (in ms) and Percentage of Error Rates in Experiment 2

	Mean latency		*Error rate (%)*	
	Lm	*Hm*	*Lm*	*Hm*
Opaque				
Subjects				
M	516.10	539.05	2.94	5.88
SD	41.08	52.10	4.70	7.12
Items				
M	517.90	542.36	2.94	5.88
SD	38.10	53.94	5.88	7.14
Transparent				
Subjects				
M	552.79	506.10	4.12	2.06
SD	61.19	43.13	3.64	3.56
Items				
M	554.43	507.84	4.12	2.06
SD	48.53	39.72	5.76	5.49

Note. Lm: low cumulative morpheme frequency; Hm: high cumulative morpheme frequency.

Two-by-two ANOVAs were conducted for both mean latency and error rate. A significant interaction was obtained between cumulative frequency and transparency for mean response latency by "subjects" analysis (MSE = 20613.31, $F(1, 16) = 39.75$, $P = .000$) and items analysis (MSE = 2073.49, $F(1, 76) = 12.17$, $P = .001$). No main effect for cumulative frequency or transparency was obtained for average latencies. The simple effect analysis demonstrated that cumulative frequency was significant for transparent words (by subjects: MSE = 18531.58, $F(1, 16) = 36.93$, $P = .000$; by items: MSE = 2046.82, $F(1, 76) = 10.61$, $P = .002$) and for opaque words the simple effect analysis was significant by subjects (*MSE* = 4477.30, $F(1, 16) = 18.67$, $P = .001$), and approached an acceptable level of significance by items (*MSE* = 2046.82, $F(1, 76) = 2.92$, $P = .091$). We must note, however, that the effects of cumulative morpheme frequency were the reverse. In the condition of transparent words, average response latency of Hm words was 47 ms faster than that of Lm words, whereas in the condition of opaque words, Hm latencies were slower by 24 ms than Lm words.

The results of error rates showed the same trend as reaction time. There was no main effect of cumulative frequency or transparency on error rates. A significant interaction effect was obtained for mean error rates by subjects (*MSE* = 4.25, $F(1, 16) = 4.86$, $P = .043$). The simple effects analysis demonstrated a significant effect for opaque word condition (*MSE* = 2.94, $F(1, 16) = 4.68$, $P = .046$) and an effect approached significance for the cumulative effect in the transparent word condition (*MSE* = 1.44, $F(1, 16) = 3.27$, $P = .090$). Mean error rate of Hm words under transparent word conditions was 2.06% smaller than for Lm words, whereas with opaque words, Hm words were 2.94% greater than for Lm words. Although the interacted effect approached significance by items (*MSE* = 1.08, $F(1, 76) = 3.36$, $P = .071$), no main effects or simple effects could be identified ($P > .10$).

Discussion

Both the response latencies and error rates demonstrate reliable CMFEs. We should note that in the conditions of transparent and opaque words, these effects were reversed. The reversal effects of cumulative morpheme frequency obtained here was not as strange as they seem to be. In a similar experiment, Taft et al. (1994) found that the response of LL words was as fast as that of HH words. They attributed this unusual phenomenon to the nature of the LL words used, most of which consisted of monomorphemic words. This is not the case in this experiment. All words were polymorphemic.

How can we account for the reversed effect for the transparent and opaque words found in the present experiment? As previously mentioned, we regard access representation as a mixed one, with both words and morphemes existing at the same level of lexical access. There are connections between word units and their morpheme units in the access level. From the results obtained here we can conclude that the connections between transparent word units and their morpheme units are positive. However, those between opaque word units and their morpheme units are negative. As a result, transparent words with high cumulative morpheme frequency are recognized rapidly, whereas the processing of opaque words with high cumulative morpheme frequencies is delayed.

We must ask the question: How are words and morphemes represented in access representation? Two hypotheses have been proposed. One considers that words and morphemes are mixed at the same level; that is, whole word units and morpheme units are at the same level, rather than at different levels (AAM, Caramazza et al., 1988). The other hypothesis considers them to be at different levels, namely a word level that is distinct from the morpheme levels (Taft, 1994; Zhou & Marslen-Wilson, 1995). However, these distinctions are not essential because the two hypotheses cannot be discriminated when we consider the connection between levels as identical to that within levels.

A key factor by which we judge models is whether there is a direct connection regardless of the assumed structure of the model. In the AAM model, the words and morpheme are at the same level and the inputs from orthography map onto them in a direct fashion. In a multilevel activation model (Taft, 1994), however, the orthographic level is connected to the morphemic level, which, in turn, is connected to the word level. For this model, there is no direct connection between word and orthographic levels.

Are there two sequential levels of processing or a single mixed level? If access representation consists of two sequential levels, namely, word and morphemic levels, both surface frequency of words and cumulative morpheme frequency can influence their processing. Therefore, whether they are transparent or not, the higher the cumulative morpheme frequency, the faster word recognition.

The results of the present experiment do not allow for this interpretation. The slower responses to opaque HH words require us to reconsider the structure of the access representation. A single mixed level may be more flexible in accounting for the results. The inputs of visual words from orthography activates corresponding morphemes and words. These morphemes and words interact with each other by way of the connection pattern. If the words are transparent, the connections are positive; if the words are opaque, the connections are negative. That is, transparent words and their morphemes activate each other, whereas opaque words and their morphemes inhibit each other. In Experiment 3, we focus on this assumption by using a repeated priming paradigm.

EXPERIMENT 3

Method

A repetition priming experiment was conducted to examine the morpheme access of transparent and opaque compound words.

Subjects. Thirty students from Beijing Normal University, paid for their participation, took part in the experiment.

Design. A two (opaque vs. transparent word priming) by two (repetition vs. no repetition) design was used. Transparency was only manipulated in the primes. The target words were transparent or semitransparent, even in opaque prime condition. The average transparency scores of targets are larger than 6 in both opaque and transparent prime conditions. A within-subject design was utilized.

Materials. Two priming conditions were used, namely, first morpheme repetition versus no repetition. The targets of these two conditions were the same words. Each condition included 20 pairs of two-morpheme words (see Appendix C). The priming words of the repetition and control condition were well matched for word frequency, character frequency, and strokes per character. The frequency of target in opaque condition was 25.5, in transparent condition 36.5. The difference was not statistically significant ($t = .62$, $p = .542$). Nonwords consisted with real characters were used as negative stimuli.

There was no semantic relation between primes and the target words in all four groups. Two types of primes were used: transparent (安宁-安装 quiet primes install) and opaque (快活-快速, happy primes rapidly). Words where transparency was bigger than 7 were identified as transparent and those with scores smaller than 3 as opaque (see Table 4.6).

Procedure. Stimuli were presented on a CRT as white characters against a black background. Presentation of a prime was preceded by a 500 ms masking stimulus. The priming word was then presented for 56 ms. The target word immediately followed presentation of the prime and remained on the screen until the subject responded. In order to exclude immediately

TABLE 4.6
The Materials Data in Experiment 3 (word frequency and character frequency for opaque priming)

	WFre	*1Fre*	*1Strk*	*2Fre*	*2Strk*
Opaque prime	64.8	0.995735	6.75	1.28196	8.6
Control prime	64.7	1.373325	8.3	1.251415	7.9
Target	25.7	0.995735	6.75	0.82307	8.15
Transparent prime	76	0.886205	9.4	0.762285	9.2
Control prime	74.1	0.92422	7.4	1.218895	7.55
Target	34.5	0.886205	9.4	1.08774	8.7

Note. WFre: The means of word frequency (token number in a corps of 1.31 million words, Beijing Language Institute Press, 1986), first and second character's cumulative frequency (thousandths), first and second character's stroke numbers for prime in experiment group, prime in control group and target words.

form repetition effect, the prime words were presented in KAI font style and the target words in SONG font. The subjects can not see the prime clearly because of the short SOA and mask. The influence of criteria can be reduced with this technique. All subjects were fully practiced on these procedures before the formal experiment was started. Presentation sequence was randomized for each subject.

Results

Only data for correct responses were included in the RT analyses. As in the other experiments, RTs that fell more than two standard deviations above or below the mean RT were replaced by RTs exactly two standard deviations above or below that mean. Fewer than 2% of RTs in any one condition had to be replaced. Unless otherwise noted, all tests of significance are two-tailed with acceptable significance level at $p < 0.05$.

Table 4.7 presents the mean RTs and error data. A 2 × 2 ANOVA was performed on the RTs and errors. In the within-subject analysis, the priming effect was significant for RTs, $F(1, 29) = 8.96$, $MSE = 8267.16$, $p < .01$, and not for errors, $F(1, 29) = .63$, $MSE = 0$. RT was significant in transparent priming condition, $F(1, 29) = 7.95$, $MSE = 6359.49$, $p < .01$. Interactive effects were not significant for either RTs or errors, $F(1, 29) = .50$, $MSE = 477.62$, $p > .05$, and $F(1, 29) = 1.11$, $MSE = .01$, $p > .05$. No significant effect was obtained under the opaque condition, $F(1, 29) = .06$, $MSE = .00$, $p = .801$. No significant effect attributable to items was obtained.

Discussion

Experiment 3 examined the repetition effect of opaque and transparent primes. We obtained evidence that morphemes are accessed in transparent but not in opaque compound words. Facilitation of recognition time was significant, although the meaning of the first component morpheme, the prime, was quite different from that of the target word. It appears that transparent compounds are decomposed during information processing. No

TABLE 4.7
Mean Response Times (RT's; in ms), and Percentage Errors (%E), and Priming Effects in Experiment 3

56 ms prime	*Opaque*		*Transparent*	
Morpheme repetition	626	(8.7)	600	(2.3)
No repetition	639	(8.0)	621	(6.3)
Effect	13	(−0.7)	21	(4)

such effect was obtained when the primes were associated with opaque compound words. Because we failed to find a significant interaction effect between transparency and repetition, we cannot determine whether there are differences in the decomposition of opaque and transparent compound words. However, we can be assured that morphemes in opaque compound words are activated to a smaller extent than is true for transparent compound words. Zhou and Marslen-Wilson (1994) failed to find such an effect in an experiment similar to this one. The results for that study approached significance for the transparent condition. Betin and Feldman (1990) also observed significant facilitation effects in their morphological condition where there was no semantic relationship between the prime and the target word. Their data, using Hebrew writing, appear to match our results with Chinese characters. A crucial issue is how to explain the repetition effect in transparent prime condition. Strong evidence against an orthographic or phonological interpretation to account for facilitation in repetition priming comes from the results of studies using Serbian (Feldman, 1992; Feldman & Moskovljević, 1987) and English words (Fowler, Napps, & Feldman, 1985; Stolz and Feldman, 1995). General conclusions were drawn from above experiments. In our study, the effect cannot be attributed to the repetition of form because a mask was presented prior to the prime and fonts were different for the prime and target stimuli. Priming effects must be attributed to the nature of the repeated morpheme because there was no semantic relationship between the prime and the compound word. We conclude that morphemes do exist as independent units in our mental lexicon.

GENERAL DISCUSSION

Experiment 1 provided evidence that both words and morphemes exist in access representation and that there are demonstrable connections between high-frequency words and their morphemes. Experiment 2 demonstrated that such connections were different for transparent and opaque words. These differences were confirmed in Experiment 3; even the interaction is not significant. From these three experiments and other empirical data, we favor an Inter/Intra Connection (IIC) model, one in which the access representation is mixed (see Fig. 4.1). Both whole words and morphemes are represented at the same level. There are some intralevel connections between morpheme and whole word units. The strength of connections is determined by the surface frequency and the relationship between the meaning of morphemes and words of which they are a part, namely transparency. Strong semantic relationships between word and morpheme (i.e., a transparent word) provides strong positive connections. Weak

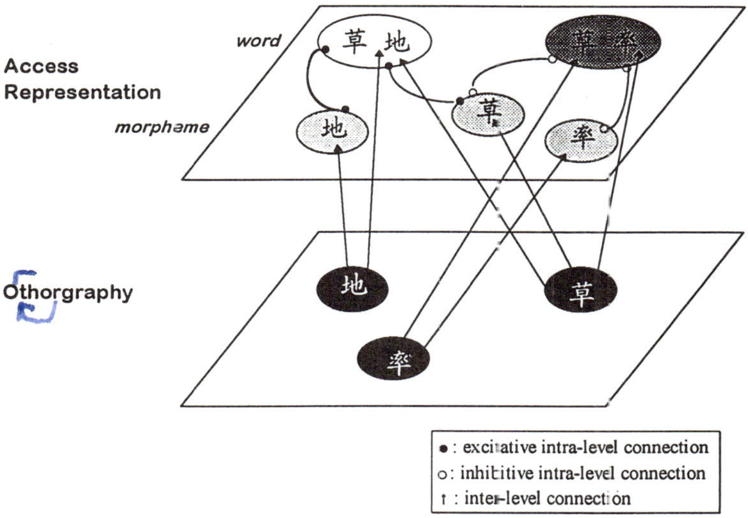

FIG. 4.1. A frame of Inter/Intra Connection model. In the access representation, "草," "地," and "率" are morphemes, "草地" is a transparent word and "草率" is an opaque word. "草" means grass, "地" means land, "草地" means meadow, and "草率" means careless.

semantic relationships (opaque words) generate weak positive connections, and where there is no relationship, there sometimes are negative connections. The inputs from orthography maps directly onto access representation via interlevel connections where both words and morphemes may be activated.

The access representation suggested by AAM model is also mixed (Caramazza et al., 1988). Word frequency determines whether a two-morpheme word is represented as a word or whether it is decomposed during access. That is, high surface frequency words tend to be represented as a whole, whereas low surface frequency words tend to be represented in a decomposed fashion. However, Experiment 1 found a cumulative morpheme frequency effect for high rather than low surface frequency words. Based on our model, recognition of transparent words with high surface frequency may be facilitated by their morphemes. If their morphemes are high-frequency ones, response latency may be short. For low surface frequency words, on the other hand, little excitation is generated by the morphemes; therefore, there is no cumulative morpheme frequency effect.

In Taft's interactive activation model (1994), one finds both whole word and morpheme units. Words and morphemes are considered to be at two different levels of representation with some connections between levels. In reality, there is no intrinsic difference between a two-level or a one-level conceptual model. The important difference is in the relationship between words, morphemes, and orthographic units. Taft's model is a bottom-up model in which the orthography activates morphemes that in turn activate word units. In this model, all connections are positive in nature. Therefore, the response to HH words must be faster than that to LL words.

In Experiment 2, we found that when words are opaque, the processing of LL words was faster than HH words. In fact, Taft also found that, in some conditions, there was no difference in access speed between LL and HH words. We assume the existence of negative connections between opaque words and their morphemes and that this connection leads to the slower responses to HH words. From this hypothesis, we infer that the processing of transparent words must be faster than that for opaque words. Unfortunately, Experiment 2 demonstrated that under conditions of low surface frequency, the mean response latency to transparent words was larger than that to opaque words, although mean response latency to transparent words was shorter than that to opaque words under conditions of high surface frequency.

We conclude that morpheme level cannot be identified as a level that is distinct from the word level. They must be at the same level. That is, the inputs from orthography map onto both morphemes and words. The strong connections between high-frequency words and their morphemes led to the cumulative morpheme frequency effects obtained in Experiment 1. The negative connection between opaque words and their morphemes led to the slower response time for HH words.

In repetition priming experiments, morphemes are always activated when a transparent prime is presented. Because of the positive connection between the morpheme and word units, such as 安 and 安宁, the morpheme unit (such as 安) will still be activated when the target word (such as 安装) is presented. The positive connection between 安 and 安装 facilitates the recognition of target 安装. As a result, priming effects were obtained even when there was no semantic relationship between 安宁 and 安装 .

When the primes are opaque compounds, inhibition from words to morphemes depresses activation of the component morpheme. The target obtains no facilitation from the preactivation of a morpheme even when the target is transparent. That is, we believe, the reason for the absence of a repetition effect when the primes are opaque.

ACKNOWLEDGMENTS

The research reported here was supported by the Chinese Natural Science Foundation (CNSF,39470250 and 397711120158). All experiments were conducted in the Cognitive Science Lab in Beijing Normal University.

APPENDIX A

The Materials of Experiment 1

Lm & Ls	*Hm & Ls*	*Lm & Hs*	*Hm & Hs*
慈悲	精辟	麻烦	把握
荣幸	打听	牺牲	检讨
碰巧	承包	扫荡	青春
荒唐	死板	宝贝	推测
肤浅	表率	毕竟	目标
挑拨	底细	介绍	快活
勒索	参与	漂亮	树立
拍卖	条理	操纵	调解
典礼	较量	讽刺	交代
逢迎	言行	玩笑	程度
申请	职责	束缚	分裂
抹杀	争执	夸奖	雄伟
俘虏	过错	驾驶	模范
崩溃	零散	腐朽	爱护
锋芒	培植	尖锐	讲述
惊恐	品德	辽阔	缺少
怨恨	边际	飞跃	坚强
忙乱	更改	暗淡	观看
埋藏	评论	欺骗	美好
驱逐	场所	欢乐	才能

Note. Ls: low surface frequency; Hs: high surface frequency; Lm: low cumulative morpheme frequency; Hm: high cumulative morpheme frequency.

APPENDIX B

The Materials of Experiment 2

Lm & OP	*Hm & OP*	*Lm & TR*	*Hm & TR*
慈悲	把握	俘虏	职责
麻烦	检讨	束缚	争执
荣幸	精辟	夸奖	过错
纯粹	打听	崩溃	零散
牺牲	青春	驾驶	分裂
扫荡	承包	腐朽	雄伟
宝贝	死板	锋芒	模范
荒唐	推测	惊恐	爱护
肤浅	目标	怨恨	培植
挑拨	底细	尖锐	节省

(Continued)

APPENDIX B (Continued)

Lm & OP	*Hm & OP*	*Lm & TR*	*Hm & TR*
毕竟	罗列	辽阔	品德
勒索	快活	飞跃	讲述
拍卖	树立	暗淡	边际
漂亮	调解	欺骗	更改
典礼	交代	诬陷	缺少
操纵	参与	欢乐	坚强
讽刺	条理	忙乱	观看
玩笑	较量	埋藏	评论
申请	程度	软弱	美好
抹杀	言行	驱逐	才能

Note. Lm: low cumulative morpheme frequency; Hm: high cumulative morpheme frequency; OP: opaque; TR: transparent.

APPENDIX C

The Materials of Experiment 3

Opaque primes	control primes	target words		transparent primes	control primes	target words
好比	过分	好感		摆动	王国	摆设
刻薄	星系	刻画		冤枉	衬衫	冤家
风光	公平	风向		欺骗	姿态	欺压
扫兴	薄膜	扫除		覆盖	酱油	覆灭
幽默	佩带	幽静		沉重	停止	沉没
快活	表情	快速		牢固	幼年	牢房
死板	红茶	死刑		派遣	压抑	派别
体面	代价	体温		能够	看见	能量
果断	再次	果实		合并	利害	合理
打量	力求	打击		明亮	电路	明天
难过	数量	难点		等待	气温	等级
条件	其他	条纹		容纳	深处	容貌
虚心	飘扬	虚弱		纯洁	浓度	纯熟
挖苦	喷气	挖掘		解答	加热	解散
封建	旁边	封闭		果实	光景	果断
马上	统治	马匹		制造	门口	制约
生意	看来	生前		平坦	书本	平安
天真	动摇	天色		安宁	落伍	安装
出色	主权	出发		快速	总务	快乐
毕竟	残酷	毕业		道路	事实	道理

REFERENCES

Andrews, S. (1986). Morphological influences on lexical access: lexical or nonlexical effects? *Journal of Memory and Language, 25*, 726–740.

Beijing University of Aeronautics and Astronautics. (1987). *Dictionary of usage frequency of modern Chinese words*. Beijing: The Aeronautics and Astronautics Press.

Bentins, S., & Feldman, L. B. (1990). The contribution of morphological and semantic relatedness to repetition priming at short and long lags: evidence from Hebrew. *The Quarterly Journal of Experimental Psychology, 42A*, 693–711.

Burani, C. (1993). What determines morphological relatedness in the lexicon. In G. T. M. Altmann & R. C. Schillcock (Eds.), *Cognitive models of speech processing: The Sperlonga Meeting II*. Hove, England: Lawrence Erlbaum Associates.

Burani, C., & Caramazza, A. (1987). Representation and processing of derived words. *Language and Cognitive Processes, 2*, 217–222.

Burani, C., & Laudanna, A. (1992). Units of representation for derived words in the lexicon. In R. Frost & L. Katz (Eds.), *Orthography, phonology, morphology, and meaning* (pp. 361–376). Amsterdam: Elsevier.

Burani, C., Salmaso, D., & Caramazza, A. (1984). Morphological structure and lexical access. *Visible Language, XVIII, 4*, 342–352.

Caramazza, A., Laudanna, A., & Romani, C. (1988). Lexical access and inflectional morphology. *Cognition, 28*, 297–332.

Colé, P., Beauvillain, C., & Segui, J. (1989). On the representation and processing of prefixed and suffixed derived words: A differential frequency effect. *Journal of Memory and Language, 28*, 1–13.

Feldman, L. B. (1991). Morphological relationships revealed through the repetition priming task. In M. Noonan, P. Downing & S. Lima (Eds.), *Linguistics and literacy* (pp. 239–254). Amsterdam/Philadelphia: John Benjamins.

Feldman, L. B. (1992). Morphological relationships revealed through the repetition priming task. In M. Noonan, P. Downing, & S. Lima (Eds.), *Literacy and linguistics*. Amsterdam/Philadelphia: John Benjamins.

Feldman, L. B. (1994). Beyond orthography and phonology: Difference between inflections and derivation. *Journal of Memory and Language, 33*, 442–470.

Feldman, L. B., & Andjelković, D. (1992). Morphological analysis in word recognition. In R. Frost & L. Katz (Eds.), *Orthography, phonology, morphology, and meaning* (pp. 343–360). Amsterdam: Elsevier.

Feldman, L. B., Frost, R., & Pnini, T. (1995). Decomposing words into their constituent morphemes: Evidence from English and Hebrew. *Journal of Experiment Psychology: Learning, Memory, and Cognition, 21*(4), 947–960.

Feldman, L. B., & Moskovljević, J. (1987). Repetition priming is not purely episodic in origin. *Journal of Experimental Psychology: Learning, Memory, and Cognition, 13*, 573–581.

Fowler, C. A., Napps, S. E., & Feldman, L. B. (1985). Relation among regular and irregular morphologically related words in the lexicon as revealed by repetition priming. *Memory & Cognition, 13*, 241–255.

Frauenfelder, U. H., & Schreuder, R. (1992). Constraining psycholinguistic models of morphological processing and representation: The role of productivity. In G. Booij & J. van Marle (Eds.), *Yearbook of morphology 1991*. Dordrecht, The Netherlands: Kluwer.

Frisson, S., & Sandra, D. (1992, July). *Whole-word representation of Dutch compounds*. Paper presented at the workshop on Psycholinguistic Approaches to the Lexical Representation of Morphological Structure, Krems, Austria.

Grainger, J., Colé, P., & Segui, J. (1991). Masked morphological priming in visual word recognition. *Journal of Memory and Language, 30*, 370–384.

Laudanna, A., Burani, C., & Cermele, A. (1994). Prefixes as processing units. *Language and Cognitive Process, 9*(3), 295–316.

Lima, S. D., & Pollatsek, A. (1983). The Basic Orthographic Syllabic (BOSS) reconsidered. *Journal of Verbal Learning and Verbal Behavior, 22*, 310–332.

Marslen-Wilson, W., Tyler, L. K., Waksler, R., & Older, L. (1994). Morphology and meaning in the English mental lexicon. *Psychological Review, 101*, 3–33.

Monsell, S. (1985). Repetition and the lexicon. In A. W. Ellis (Ed.), *Progress in the psychology of language* (Vol. 2, pp. 147–196). Hillsdale, NJ: Lawrence Erlbaum Associates.

Napps, S. E. (1989). Morphemic relationships in the lexicon: Are they distinct from semantic and formal relationships? *Memory & Cognition, 17*, 729–739.

Napps, S. E., & Fowler, C. A. (1987). Formal relationships among words and the organization of the mental lexicon. *Journal of Psycholinguistic Research, 16*, 257–272.

Osgood, C. E., & Hoosain, R. (1974). Salience of the word as a unit in the perception of language. *Perception & Psychophysics, 15*, 168–192.

Peng, D.-L., Li, Y.-P., & Liu, Z. Z. (1994). Identification of the Chinese two character word under repetition priming condition. *ACTA Psychological Sinica, 26*, 393–400.

Peng, D.-L., Zhang, B. Y., Liu, Z. Z. (1993). Lexical decomposition and whole word storage of Chinese coordination two-character word. In W. Su (Ed.), *Proceedings of the Second Afro-Asian Psychology Congress* (pp. 92–97). Beijing: Beijing University Press.

Sandra, D. (1990). On the representation and processing of compound words: Automatic access to constituent morpheme does not occur. *The Quarterly Journal of Experimental Psychology, 42*(A), 529–567.

Sandra, D. (1994). The morphology of the mental lexicon: Internal word structure viewed from a psycholinguistics perspective. *Language and Cognitive Processes, 9*(3), 227–269.

Seidenberg, M. S. (1987). Sublexical structures in visual word recognition: Access units or orthographic redundancy? In M. Coltheart (Ed.), *Attention and performance XII* (pp. 245–263). Hillsdale, NJ: Lawrence Erlbaum Associates.

Seidenberg, M. S. (1989). Reading complex words. In G. N. Carslon & M. K. Tanenhaus (Eds.), *Linguistic structure in language processing*. Dordrecht, The Netherlands: Kluwer.

Seidenberg, M. S., & McClelland, J. L. (1989). A distributed, developmental model of word recognition and naming. *Psychology Review, 96*, 523–568.

Stolz, J. A., & Feldman, L. B. (1995). The role of orthographic and semantic transparency of the base morpheme in morphological processing. In L. B. Feldman (Ed.), *Morphological aspects of language processing* (pp. 109–130). Hillsdale, NJ: Lawrence Erlbaum Associates.

Taft, M. (1979a). Lexical access via an orthographic code: The Basic Orthographic Syllabic Structure (BOSS). *Journal of Verbal Learning and Verbal Behavior, 18*, 21–39.

Taft, M. (1979b). Recognition of affixed words and the word frequency effect. *Memory & Cognition, 7*, 263–272.

Taft, M. (1985). The decoding of words in lexical access; A review of the morphographic approach. In D. Besner, T. G. Waller, & G. E. MacKinnon (Eds.), *Reading research: Advances in theory and practice* (Vol. V). New York: Academic Press.

Taft, M. (1991). *Reading and the mental lexicon*. Hillsdale, NJ: Lawrence Erlbaum Associates.

Taft, M. (1994). Interactive-activation as a framework for understanding morphological processing. *Language and Cognitive Processes, 9*(3), 271–294.

Taft, M., Forster, K. I. (1975). Lexical storage and retrieval of prefixed words. *Journal of Verbal Learning and Verbal Behavior, 14*, 638–647.

Taft, M., & Forster, K. I. (1976). Lexical storage and retrieval of polymophemic and polysyllabic words. *Journal of Verbal Learning and Verbal Behavior, 15*, 607–620.

Taft, M., Huang, J., & Zhu, X. (1994). The influence of character frequency on word recognition responses in Chinese. In *Advances in the Chinese language processing* (Vol. 1, pp. 59–73). Department of Psychology, National Taiwan University.

Taft, M., & Zhu, X. (1995).The representation of bound morphemes in the lexicon: A Chinese study. In I. B. Feldman (Ed.), *Morphological aspects of language processing* (pp. 293–316). Hillsdale, NJ: Lawrence Erlbaum Associates.

Taft, M., & Zhu, X. (1997). The model of lexical processing: orthographic, phonology and morpheme. In D.-L. Peng, H. Shu, & H.-C. Chen (Eds.), *The cognitive research of Chinese* (pp. 85–99). Shandong: Shandong Educational Publisher.

van Jaarsfeld, H. J., & Rattink, G. E. (1988). Frequency effect in the processing of lexicalized and novel nominal compounds. *Journal of Psycholinguistic Research, 17*, 447–473.

Zhang, B. Y., & Peng, D.-L. (1992). Decomposed storage in the Chinese lexicon. In H.-C. Chen & O. J. L. Tzeng (Eds.), *Language processing in Chinese*. Amsterdam: North-Holland.

Zhou, X. (1992). *The mental representation of Chinese disyllabic words*. Unpublished doctoral dissertation, University of Cambridge, England.

Zhou, X., & Marslen-Wilson, W. (1992, July). *The morphological priming effects in Chinese compound words*. Paper presented at the 25th International Congress of Psychology, Brussels.

Zhou, X., & Marslen-Wilson, W. (1994). Words, morphemes and syllables in the Chinese mental lexicon. *Language and Cognitive Processes, 9*(3), 341–368.

Zhou, X., & Marslen-Wilson, W. (1995). Morphological structure in the Chinese mental lexicon. *Language and Cognitive Processes, 10*(6), 545–600.

Zwitserlood, P. (1994). The role of semantic transparency in the processing and representation of Dutch compounds. *Language and Cognitive Processes, 9*(3), 341–368.

CHAPTER

5

Morphemic Processing in Reading Chinese

Marcus Taft
University of New South Wales, Australia

Ying Liu
Beijing Normal University, China

Xiaoping Zhu
University of New South Wales, Australia

Written words are not composed merely of a random assignment of lines and dots. In all languages, words have a systematic internal structure. Some of these structures are purely form-based, as with the combination of lines that create letters in alphabetic languages (like the *h*, *a*, and *t* of *hat*), the grouping of letters that create syllables (like the *chim* and *ney* of *chimney*), and the recurrent combination of strokes found in Chinese (like the 彐 of 非 and 诽, and the 昷 of 温 and 愠). However, other structures, called *morphemes*, involve the interaction of form and meaning, and these are the focus of this chapter.

DEFINING THE MORPHEME

The traditional definition of a morpheme is the smallest unit of language that can be associated with meaning and grammatical function (e.g., Hockett, 1958; Lyons, 1969). Thus, the word *unfriendliness* is composed of four morphemes, the prefix *un*, the stem *friend*, and the suffixes *li* (=*ly*) and *ness*. Although this traditional definition is unable to handle words whose putative morphemes have no meaning (e.g., *receive*, *deceive*, *refer*, *defer*; see Aronoff, 1976), most psycholinguistic studies focus on those cases where the semantic motivation for the morphemic structure is clear (though see Emmorey, 1989; Taft & Forster, 1975), or they directly contrast those cases where the word and its component morphemes have a clear semantic relationship (i.e., "transparent" cases) with those where it does not (i.e., "opaque" cases).

The morphemic structure that is of most relevance to Chinese is where two stem morphemes combine to form a compound word (like *teapot*, *handgun*, and *nightmare* in English). Despite some linguists' description of Chinese as an "isolating" language where most words are monomorphemic (e.g., Bybee, 1995), according to statistical analyses of the language (e.g., Lin, 1971), around 80% of Chinese words are actually polymorphemic. A typical example is 手枪 (*handgun*), where the character 手 means *hand* and 枪 means *gun*, and both can combine with other characters to form different words, like 手表 (*wristwatch*), and 机枪 (*machine-gun*). In printed Chinese, then, each character typically corresponds to a morpheme (e.g., Chao, 1968; Chen, 1992; Hoosain, 1992; Li & Thompson, 1981; Zhang & Peng, 1992). This is obviously so when the character can be used as a word in its own right (i.e., as a "free" morpheme like 手 or 枪), but what about when it cannot? Is it still a morpheme (a"bound morpheme"), or is it merely a form-based unit (the orthographic representation of a syllable)?

Taft and Zhu (1995a) suggest that a character that cannot stand alone is a bound morpheme when it (a) can be found in at least two semantically related words, like the 浩 of 浩大 (*huge*), 浩荡 (mighty), and 浩瀚 (vast), or (b) only occurs in one word, but the other character is related in meaning to that word, like the 芹 of 芹菜 (*celery*) where 菜 means *vegetable* (see discussion by Aronoff, 1976, of English examples like the *cran* of *cranberry*). In the former case, the meaning of the character can be abstracted out as the intersection of the meaning of the words in which it occurs (e.g., 浩 must mean something like *very big*), whereas in the latter case, the meaning of the character takes on the qualities of the meaning of the whole word (e.g., 芹 means *the quality of a vegetable that makes it a celery*). In both cases, this meaning might be quite abstract and maybe cannot even be expressed in words, as with the meaning of the 凌 of 凌辱 (*humiliate*), 凌空 (*soar*), 凌乱 (*messy*), etc. (or the *fer* of *refer*, *defer*, *infer*, *confer*, *prefer*, etc., in English).

On the other hand, a character has no morphemic status when it can only combine with one other character to create a word and where that other character also occurs only in that one word. Such words are called "binding words" (连绵词). For example, 蚯 only occurs with 蚓 and 蚓 only occurs with 蚯. The word they create when put together, 蚯蚓, means *earthworm*. Neither character alone has any meaning because there are no other words in which they occur that allows one to abstract out that meaning. An English example would be *zigzag*, where *zig* and *zag* alone do not have any meaning except, jocularly, "half a zigzag." Certainly, 蚯 and 蚓 do not each refer to "half an earthworm"!

Another way in which two-character monomorphemic words occur is when a word is copied from another language and the individual characters have been chosen purely so that the sound of the Chinese word will approximate its pronunciation in the original language. For example, 沙 is pronounced *shā* and means *sand*, whereas 发 is pronounced *fā* and means *develop*. The two characters are combined together to create the word for *sofa* because of the phonological similarity of *shāfā* to *sofa*; 沙发 has nothing to do with "sand development." Thus, 沙 and 发 are purely syllabic, and not morphemic, when used in the word 沙发.

There are other two-character words, however, whose constituents have a dubious morphemic status inasmuch as their meanings can be combined together to form a coherent concept, but this concept is not the actual meaning of the word. These are called "opaque" words. For example, 小 means *little* and 心 means *heart* or *spirit*, but 小心 means *beware*, and not *faint-hearted* or *mean-spirited* as one might expect. Examples in English are *cocktail* and *butterfly*. It can be seen from these examples that there is actually a range of opacity; for example, *cocktail* has nothing to do with *cocks* or *tails*, whereas *butterfly* is at least something that *flies*. It is hard to know whether opaque words should be considered polymorphemic or monomorphemic, but what can be said is that if there is no difference observed in the processing of opaque and nonopaque (i.e., "transparent") words in reading, the units of processing would seem to be generated more on the basis of the physical structure of the word (i.e., its form) than on the basis of its semantic structure.

Finally, if we are focusing on the morphemic structure of the orthographic representations of words, we need to consider something else when looking at Chinese. Most characters are composed of more than one "radical" (e.g., the 氵 and 永 of 泳), and often one of these radicals conveys meaning to the whole character (usually the left-hand one in a horizontally structured character). An example of such a *semantic radical* is the 钅 of 铁 (*iron*), 银 (*silver*), and 铜 (*copper*), where 钅 clearly means *metal*. Thus, the semantic radical has many of the qualities of a morpheme: The characters in which it occurs can be highly related to each other so that a clear meaning for the common radical can be extracted (thus, 钅 has the same relationship with 铁, 银, and 铜 as 浩 has with 浩大, 浩荡, and 浩瀚) or else, like 组 (*group*), 纺 (*weave*), and 红 (*red*), the relationship between the characters can be quite opaque (thus, 纟 has the same relationship with 组, 纺, and 红 as 凌 has with 凌辱, 凌空, and 凌乱).

Given these parallels, should the semantic radical therefore be considered to be a morpheme? Linguists would certainly respond negatively to this question even if only on the grounds that the semantic radical is an orthographic unit and orthography is a culturally imposed convention that does not fall within the scope of linguistic analysis. The only form of a word that warrants attention is its phonological form, according to most linguists.

This, however, is not a satisfactory answer for a cognitive psychologist interested in the processing of orthographic structure in reading.

What is probably more important in deciding whether the semantic radical should be considered to be a morpheme is the fact that the rest of the character (e.g., the 永 that remains after the semantic radical 氵 is removed from 泳) does not behave like a morpheme. So, the only function that the 永 of 泳 has is phonological (in that both 永 and 泳 are pronounced *yǒng*), and even the phonological function is unreliable given that the vast majority of nonsemantic radicals are not homophonic with the character in which they occur (Fan, Gao, & Ao, 1984). Such a situation means that a semantic radical should probably not be termed a morpheme, just as the *-ash* of *clash, dash, smash, crash, flash, bash*, etc. is not considered to be a morpheme in English despite the common sense of "sharp movement" among these words: The remaining letters *cl-, d-, sm-*, etc. are not morphemic in nature. For this reason, we do not discuss processing of the radical (though see Taft & Zhu, 1997a; Zhu & Taft, 1994, and the chapters in this volume by Zhou & Marslen-Wilson and by Feldman & Siok).

Later in this chapter we consider whether what we do or do not label as a morpheme is actually important in relation to lexical processing, but first we consider how morphemes might be lexically represented in relation to their orthography, meaning, and phonology.

THE MULTILEVEL INTERACTIVE-ACTIVATION FRAMEWORK

A useful framework for thinking about morphemic processing is a multilevel interactive-activation system where activation units are hierarchically organized in levels corresponding to words, morphemes, and submorphemic components. Taft (1994) presents an exploration of how this framework might be used in relation to English; Taft and Zhu (1995, 1997a, 1997b) and Zhu and Taft (1994) offer a Chinese version. The latter is depicted in Fig. 5.1.

The figure depicts orthographic units on the left, phonological units on the right, and meaning units at the top. "Stroke" refers to the lines and dots that make up a Chinese character (these may alternatively be seen as "features"). Stroke units feed their activation to the units representing radicals that contain those strokes; these feed their activation to the units representing the morphemes that contain those radicals; and these feed their activation to the units representing whole-word (polymorphemic) words that contain those morphemes. The units at the "meaning" level are merely a summary of what are likely to be a large number of semantic feature units discussed later. These can be linked to units at all but the stroke level because radicals, morphemes, and polymorphemic words can all be asso-

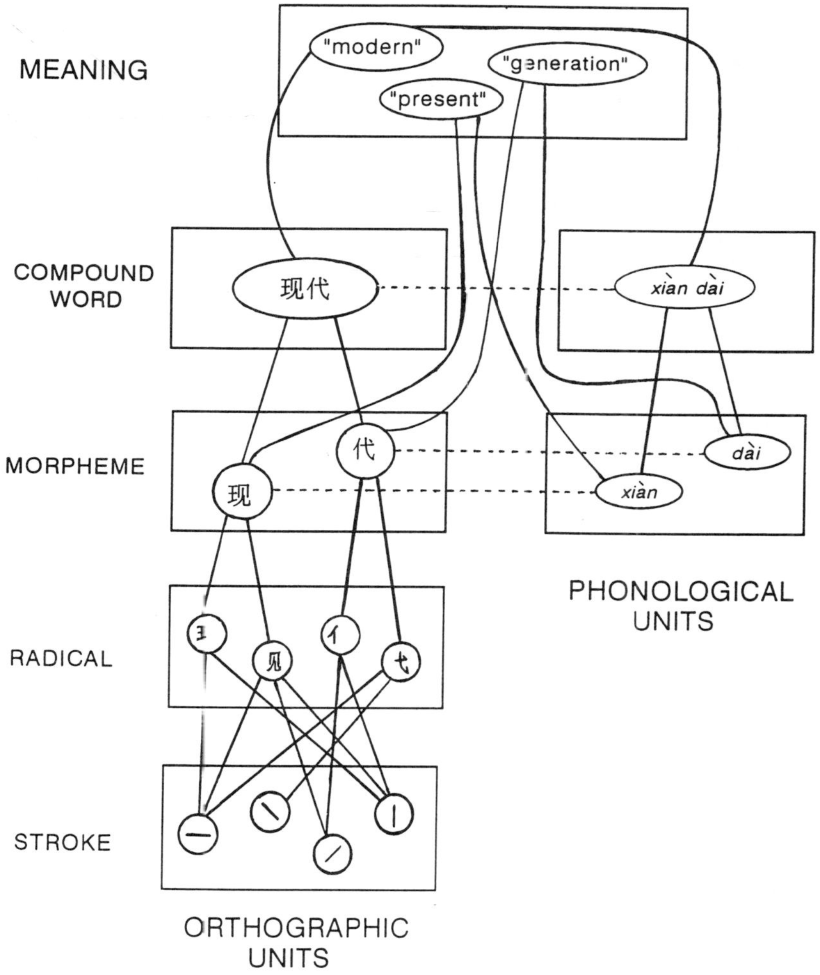

FIG. 5.1. The multilevel interactive-activation architecture using the example of 现代 to illustrate the pathways linking orthographic, phonological, and semantic units.

ciated with a meaning (though links from the radical level are not actually shown in Fig. 5.1).

At the morpheme and whole-word levels, there are both orthographic and phonological representations. Phonological links may conceivably exist at the radical level also, but these are not incorporated into the system as depicted in Fig. 5.1 because the pronunciation of a radical is such an unreliable guide to the pronunciation of the morpheme (Fan et al., 1984) that phonological units at the radical level could not be systematically linked to the phonological units at the morpheme level. After all, the radical is an orthographic construct, not a phonological one.

When a word is visually presented, activation passes up through the stroke level to the meaning via the other orthographic levels and also, potentially, via the associated phonological units (e.g., Perfetti & Zhang, 1995; Tan, Hoosain, & Peng, 1995). The strength of connection between units would be influenced by the frequency with which that connection is used and, therefore, high-frequency (i.e., common) words are activated via stronger links, and hence more quickly, than low-frequency words.

A response to the word could be made at different points within the processing system depending on the task requirements. If a lexical decision response is required (i.e., "Is the stimulus item a real word or not?"), the response could be logically based solely on whether or not there exists an orthographic representation at the word level, but it is also possible that it will be based on whether or not a semantic representation exists. It is also possible that phonological information becomes activated in the course of this process, though the lexical decision response cannot be made solely on the existence of such phonological information or else a nonsense word that is pronounced homophonically with a real word would be judged to be that word. On the other hand, phonological information will be the basis for the response if the task requires that the stimulus be named.

The issues we consider in relation to this multilevel interactive-activation framework are the existence of the morpheme level, and the role that semantics and phonology play in morphemic processing. The involvement of a morpheme level of analysis in the recognition of compound words has been examined in English and Dutch as well as Chinese. There have been a number of different approaches taken, and we describe these next.

COMPOUND NONWORDS

If compound words are recognized via a morpheme level of analysis, then one can make the assumption that compound nonwords are also processed in this way. If so, one can assess the relative importance of the first or second constituent in the recognition of a compound word by manipulating the nature of the putative morphemes of a compound nonword. Taft and Forster (1976) did this in an English lexical decision task by manipulating whether the first and/or second constituent of a nonword was a real word or not. When the nonword was composed of two real words (RR, e.g., *toastpull*) or of a real word followed by a nonword (RN, e.g., *spellcung*) response times were longer than when they were composed of a nonword followed by a word (NR, e.g., *flurbpair*) or of two nonwords (NN, e.g., *thrimnade*). That is, the lexical status (word or nonword) of the first constituent affected the responses. On the other hand, the lexical status of the second constituent had no effect (i.e., RR = RN, and NR = NN).

Subsequent work failed to replicate this pattern, however. Experiments by Lima and Pollatsek (1983) and Monsell (1985) in English, van Jaarsfeld and Rattink (1988) in Dutch, and Taft, Huang, and Zhu (1994) in Chinese suggested that the lexical status of both the first and second constituents were relevant; that is, there appeared to be a gradation of response times from RR to RN to NR to NN. Within the multilevel interactive-activation framework, this would be interpreted in terms of activation happening at the morpheme level within the system and therefore partially activating word-level information that renders the "nonword" response more difficult. Representations at the morpheme level of both the first and second constituent are accessed, though the first must be providing greater activation (given the difference between RN and NR).

While these findings might imply that nonwords are processed via analysis of both constituents, they do not entail that the same is true of real compound words. It is possible that compounds are normally processed as whole words and are only analyzed into their constituents when no whole-word representation is accessed, that is, when the compound is a nonword. Therefore, results with nonword compounds are restricted in what they can tell us about normal reading.

FREQUENCY MANIPULATIONS

Instead of examining the lexical status of the constituents of nonword compounds, then, it is preferable to manipulate characteristics of the constituents of real compounds. The characteristic of a constituent that is most likely to reveal an effect is its frequency of occurrence. Again, this has been examined in English (Andrews, 1986; Taft & Forster, 1976), Dutch (van Jaarsfeld & Rattink, 1988) and Chinese (Zhang & Peng, 1992; Taft, Huang, & Zhu, 1994). The Chinese experiments are described next.

Zhang and Peng (1992) found that lexical decision responses to words of the same frequency were affected by the frequency of both the first and second character. They established this by comparing words composed of two high-frequency characters (HH) with those composed of one high- and one low-frequency character (HL or LH). In fact, Zhang and Peng claimed that the HH/LH difference only held true for "coordinative" compounds, where the meanings of both morphemes contribute equally to the meaning of the whole word (e.g., 城镇 meaning *urban area*; 城 means *city* and 镇 means *town*), and not for "modifier" compounds, where the meaning of the first character merely serves to modify the meaning of the second (e.g., 金鱼 meaning *goldfish*; 金 means *gold* and 鱼 means *fish*). The HH/HL difference held for both types of compound.

If it is true that the frequency of a character only has an effect when it represents a morpheme that is semantically central to the meaning of the word (i.e., the second character of a modifier compound and both characters of a coordinative compound), it would imply that semantic factors have a major impact on the storage and access of compound words. The involvement of semantics in morphemic processing is considered later, but it must be pointed out here that Taft, Huang, and Zhu (1994) cast doubt on the processing distinction between coordinative and modifier compounds, both in relation to their own results and in their reassessment of the data obtained by Zhang and Peng themselves. Taft et al. failed to find any difference between coordinative and modifier compounds in a post hoc analysis of their data, and suggested that the difference observed by Zhang and Peng could be explained on methodological grounds.

In fact, the results obtained in relation to character frequency by Taft et al. (1994) were somewhat surprising. Like Zhang and Peng, they found that lexical decision was easier with HH words (e.g., 城关) than HL words (e.g., 家眷) and LH words (e.g., 桉树), but additionally obtained the unexpected result that LL words (e.g., 惆怅) were just as easy as HH ones. Many of the LL stimuli were actually binding words, and this could explain the lack of difficulty with these words. If it is morpheme frequency rather than character frequency that is important, the frequency of the individual characters of a binding word will have no impact because they are not morphemes at all. Thus, responses to LL words could turn out to be relatively fast. Taft et al. (1994) tested this possibility by comparing LH and LL words where the LL words were specifically designed to be either genuine compound words or monomorphemic binding words. Only the former showed a morpheme frequency effect, supporting the suggestion that it is morpheme frequency, not character frequency, that is important in word recognition. Such a result supports the existence of a distinctively morphemic level of representation.

One factor that might affect response times to compound words is the predictability of the first constituent. Obviously, if one anticipates the identity of a word after processing the first character, recognition of the word should be facilitated. In processing terms, if a candidate set is established on the basis of the initially processed information (i.e., the first character), then the smaller the set size, the easier it should be to recognize the stimulus as being a member of it. It is interesting to note, though, that predictability per se cannot explain the pattern of responses that was observed by Taft et al. (1994). Predictability was measured in terms of the percentage of occasions an independent group of subjects thought of the stimulus word when given the first character only. The stimulus words beginning with a common character (i.e., the HH and HL words) were naturally far less predictable than those beginning with a rare one (i.e., the LH and LL words): 6.7% versus 59.5%. Despite the huge difference in predictability of LH words over HL words, however, the former was certainly not easier to recognize than the latter; in fact, if anything, the reverse was true. Such a result argues against any theoretical account of compound word processing that centers on the activation of a candidate set via access to a representation of the first character.

Using Japanese Kanji compound words (equivalent to Chinese compound words in structure), Tamaoka and Hatsuzuka (1995) found a somewhat different pattern of data from Taft et al.'s (1994), but came to a similar conclusion. Lexical decision times showed an effect of frequency of the second character but not the first, whereas the reverse was true for naming times. Tamaoka and Hatsuzuka concluded that both characters are involved in activating the whole-word representation, but in lexical decision, the effect of the first character is obscured by the time spent processing the second character, whereas a naming response can commence on the basis of the first character prior to the second character being fully processed.

These results, then, can be taken as general support for the idea that a morpheme level of processing exists, and this is captured in the multilevel interactive-activation framework as a distinct morpheme level of representation.

MORPHEME PRIMING

Another approach adopted to examine morphemic analysis in reading is the priming technique. Several studies, again in English and Dutch, looked at the effect on lexical decision responses to a compound word of the prior presentation of one of the constituents of that word, relative to an unrelated control prime. Monsell (1985) found that responses were facilitated by the prior presentation of both the first and second constituents regardless of the semantic transparency of the compound (e.g., *day* or *dream* priming *daydream*; *cock* or *tail* priming *cocktail*). However, such an effect was shown to be purely orthographic because facilitation was also obtained with monomorphemic *pseudocompounds* (e.g., *car* or *pet* priming *carpet*).

Zwitserlood (1994) reversed the prime and target (e.g., *cocktail* preceding *cock*) and again found a priming effect for both constituents regardless of semantic transparency, but was able to discriminate this from a purely orthographic effect by showing that parts of a compound that are not morphemic constituents produced inhibitory effects when used as primes (e.g., prior presentation of *patchwork* and *cufflink* would inhibit responses to *pat* and *ink*, respectively).[1] An indication that these effects were morphemic, rather than purely syllabic, came from the fact that the amount of priming of the second constituent was greater than that of the first constituent in the case of transparent words but not opaque words. Given that Dutch

[1]The materials used by Zwitserlood were Dutch.

compounds tend to be of the "modifier" type, these results are consistent with the idea that the relationship between the meaning of the word and its constituents has some impact on processing and, therefore, Zwitserlood argues for the existence of a specifically morphemic level of analysis.

This conclusion was reinforced by a further experiment where the relationship between prime and target was based on meaning rather than form. Priming with a transparent word (e.g., *daydream*) facilitated responses to target words that were semantically related to either its first constituent (e.g., *night*) or its second constituent (e.g., *sleep*), though more so for the latter. On the other hand, priming with an opaque word (e.g., *cocktail*) did not facilitate responses to words related in meaning to either its first constituent (e.g., *hen*) or second constituent (e.g., *wag*), but did facilitate responses to words related to the meaning of the compound (e.g., *gin*). Opaque words were treated in exactly the same way as monomorphemic pseudocompounds (e.g., *carpet* did not prime *vehicle* or *dog*, but did prime *rug*). In fact, it is unclear how one might define the difference between a truly opaque word and a pseudocompound,[2] so it is not surprising that they were responded to in the same manner.

The conclusion from Zwitserlood's experiments seems clear. Form-based morpheme structure is represented in lexical memory as evidenced by the priming effect observed in lexical decision with opaque words (though this should perhaps be seen as a syllable effect), but the pattern of semantic priming suggests that activation of the links from these morpheme representations to the semantic system is only present for semantically transparent constituents.

An experiment by Sandra (1990) failed to find any semantic priming from the opaque constituents of "partially opaque" words, like the *butter* of *butterfly*. That is, *bread* did not prime *butterfly*. On the other hand, Zwitserlood (1994) found facilitatory semantic priming with partially opaque words, though the prime and target conditions were reversed. That is, *butterfly* did prime *bread*. Such a conflict in results could be interpreted if it is supposed that the meaning of the constituent becomes available only after the meaning of the whole word does, and therefore the meaning of *butter* is only extracted after *butterfly* has been recognized. In this way, the meaning of *butter* will be available by the time *bread* is presented, but will play no role in the recognition of *butterfly* even when preceded by *bread*.

We turn now to similar research in Chinese. Liu and Peng (1995), using a brief SOA (43 ms) between prime and target, found that there was no facilitation of responses to a target word that was semantically related to one of the constituents of an opaque prime word, but there was such facili-

[2]If there really is a difference, Zwitserlood does not adequately define it. In fact, she uses *mandrill* as an example of a pseudocompound, even though it is historically composed of *man* and *drill* and therefore would be better described as opaque.

tation for transparent prime words. For example, 马 means *horse* and 上 means *up*, which when put together make an opaque word meaning *immediately*. Presentation of 马上 did not prime responses to 绵羊 (*sheep*). On the other hand, the constituents of the word 背叛 (*betray*) are transparently related to its meaning (背 meaning *back* and 叛 meaning *rebel*), and 背叛 did facilitate responses to 覆盖 (*cover*), a target that is not semantically related to the whole prime word 背叛, but is related to the constituent 背. Such results replicate those of Zwitserlood in Dutch (though she used an SOA of 300 ms).

In a previously unreported experiment, Liu, Peng, and Taft also used the partial repetition priming paradigm to compare transparent and opaque words. This was different from the partial repetition paradigm used by Monsell (1985) and by Zwitserlood (1994) in that the prime was not merely a constituent of the target word or vice versa (e.g., *day* preceding *daydream*, or *daydream* preceding *day*), but rather, shared their first constituent with the target (like *daydream* preceding *daybreak*). Another difference was that the prime was masked from conscious awareness by using the masked priming technique pioneered by Forster and Davis (1984) where guessing strategies are eliminated. In this experiment, it was the prime word that could be either transparent or opaque (with the target always being transparent). The transparent word primes were designed so that their first constituent had two distinct meanings (e.g., 安, pronounced *ān*, means both *quiet* and *install*), and the prime and target word each made use of a different meaning. For example, 安宁 (*peaceful*) was used as a prime and 安装 (*install*) was its target. Therefore, the 安 of 安宁 is as different a morpheme to the 安 of 安装 as the 马 of the opaque word 马上 (*immediately*) is to the 马 of 马掌 (*horseshoe*). Even so, facilitation was found with the transparent primes, but not with the opaque ones. How can one explain the pattern of results in Chinese and other languages?

It seems that any explanation for the differential priming of transparent and opaque words (in both the partial repetition and semantic relatedness tasks) requires the postulation of a morpheme level of representation where each morpheme is associated with a meaning. An explanation in terms of semantic priming at the word level will not suffice because the whole-word meaning of the prime and target were no more related for the transparent words than they were for the opaque ones (as determined by ratings).

As was suggested earlier, the representation of meaning may be best conceived of as a set of units, each of which represents a semantic primitive. These primitives need not necessarily be describable in words. A compound word will activate a constellation of such semantic units, as will its component constituents (even if they are bound morphemes). The transparency of a compound word can actually be characterized in terms of the overlap of the semantic units associated with that word and those associated with its constituents: The greater the overlap, the greater the transparency. That is, an opaque word is one whose meaning is unrelated to the meaning of its constituents, while a fully transparent word is one whose meaning is entirely contained within the combination of the meanings of its constituents.

Now, when a compound word is presented, it activates the representation of the whole word via the representation of its constituent morphemes. The meaning associated with each of these representations becomes activated as a result. If there is no overlap found between the word meaning and morpheme meaning (i.e., if the word is opaque), activation in the morpheme unit is reset to baseline because the meaning units activated by it only serve to compete with the meaning units associated with the whole word. Thus, there is no facilitation observed with opaque words. For example, the whole-word representation for the opaque word 马上 will be activated via the units representing 马 and 上, but the activation within these morpheme units will be reset when the semantic units associated with 马上 provide competing activation to those associated with 马 and 上. Therefore, when 马掌 is subsequently presented, there will be no facilitation arising from the prior processing of 马.

In the case of transparent words, the overlap found to exist between the semantic units associated with the whole word and that associated with at least one of the meanings of its constituent morphemes leads to a maintenance of the activation set up in the morpheme units. For example, when the transparent word 安宁 is presented, it activates the morpheme representation for 安 and 宁, which together activate the whole-word representation 安宁. Activation of 安 in turn starts to activate the set of semantic units representing the meaning *quiet*, as well as the set of units representing the meaning *install*, whereas activation of 安宁 starts to activate the set of units representing the meaning *peaceful*, which overlap with the set representing *quiet*. This overlap prevents any resetting in the activation that exists in the unit for 安 because the overlap confirms that 安 is a useful contributor to the understanding of 安宁. In this way, responses to a subsequently presented word are facilitated if it contains the constituent 安, regardless of the meaning of 安 in that word (e.g., 安装).

The same explanation can be used to explain the finding of semantic priming with transparent words and not with opaque (Liu & Peng, 1995; Zwitserlood, 1994). Simply, the resetting of activation for the constituents of opaque words means that the semantic units associated with those morphemes are no longer active when the target word is presented. For transparent words, they are.

What is not explained, however, is the facilitation observed by Zwitserlood (1994) when the target is a constituent of a previously presented opaque word (e.g., *cocktail* priming *cock*). If activation in the morpheme unit *cock* is reset after it is established that the meaning of the first constituent *cock* is unrelated to the meaning of *cocktail*, obviously there should be no facilitation of responses to *cock*. There is clearly a conflict here with the lack of facilitation in Chinese when the target and opaque prime share a constituent (e.g., 马上 preceding 马掌). It is hard to know what the relevant difference is between the two studies (Chinese vs. Dutch; compound word target vs. monomorphemic target; masked priming vs. unmasked priming), but logically, awareness of the prime seems to be the most likely candidate. When the prime is not masked, subjects can adopt experiment-specific strategies. For example, when *cocktail* is presented and is available to consciousness, the subject can anticipate that *cock* (or *tail*) might follow as the target. In other words, the subject can either ignore the semantic information activated by *cocktail* or else override the resetting of activation in the morpheme unit for *cock*. Experiment-specific strategies are a potential problem for any experiment where the prime is not masked.

The difference in response patterns between the masked and nonmasked printing paradigms is borne out in research by Zhou, Taft, and Shu (1995). They also looked at cases where the first character of the target was orthographically the same, but morphologically different from the first character of the prime. In one condition, this was achieved by using targets that were borrowed words (e.g., 沙发, *shāfā*, meaning *sofa*) preceded by transparent primes (e.g., 沙滩, *shātān*, meaning *sandy beach*). When the prime was masked, facilitation of responses was observed, whereas when the prime was presented in clear view, there was no facilitation. We argue (as does Forster & Davis, 1984) that the masked paradigm is the more reliable guide to strategy-free processing, given that subjects are totally unable to report what the prime word is. Note that, even though a borrowed word is the most extreme case of opacity, the facilitation observed with such words as targets in a masked priming task is consistent with the explanation we gave for the lack of facilitation observed by Liu et al. (unpublished) with opaque primes. The resetting of activation levels in the morpheme representation will only be detected when it is the prime that is opaque. When the prime is a transparent compound, its morpheme representations remain in an activated state and serve to facilitate activation of the whole-word representation of any target word that contains those morphemes, even if it is opaque.

WHERE IS PHONOLOGY?

We suggest, then, that semantic units are associated with the representations of both morphemes and whole words. Is the same true of phonology? There is considerable redundancy in activating the phonology of a compound word via both its morphemes and its whole-word representation. It is only in those few cases where the morpheme has more than one pronunciation (like the 会 of 会计, *kuàijì*, and 会议, *huìyì*) that the whole word provides phonological information beyond that provided by the morpheme alone. So, is there evidence that phonology can be activated at the morpheme level as well as at the word level?

If phonology is only represented at the whole-word level and not at the morpheme level, then naming a character that is only ever found in the second position of a word should take longer than one that is only ever found in the first position. Take, for example, the bound morpheme 缭, which only ever occurs in first position (e.g., 缭乱, *liáoluàn*, meaning *confused*; 缭绕, *liáorào*, meaning *wind around*), and the bound morpheme 疚, which only ever occurs in second position (e.g., 内疚, *nèijiū*, meaning *guilty feeling*; 负疚, *fùjiū*, meaning *feel guilty*). If the only way to pronounce 缭 is to think of the pronunciation of the whole word 缭乱, for example, and the only way to pronounce 疚 is to think of the pronunciation of 负疚, then it should take longer to name 疚 than 缭 because it is the second half of the word that needs to be pronounced. That is, it should take time to skip over the first syllable of the whole word in order to pronounce the second syllable, as opposed to merely pronouncing the first syllable.

The logic of this was confirmed by Taft and Zhu (1995) using binding words whose morpheme level representation is equivalent to their whole-word form (i.e., is made up of both syllables) and, therefore, should show the effect of having to skip a syllable in pronouncing the second character. Indeed, characters that only occur in the second position of binding words (e.g., the 蚓, *yǐn*, of 蚯蚓, *qiūyǐn*) took longer to name than those that only occur in the first position (e.g., 蚯, *qiū*). When it came to bound morphemes, however, Taft and Zhu (1995) found no difference between naming responses to second position characters (e.g., 疚, *jiū*) and first position characters (e.g., 缭, *liáo*). This result seems to indicate that individual morphemes, whether they be free or bound, monosyllabic or bisyllabic, are associated with their phonology such that activation of the orthographic form of the morpheme provides a direct link to the pronunciation of that morpheme, as depicted in Fig. 5.1.

The question remains, though, whether phonology is associated with the whole word as well as the morpheme. Logically, such whole-word representations would only be required for words like 会计 and 会议 that have an ambiguously pronounced morpheme because, otherwise, the whole-word pronunciation need only be a concatenation of the pronunciation of its component morphemes.

An attempt was made by Taft and Zhu (1997b) to determine if whole-word pronunciations are represented. Using the logic that the first character of a compound word could be named by pronouncing the first half of a whole-word representation if it existed, Taft and Zhu examined whether such a naming response could be facilitated when the whole-word representation for that compound was preactivated. For example, the target word 水果 was presented, and naming responses were measured for 水 (*shuǐ*) when the target had been preactivated by a masked prime. The prime that was used was not the whole word (水果) because any facilitation that might have been found in naming the first character could then have arisen from preactivation of the morpheme-level representation of that character. Instead, the whole word was preactivated by using its second character (e.g., 果, *guǒ*) as the prime, and a facilitatory effect was observed. That is, naming of the first constituent of a compound was facilitated by the prior presentation of the second constituent of that word. This suggests that 水 can be pronounced on the basis of the whole-word representation *shuǐguǒ* because the whole word 水果 can be partly activated by the prime 果. If it were only possible to pronounce 水 at the morpheme level, then priming with 果 would have been of no help and may have even interfered with the naming of 水 when 水果 was presented.

So, our experiments suggest that phonology can be derived at both the morpheme and whole-word levels.

PROBLEMS FOR THE MULTILEVEL INTERACTIVE-ACTIVATION FRAMEWORK

The framework we used is able to account for all of the experimental data presented here. The essential feature of this framework in the current context is the fact that there are separate morpheme and whole-word levels of orthographic representation, and each of these can be associated with both semantic and phonological representations. There are (at least) two features of this conceptualization of the system, however, that could be seen as being unsatisfactory.

First, there is redundancy in both the orthographic and phonological representations. In both cases, compound words are fully represented even though they could be directly constructed from their constituents. For example, if both 手 and 枪 have orthographic representations, why have one for 手枪 as well? Even though the meaning of 小心 is not captured by the combination of the individual meanings of 小 and 心, there is no need for 小心 to be explicitly represented. Simply, 小 could activate one set of semantic units when on its own, and 心 another set, whereas they activate a different set of semantic units when activated together (i.e., the set appropriate for 小心). The same argument holds for the phonological representations.

Second, if the orthographic representations below the whole-word level are actually morphemic in nature, then homographic morphemes should really have separate representations. That is, even though *ān* meaning *quiet* and *ān* meaning *install* are both written as 安, they are different morphemes (by definition) and, therefore, at a morpheme level of representation they should be differentiated. Yet in order to explain why 安宁 (*peaceful*) primes 安装 (*install*), it is necessary to say that there is a single representation for 安 at the morpheme level. If they were separate, the *install* version of 安 would have its activation level reset after 安宁 was presented as a prime in exactly the same way that 马 (*horse*) has its activation reset after 马上 (*immediately*) is presented. In both cases, the meaning of the morpheme is incompatible with the meaning of the whole word and hence there would be no priming. So, in the current instantiation of the model, one needs to account for the priming that was observed by saying that homographic morphemes have a single representation at the morpheme level of orthography. Yet if this level is truly morphemic, all morphemes should be separately represented regardless of their homography.

A MODIFICATION TO THE FRAMEWORK

In order to handle these concerns, a modification to the multilevel interactive-activation framework is suggested. Instead of having a whole-word level for orthography and for phonology, there could be a set of modality-free units that link together orthographic character units, phonological syllable units, and semantic units. Abstract units of this type can be called "lemmas" (e.g., Levelt, 1989). These lemmas are way-stations that develop when the same semantic information repeatedly co-occurs with the same form information (either orthographic, phonological, or both). That is, these units capture the correlation between meaning and form. Such lemmas are of the type proposed by Schreuder and Baayen (1995), though they call them "concept nodes." Similar ideas about units emerging through frequent co-occurrence are expressed in distributed connectionist models of lexical processing such as that of Seidenberg and McClelland (1989) where there are "hidden units" that intervene between input and output units (e.g., between orthographic and semantic units) and whose activation pattern developes on the basis of the statistical characteristics of the co-occurrence of input and output information. Figure 5.2 illustrates the architecture of the lemma system.

Here we see orthographic units (called "access representations" by Schreuder and Baayen), each of which represents a character. These units can be seen as being at the highest level of an orthographic module, below which are radical and stroke levels as depicted in Fig. 5.1. The character unit 马 sends activation to a lemma unit representing the free form that is pronounced *mǎ* and means *horse*, as well as to all other lemma units that represent 马 in combination with another character, like the lemma that is activated by the combination of 马 and 掌 (*zhǎng*). This latter "compound" lemma is also linked to the semantic units representing *horse-related* things and the relevant semantic units associated with 掌 (summarized here as *bottom of foot*), as well as any other semantic units that distinguish 马掌

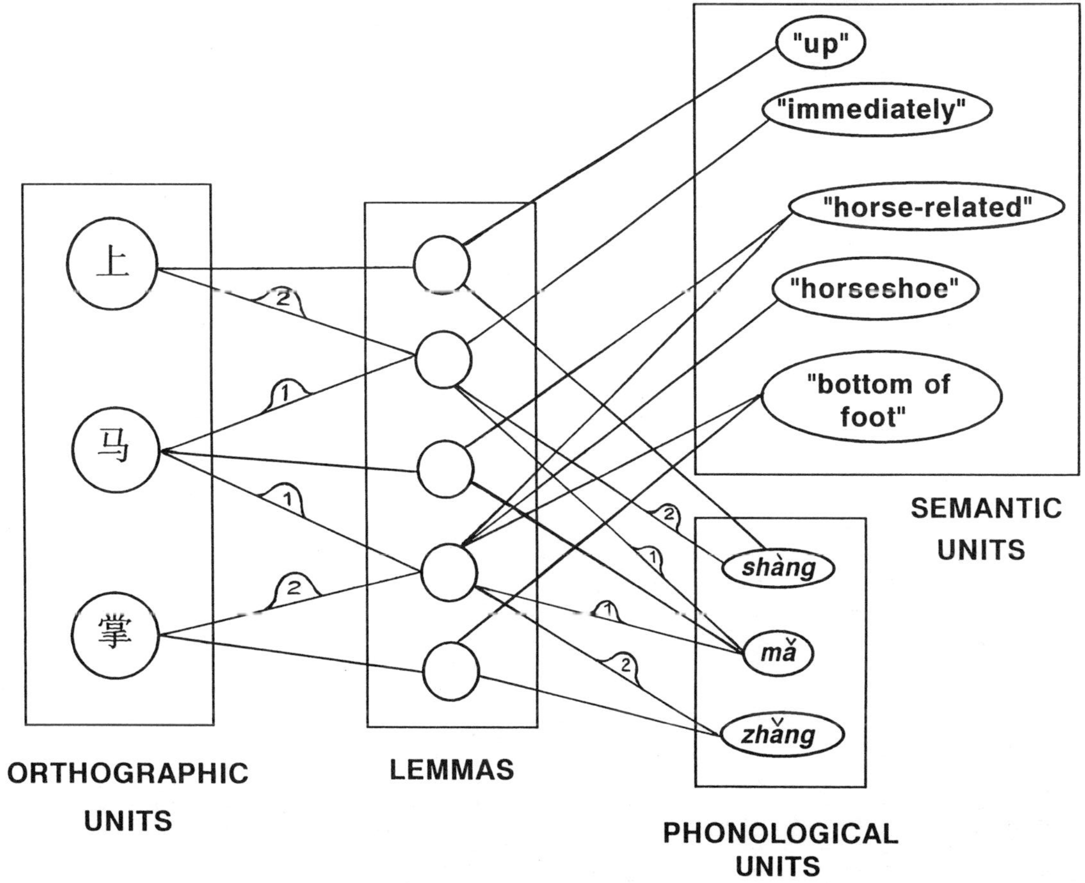

FIG. 5.2. A modification to the interactive-activation model whereby lemma levels exist and there are no orthographic or phonological representations of complete compound words.

from the mere concatenation of 马 and 掌 (like it being an object of a particular shape and function, summarized here as *horseshoe*). This lemma unit is also linked to the phonological units for *mǎ* and *zhǎng*. When there are two units at one level linked to one unit at another, the linkages need to be ordered in some way so that the nonword 掌马, for example, is not recognized as 马掌, or so that 马掌 is pronounced *mǎzhǎng* and not *zhǎngmǎ*. This ordering is depicted arbitrarily in Fig. 5.2 as numbered linkages.

Such a modification to the architecture of the theoretical framework eliminates the problems raised earlier. There is no longer the redundancy of representing the orthography and phonology of both a compound word and its constituents. In addition, there is no morphemic level of orthographic representation, so that a pair of homographic morphemes (like 安 meaning *quiet* and 安 meaning *install*) are represented together in terms of orthography, but separately in terms of function. This means that the resetting that we proposed took place in the masked priming task occurs at the lemma level. When the meaning of a component of a compound word does not coincide with the meaning of the whole word (as with the *install* meaning of the 安 of 安宁 *peaceful*, and the 马 of 马上 *immediately*), the lemma associated with that meaning will be reset. In the case of 安宁, however, the lemma representing the *quiet* meaning of 安 remains active. If we assume that an orthographic character unit remains active as long as there is an active lemma associated with it, then 安宁 will prime 安装 (*install*) via their shared orthographic unit, while 马上 will not prime 马掌 (*horseshoe*) because the orthographic unit for 马 will no longer be active. Without the postulation of lemma units that are distinct from the semantic and orthographic units, this pattern of priming results cannot be explained (unless one accepts that homographic morphemes can share the same orthographic unit despite this unit being defined morphemically).

Note that the lemma units are not equivalent to morpheme units. As can be seen from Fig. 5.2, lemmas can represent polymorphemic words as well as morphemes. So, does this mean that lemma units are equivalent to word units? Not necessarily. If lemmas are simply way-stations that develop when semantic information and form information regularly co-occur, they can develop as much for bound morphemes as for free-standing words. For example, the character 浩 will have its own orthographic representation, but since it only occurs in combination with other characters (as in 浩大 *huge*, 浩荡 *mighty*, and 浩瀚 *vast*), one might suppose that it does not have its own lemma node. However, because of the repeated activation of the semantic units representing *very big* whenever the form 浩 appears, we assume that a lemma automatically develops as a link between the character unit 浩 and the semantic units for *very big*. In fact, we assume that the lemmas for free morphemes develop in the same way. That is, because the appearance of 马 co-occurs with the meaning of *horse* when it is encountered in different contexts, it develops a lemma linking the unit 马 to *horse*. On the other hand, no lemma unit will develop for the individual characters of a binding word (e.g., 蚯 and 蚓) because they only ever occur together. Thus, there is nothing to be abstracted out for each of the individual characters. The meaning *earthworm* co-occurs with the form 蚯蚓 and cannot be decomposed any further, so there is no meaning that can be ascribed to 蚯 and 蚓 alone.

The existence of lemma nodes for bound morphemes, and not for the characters of binding words, allows one to explain the results described earlier in relation to the naming of characters that only ever occur in the second position (Taft & Zhu, 1995). When that character is a bound morpheme (e.g., 疚) it takes no longer to name than one that only appears in the first position (e.g., 缭), whereas when it is the second part of a binding word (e.g., 蚓), it does take longer to name than a first position character (e.g., 蚓). The latter occurs because, when naming 蚯, one needs to ignore the "number 1" link from lemma to phonology (i.e., that between the lemma for 蚯蚓 and the pronunciation *qiū*) and to use the "number 2" link (i.e., that between the lemma for 蚯蚓 and the pronunciation *yǐn*). For bound morphemes, however, the phonology is linked directly to the lemma for each character, so it does not matter where the character typically appears within a compound word.

The main problem faced when conceptualizing the system in this way is to explain how real words (e.g., 握) can be distinguished from bound morphemes (e.g., 浩). If a lexical decision response is made on the basis of lemma activation, a bound morpheme will presumably be treated as a real word. Although it may at times be difficult for the average Chinese speaker to decide that a bound morpheme is not a word (not a 词, *cí*), it is certainly not impossible to do so. This suggests that semantic information is also involved in discriminating between a bound and a free morpheme. The status of the bound morpheme might be determined on the basis of the fact that the set of semantic units activated by its lemma is entirely contained within the set of semantic units associated with other lemmas. That is, a free morpheme will be associated with some semantic units that are unique to that morpheme, while a bound morpheme will not. For example, there will be units associated with 马 that are not associated with the compounds that include 马, namely all the things that uniquely make a horse a horse as opposed to a pony, a stable, or a horseshoe. So there actually is information available in the system that allows one to distinguish free from bound morphemes, though it is clearly not an easy thing to do (see Taft & Forster, 1975, for an example of the difficulties involved in classifying bound morphemes as nonwords in English).

REVISITING THE DEFINITION OF A MORPHEME

If we adopt the lemma version of the interactive-activation model just proposed, we eliminate the need to define what a morpheme is for the purposes of word recognition. The units of analysis that develop are either based on physical form (i.e., characters and syllables) or ones that simply capture the co-occurrence of those form units with meaning. When an orthographic form is repeatedly associated with a particular meaning, this association becomes hardwired via a linking way-station, namely, a lemma. Some of these lemmas represent what a linguist would define as a morpheme, and some represent a polymorphemic word, but the definition does not really matter. All that matters is whether there is a correlation between form and meaning.

One remaining question, though, concerns bound morphemes. How much correlation between form and meaning is required before a lemma develops for these characters? Presumably, the appearance of a lemma is not an all-or-none matter, but rather a gradual development with an increasing representation strength (see Schreuder & Baayen, 1995). If the correlation between form and meaning increases with exposure to that character in context, so will the representation strength of the lemma.[3] Thus, there is no explicit point at which one can definitely say that a character is a bound morpheme as opposed to merely being a visual label for a meaningless syllable, and therefore we do not need to define what is and what is not a bound morpheme. It is a fuzzy boundary that depends on the quantity and quality of the exposure one has to the words of one's language.

Finally, we must consider where radicals fit into the modified interactive-activation system. It was pointed out earlier that radicals have characteristics that are typical of a morpheme, in that many are consistently associated with a particular meaning. For example, the characters in which 氵 occurs are often related to *water* (e.g., 泳 *swimming*, 海 *sea*) and the characters in which 钅 occurs are usually related to *metal* (e.g., 铁 *iron*, 银 *silver*). There are at least two ways in which this radical semantic information could be captured in the model.

First, the interface between the orthographic module and the lemma units need not be restricted to the character level, even though this is the highest level within the orthographic module. That is, the consistent co-occurrence of a radical with a particular meaning could lead to the development of a lemma unit linking that radical to a set of semantic units. Such an idea means that there are lemmas that have no links to the phonological module, since

[3]Strength of lemma representation is another cue for deciding whether a character is a word (词) or not. If lemma strength is reasonably low, then it is likely that the character is not a free-standing word because the correlation between form and meaning for free-standing words should be very high. That is, a word is a word because of its highly consistent meaning in the face of different contexts.

any semantic radical that is not a character in its own right (e.g., 氵) has no pronunciation. Alternatively, it may be that lemmas are only associated with characters and not with radicals, and that one only knows about the meaning of a radical by attending to the semantic units that become strongly activated via the characters in which the radical occurs. That is, one knows that 氵 is related to *water* by virtue of the fact that many of the character units that are partly activated by it (e.g., 泳, 海) all activate the semantic units for *water*.

Whichever explanation is given, it remains the case that whether one considers a radical to be a morpheme or not is purely an academic point. It is unimportant for the purposes of processing characters in reading.

CONCLUSION

We presented evidence here that Chinese compound words are recognized via a processing of their characters. It appears that this processing is sensitive to the contribution that those characters make to the meaning of the compound word in which they occur, in that there are differences observed between transparent and opaque compounds. A framework was presented for thinking about this analytic processing whereby morphemic units are explicitly represented (cf. Fig. 5.1). On considering some conceptual problems with this framework, however, a modification was suggested in which there are no explicit morphemic units, but rather a set of abstract units linking orthographically based character units with semantic units (and phonologically based syllable units). These lemma units simply emerge from the co-occurrence of orthographic forms with meaning (cf. Fig. 5.2). The empirical evidence for our conceptualization of the recognition system for compound words is still quite scanty and clearly more is needed to confirm the ideas proposed in this chapter.

ACKNOWLEDGMENTS

Preparation of this chapter and the research conducted by the senior author, which is referred to therein, was supported by a grant from the Australian Research Council.

REFERENCES

Andrews, S. (1986). Morphological influences on lexical access: Lexical or nonlexical effects? *Journal of Memory and Language, 25*, 726–740.

Aronoff, M. H. (1976). *Word formation in generative grammar*. Cambridge, MA: MIT Press.

Bybee, J. (1995). Diachronic and typological properties of morphology and their implications for representation. In L. Feldman (Ed.), *Morphological aspects of language processing* (pp. 225–246). Hillsdale, NJ: Lawrence Erlbaum Associates.

Chao, Y. R. (1968). *A grammar of spoken Chinese*. Berkeley: University of California Press.

Chen, H.-C. (1992). Reading comprehension in Chinese: Implication for character reading times. In H.-C. Chen & O. J. L. Tzeng (Eds.), *Language processing in Chinese* (pp. 175–205). Amsterdam: North-Holland.

Emmorey, K. D. (1989). Auditory morphological priming in the lexicon. *Language and Cognitive Processes, 4*, 73–92.

Fan, K. Y., Gao, J. L., & Ao, X. P. (1984). Pronunciation principles of Chinese characters and alphabetic script. *Chinese Character Reform, 3*, 23–27. [In Chinese]

Forster, K. I., & Davis, C. (1984). Repetition priming and frequency attenuation in lexical access. *Journal of Experimental Psychology: Learning, Memory, and Cognition, 10*, 680–698.

Hockett, C. F. (1958). *A course in modern linguistics*. New York: Macmillan.

Hoosain, R. (1992). Psychological reality of the word in Chinese. In H.-C. Chen & O. J. L. Tzeng (Eds.), *Language processing in Chinese* (pp. 111–130). Amsterdam: North-Holland.

Levelt, W. J. M. (1989). *Speaking: From intention to articulation*. Cambridge, MA: MIT Press.

Li, C., & Thompson, S. (1981). *Mandarin Chinese: A functional reference grammar*. Berkeley, CA: University of California Press.

Lima, S. D., & Pollatsek, A. (1983). Lexical access via an orthographic code? The Basic Orthographic Syllabic Structure (BOSS) reconsidered. *Journal of Verbal Learning and Verbal Behavior, 22*, 310–332.

Lin, Y. T. (1971). *Chinese-English dictionary of modern usage*. Hong Kong: Chinese University of Hong Kong.

Liu, Y., & Peng D.-L. (1995). *Time-course of Chinese compound word recognition*. Poster presented at the Seventh International Conference on the Cognitive Processing of Chinese and Other Asian Languages, Hong Kong.

Lyons, J. (1969). *Introduction to theoretical linguistics*. Cambridge, England: Cambridge University Press.

Monsell, S. (1985). Repetition and the lexicon. In A. W. Ellis (Ed.), *Progress in the psychology of language* (Vol. 2, pp. 147–195). London: Lawrence Erlbaum Associates.

Perfetti, C. A., & Zhang, S. (1995). Very early phonological activation in Chinese reading. *Journal of Experimental Psychology: Learning, Memory, and Cognition, 21*, 24–33.

Sandra, D. (1990). On the representation and processing of compound words: Automatic access to constituent morphemes does not occur. *The Quarterly Journal of Experimental Psychology, 42A*, 529–567.

Schreuder, R., & Baayen, R. H. (1995). Modeling morphological processing. In L. Feldman (Ed.), *Morphological aspects of language processing* (pp. 131–154). Hillsdale, NJ: Lawrence Erlbaum Associates.

Seidenberg, M. S., & McClelland, J. L. (1989). A distributed, developmental model of word recognition and naming. *Psychological Review, 96*, 523–568.

Taft, M. (1994). Interactive-activation as a framework for understanding morphological processing. *Language and Cognitive Processes, 9*, 271–294.

Taft, M., & Forster, K. I. (1975). Lexical storage and retrieval of prefixed words. *Journal of Verbal Learning and Verbal Behavior, 14*, 638–647.

Taft, M., & Forster, K. I. (1976). Lexical storage and retrieval of polymorphemic and polysyllabic words. *Journal of Verbal Learning and Verbal Behavior, 15*, 607–620.

Taft, M., Huang, J., & Zhu, X. (1994). The influence of character frequency on word recognition responses in Chinese (pp. 59–73). In H.-W. Chang, J.-T. Huang, C.-W. Hue, & O. Tzeng (Eds.), *Advances in the study of Chinese language processing* (Vol. 1). Taipei: Department of Psychology, National Taiwan University.

Taft, M., & Zhu, X. (1995). The representation of bound morphemes in the lexicon: A Chinese study (pp. 293–316). In L. Feldman, *Morphological aspects of language processing*. Hillsdale, NJ: Lawrence Erlbaum Associates.

Taft, M., & Zhu, X. (1997a). Sub-morphemic processing in reading Chinese. *Journal of Experimental Psychology: Learning, Memory, and Cognition, 23*, 761–775.

Taft, M., & Zhu, X. (1997b). Using masked priming to examine lexical storage of Chinese compound words. In H.-C. Chen (Ed.), *Cognitive processing of Chinese and related Asian languages*. Hong Kong: The Chinese University Press.

Tamaoka, K., & Hatsuzuka, M. (1995). The effects of Kanji printed frequency on processing Japanese two-morpheme compound words. *The Science of Reading, 39*, 121–137. [In Japanese with English abstract]

Tan, L. H., Hoosain, R., & Peng D.-L. (1995). Role of early presemantic phonological code in Chinese character identification. *Journal of Experimental Psychology: Learning, Memory, and Cognition, 21*, 43–54.

van Jaarsfeld, H. J., & Rattink, G. E. (1988). Frequency effects in the processing of lexicalized and novel nominal compounds. *Journal of Psycholinguistic Research, 17*, 447–473.

Zhang, B. Y., & Peng D.-L. (1992). Decomposed storage in the Chinese lexicon (pp. 131–149). In H.-C. Chen, & O. J. L. Tzeng (Eds.), *Language processing in Chinese*. Amsterdam: North-Holland.

Zhou, X., Taft, M., & Shu, H. (1995). *Orthographic and phonological priming in Chinese*. Paper presented at the Seventh International Conference on the Cognitive Processing of Chinese and Other Asian Languages, Hong Kong.

Zhu, X., & Taft, M. (1994). The influence of perceptual experience on Chinese character processing (pp. 85–99). In H.-W. Chang, J.-T. Huang, C.-W. Hue, & O. Tzeng (Eds.), *Advances in the study of Chinese language processing* (Vol. 1). Taipei: Department of Psychology, National Taiwan University.

Zwitserlood, P. (1994). The role of semantic transparency in the processing and representation of Dutch compounds. *Language and Cognitive Processes, 9*, 341–368.

CHAPTER

6

The Constituency Model of Chinese Word Identification

Charles A. Perfetti
Li Hai Tan
University of Pittsburgh

Our goal in this chapter is to provide an account for Chinese word identification, informed by recent research and reflecting principles that seem to underlie the results of this research. The first section discusses basic issues arising in an analysis of Chinese writing, and introduces the descriptive and theoretical vocabulary required for clear discussion. The second section summarizes the results of research that has pointed to a significant role for phonology in Chinese word reading. The third section describes a theory of word identification, developing further ideas published in Tan and Perfetti (1997) and Perfetti and Tan (1998).

WORDS, CHARACTERS, AND OTHER GRAPHOLINGUISTIC OBJECTS

We begin with a brief discussion of the components of Chinese writing because the richness of the system entails correspondingly rich possibilities for reading and because it is so easy to mislead, even with careful use of terms. It is, after all, important for a model of word identification to be clear on what a word is. In fact, it may not be possible to be completely clear, but it is necessary to try.

Characters and Components

The basic unit of the modern Chinese writing system is the *character*, which maps onto a syllabic morpheme rather than a phoneme in the spoken language. Because our focus is on the orthography of the writing system and not its visual appearance (the script), we ignore some aspects that may be important for many purposes: the variations in script conventions (e.g., traditional vs. simplified) and the transparency of the meaning of a character (e.g., pictographs vs. abstract logographs). We follow the tradition of scholars of Chinese who have taken the character as the basic writing unit and have classified this unit in various ways (DeFrancis, 1989; Wang, 1981; Zhou, 1978). The characters are composed by basic *strokes* expressed according to strict conventions. There are 24 distinct strokes, with the four major strokes (一 丨 丿 乚) accounting for 74% of occurrence frequency (Zheng, 1983). It is the character, however, not the stroke, that is basic to the writing system. We may say that the stroke is the basic (minimal) compositional unit, whereas the character is the basic orthographic unit.

Each Chinese character is pronounced as an open syllable, which may be analyzed into initial and final segments. In present-day usage, nearly all characters are free morphemes with rich meanings. Only the very few characters lacking a separate meaning are required to combine with other characters, forming two-syllable morphemes (e.g., 玻璃, /bo1 li4/, *glass*; 葡萄, /pu2 tao2/, *grape*. The numerals following pinyin refer to tones.)

Characters may be classified into two categories in terms of structural complexity: simple characters and compound characters. Simple characters are not further divisible into distinct components. Pictographs (e.g., 刀, /dao1/, *knife*) and other transparent characters (e.g., 刃, /ren4/, *the edge of a knife*) belong to this category. Compound characters, however, contain distinct components. There are two kinds of compound characters—semantic compounds and phonetic compounds. Semantic compounds are composed of two components, both of which may hint at the meaning of compounds. For example, the semantic compound 明 (/ming2/, *bright*) contains two simple characters 日 (/ri4/, *sun*) and 月 (/yue4/, *moon*). The compound 阳 (/yang2/, *sun*) has a semantic component that is not an independent character (阝) and a simple character 日 (/ri4/, *sun*). According to our study of the *Modern Chinese Dictionary* (1995), which contains 9,641 (simplified) characters, there are 1,341 simple characters (pictographs and self-explanatory characters) and semantic compounds, which together amount to about 15% of the present-day used characters.

The second, more frequent (85% of modern characters) type of compound is usually referred to as a "phonetic compound." Phonetic compounds are composed of two functionally independent components: One component may signal the meaning category of the whole compound, whereas another may signal something about the pronunciation of the compound. The first is often called the *significate*, the second, the *phonetic*. Significates are typically, but not always, on the left half of the compound; phonetics are typically, but not always, on the right. Our own study of the *Modern Chinese Dictionary* (1995) indicates 192 significates and 1,499 phonetics. This estimate of significates is very close to that of other researchers based on other dictionaries. Our estimate for number of phonetics, however, exceeds some of the earlier estimates of 800 phonetics (Taylor & Taylor, 1983). Our study further indicates that 75% of the 8,300 phonetic compounds have their significate on the left and their phonetic on the right. Some significates are characters on their own (e.g., the significate 月 in the phonetic compound 胆 and 脍), some are not (e.g., 扌 in the compound 据 抵 拉). On the other hand, almost all phonetics are themselves real characters, with very few exceptions.

The function of the two components in a phonetic compound has been widely noted. Most significates provide a clue to the meaning, and many phonetics provide a clue to pronunciation. However, many compounds have phonetics that do not serve the whole compound character. Table 6.1 illustrates a number of phonetic compounds, varying in the degree to which their components provide phonetic or semantic information. The character for the meaning ocean 洋, known as *yang*(2), contains as its right-side component the character (the phonetic) 羊, which is also pronounced *yang*(2). The left side of the ocean character is 氵, a noncharacter significate referring

TABLE 6.1
Validity of Phonetic Compounds and Significates in Phonetic Compounds

	Phonetics on the Right		*Phonetics on the Left*	
Significates	*Valid*	*Invalid*	*Valid*	*Invalid*
Valid				
Compound character	洋	炉	削	鹅
Character pronunciation	/yang2/	/lu2/	/xiao1/	/e2/
Character meaning	ocean	stove	cut	goose
Significate	氵	火	刂	鸟
Significate meaning	water	fire	knife	bird
Phonetic component	羊	户	肖	我
Phonetic pronunciation	/yang2/	/hu4/	/xiao1/	/wo3/
Invalid				
Compound character	但	捌	歌	胡
Character pronunciation	/dan4/	/ba1/	/ge1/	/hu2/
Character meaning	but	eight	song	recklessly
Significate	亻	扌	欠	月
Significate meaning	human being	action or behavior	owe	body
Phonetic component	旦	别	哥	古
Phonetic pronunciation	/dan4/	/bie2/	/ge1/	/gu3/

to something relating to "water." Thus the reader of this character has information about both its meaning and its pronunciation. However, the character 炉 for the word corresponding to *stove* has a right-side component lacking any phonetic value for this compound. In isolation, this component is a simple character pronounced *hu*(4), and some compounds that contain this character have a pronunciation that is segmentally the same, *hu*(4); but not the compound for *stove*, which is *lu*(2). The left side of *lu*(2) is a character significate, meaning *fire*, pronounced *huo*3.

One might argue that even when the compound and the component are not identical in pronunciation, as in the last example, phonetic overlap in the vowel portion of the syllables provides useful phonetic information for the identification of the compound. This may be, and researchers who have tried to classify compounds have acknowledged partial cues (DeFrancis, 1989; Zhou, 1978); however, it is unclear to what extent such information is helpful during actual reading. If we take a strict view of phonetic overlap, requiring that the pronunciations of a compound and its component be identical in their segments (but not their tones), then the majority of compounds are pronounced differently from their components. Zhou (1978), using a complex averaging statistic, estimated that about 39% of compounds had components with a helpful pronunciation. DeFrancis (1989), with a different procedure, made a similar estimate (42%) for segmentally identical pronunciations. (However, DeFrancis' additional inclusion of partial segmental overlap (24%) led him to suggest a higher overall estimation of the frequency of helpful phonetic information.) In a sample of 300 compound characters, Perfetti, Zhang, and Berent (1992) found that 37% of compounds contained a valid (segmentally identical) phonetic, a figure very close to that of Zhou. They also found that as the frequency of the compound decreased, both its semantic component's validity and its phonetic component's validity increased. The extent to which partial phonetic overlap, as opposed to segmental identity, between a component and a compound is functionally facilitative has yet to be determined by research. It is possible that partial phonetic identity aids in "problem solving" in reading, where the reader encounters an unfamiliar character and attempts to infer its meaning, more than it does in the fluent reading of familiar characters.

In summary so far, many Chinese characters, about 85%, are phonetic compounds; the components of these compounds sometimes provide information about meaning and pronunciation, with the meaning-cueing component always different from the phonetic-cueing component. However, this component information is far from perfectly reliable. It functions well as a reader encounters a rare word–the odds increase that the reader can figure out something about the meaning and be able to infer its pronunciation. In addition, most phonetic components are characters on their own. Many significates, on the other hand, are themselves noncharacters. Thus, when a compound character consists of two components that are themselves real characters, there is a complex interaction, orthographically, between the whole character and its two constituents. How this affects word identification is a question that must be addressed eventually by any model of word identification in Chinese. We return to this question in the last section.

Other orthographic aspects of characters also require brief mention, especially the common use of *radical* to refer to a component that has semantic value. As a matter of fact, radical is a traditional neutral term that covers all components, one that ignores their possible meaning or phonetic functions. In addition to their imperfect tendency to give semantic or phonetic hints, it is possible that radicals have special roles in character reading, owing to their frequency, their high combinability, or their function in organizing dictionaries (Feldman & Siok, 1997; Taft & Zhu, 1997). Although most compounds contain two components, some have three or more components, with at least one having no phonological or semantic function.

In summary, if a character is a compound, it has two or more components, one that may or may not correspond to an element of the meaning of the compound, and one that may or may not correspond to its pronunciation. The function of components, and their frequency, combinability, even their position, may influence compound character recognition.

Single- and Multicharacter Words

Most characters are words, but are most words characters? If we follow the linguistic tradition, taking a word to be the minimum free unit capable of independent usage (Bloomfield, 1926), we conclude that most single characters can occur as words. However, in daily usage, these characters usually combine with other characters to compose compound words, that is, two-character words. According to the *Modern Chinese Frequency Dictionary* (1986), two-character words are twice as frequent in texts as one-character words, with fewer than 2% being words of three or more characters. Thus, most characters are words, but most words are not characters. There are some interesting correlates of this fact. For example, because syntactic functions are controlled by words or morphemes rather than by characters, such functions are different when a given character is a word than when it is merely part of a word. The difference between single characters and multicharacter words led Chao (1968) to distinguish the *sociological word* (zi4), corresponding to the nonlinguistic public's concept of a word, from the more technical *syntactic word* (ci2).

More generally, the fact that readers more often encounter words of two characters than words of one character has implications for a complete account of reading processes in text. For one thing, there is the semantic compounding to consider. Although the pronunciation of constituent characters of a two-character word is usually that of their pronunciation in isolation, the meanings of constituent characters may not hint at the meaning of the whole compound words. Even more basic is the text segmentation problem introduced by multicharacter words. Because normal Chinese text displays characters separated by equal spacing, irrespective of word boundaries, simply finding words can be a problem. Indeed, Hoosain (1992) reported high variability in where subjects place word boundaries in normally spaced sentences. Perfetti and Tan (1996) reported experiments suggesting that readers attempt to segment character strings into two-character components rather than assuming that each character is a word. In these experiments, subjects read lexical garden path sentences and matched control sentences. For example, sentences (1) and (2) below are identical except for the fifth character. In (1) the fifth character creates an ambiguity that allows the following character to be part of a two-character word, whereas the underlined disambiguating region requires that it be a single character word. Sentence (2) has a fifth character that does not combine with the sixth character, thus avoiding the lexical garden path.

(1) 经理同意照顾客的想法来设计产品

(2) 经理同意按顾客的想法来设计产品。

The reading times were longer for the underlined region in sentence (1), suggesting that subjects first combined the character 照 with the following character 顾, forming a two-character word 照顾, initiating a garden path. Perfetti and Tan (1996) suggested that a two-character assembly strategy is adopted in reading Chinese sentences.

These issues, although beyond our present purpose, are important in the long run if a theory of word reading in texts is to be developed. In our narrower context, the question is the definition of *word*, to set the scope for a theory of word identification. There are two possibilities. One is to take word finding as part of what a theory has to account for. We think this is more than can be accomplished at present; it also works against comparisons with the research on word identification in alphabetic writing systems, which focused on single monomorphemic words in the development of models of identification. Thus, we take the second option, which is to take single-character words as the primary target of a model of word identification. This does not exclude attention to multicharacter words. Instead we assume that identification processes occur at the character level, whether a given character turns out be a word or a morpheme participating in a multicharacter (multimorphemic) word.

Before leaving this issue, we can try to make sense of our approach on analogy with English. Models that address the orthographic and phonological units responsible for word identification are based largely on four- and five-letter words presented in isolation. This allows experimental control over both constituent orthography (letters) and phonology (phonemes, onsets, rimes, syllables). However, English readily allows a kind of compounding that is analogous to Chinese. Thus, *blackbird* is an English compound just as 美丽 (/mei3 li4/, *beautiful*) is a Chinese compound. The difference is mainly in the spacing, although one must avoid the temptation to think the spacing convention in English is arbitrary. Instead, it partly reflects linguistic differences, as the stress difference between *blackbird* (lexically defined) and *black bird* (with stress assignable according to discourse principles). English, in fact, has many noun compounds closer to the Chinese case: *sky scraper, commuter train, race horse, fly fisher* are all compounds of this type; writers are sometimes tempted to hyphenate them to reflect their one-wordness. We are not sure whether such examples, when joined with Chinese, should raise a question about the basic idea of a word, a concept that linguists have traditionally avoided in favor of morpheme. However, discovering at least some of the orthographic and phonemic units that are involved in word identification, in either English or Chinese, does not require multimorphemic compounds, as interesting as these are for other issues. Thus, we address multicharacter words to the considerable extent that single-character word principles apply to them.

CONSTITUENTS OF CHINESE WORD IDENTIFICATION

Having defined descriptive terms and discussed problems with words, we turn now to a brief review of the state of affairs provided by research on Chinese word reading. Our focus is especially on phonology as a constituent of word identification because of discoveries that phonology may have a role in identification. However, we include attention to meaning and orthographic components as well.

As part of the increased research interest in Chinese language processes, which has stimulated published international conferences for the last 10 years or so (e.g., Kao & Hoosain, 1986; Chang, Huang, Hue, & Tzeng, 1994), there has been attention to the role of phonology in reading Chinese. In our own research, results have implicated a role for phonology at the word level (Lam, Perfetti & Bell, 1991; Perfetti & Tan, 1998; Perfetti & Zhang, 1991, 1995a, 1996; Tan, Hoosain, & Peng, 1995; Tan, Hoosain, & Siok, 1996; Tan & Perfetti, 1997) and the text level (Zhang & Perfetti, 1993). The conclusions at the text level converge with earlier research (Tzeng, Hung, & Wang, 1977), while establishing a link between phonological memory and comprehension. However, at the word level, recent research departs in finding phonology where standard accounts have emphasized the morphological–semantic basis of

Chinese reading (e.g., Baron & Strawson, 1976; Chen, Yung, & Ng, 1988; Hoosain & Osgood, 1983; see Hoosain, 1991, for a review). These newer conclusions have received confirmation in other laboratories (e.g., Chua, 1995; Leck, Weekes, & Chen, 1995; Zhang, Feng, & He, 1994). Although it may be useful to review the evidence in detail for some purposes, we instead highlight only a few of the results that we see as especially informative for the present theoretical goal.

1. Phonology is involved when readers make meaning judgments. Perfetti and Zhang (1995a) found that decision times for whether two characters had the same meaning were slowed when two characters of different meaning had the same pronunciation. This homophone effect was observed when the two characters were presented with only 90 ms between the onset of the first and the onset of the second. The interference pattern was also seen in error rates, and this is the case in all other studies producing *a homophone interference effect.*

2. This homophone effect does not depend on phonetic components of the characters. In her dissertation, Zhang (1996) varied whether the two homophones that produced the effect reported in Perfetti and Zhang (1995a) had valid phonetic components. In the valid condition, the name of the character was the same as the name of its component. In the invalid condition, the name of the character was different from the name of its component. There was no effect of this manipulation on the homophone interference effect. A homophone interference occurred so long as the two characters had the same name. The pronunciation of the subcharacters did not matter.

3. This homophone effect does not depend on asynchronous presentation of characters. Zhang (1996) presented two characters simultaneously, requiring meaning judgments. She found the same homophone interference reported by Perfetti and Zhang (1995a) when characters were presented asynchronously between 90 and 310 ms.

4. A homophone effect is also observed with two-character words. Tan and Perfetti (1998) carried out a two-character word experiment parallel to the Perfetti and Zhang single-character word experiment. Very strong interference effects were found when the two-character words had the same two-syllable pronunciation with simultaneous presentation and with SOAs of 71 and 157 ms.

5. Lexical decisions to two-character words show evidence of phonology. Tan and Perfetti (1997, Experiment 2) varied whether one character of a two-character word was phonologically ambiguous, that is, whether it had more than one pronunciation. The two-character sequence was a word only with one of the pronunciations. The lexical decision was slowed in this condition, compared with a control condition in which both characters had unique unambiguous pronunciations. This consistency effect was inde-

pendent of whether the ambiguous character appeared on the right or the left of the sequence.

6. Although phonological effects were not observed in brief exposure masking with targets presented approximately at identification threshold (Perfetti & Zhang, 1991; Tan et al., 1996) nor when target durations were fixed at 50 ms (Tan et al., 1995), phonological effects were found at slightly longer (60ms) durations (Tan et al, 1995). This pattern of results suggests phonological activation accompanies the identification of a single-character word, but does not precede it.

7. In naming, phonological priming effects are observed directly by homophone primes (Perfetti & Zhang, 1991), even with primes exposed as briefly as 57 ms (Perfetti & Tan, 1998). Furthermore, phonological effects are observed indirectly in phonologically mediated priming under favorable conditions (Tan & Perfetti, 1997). In mediated priming, a homophone to a synonym of a target word facilitated that word indirectly through its phonological links to the synonym. The favorable conditions refer to the fact that this effect was found when the number of homophone characters for the prime character was relatively small (5 or fewer), but not when the number was relatively large (20 or more). This suggests that activation of a name that has to be shared among many characters is too diffuse to be effective for any single character.

8. In naming, semantic priming was influenced by the number of homophones of a semantic prime. The priming effect was greater when semantic primes had few homophones than when semantic primes had many homophones (Tan & Perfetti, 1997).

9. In no case in which an experiment could observe both semantic effects and phonological effects were semantic effects stronger or earlier occurring than phonological effects. Where the time course of direct effects was explicitly tested, phonological effects preceded semantic effects (Perfetti & Zhang, 1995a; Perfetti & Tan, 1998; Tan et al., 1995, 1996). The only situation in which a stronger semantic effect has been observed is in mediated priming, where a synonym produced greater (but not earlier) priming than a homophone of a synonym (Tan & Perfetti, 1997).

10. Meaning effects are influenced by the meaning precision of a character, both in backward masking (Tan et al., 1996) and in primed naming (Perfetti & Tan, 1998). Characters with vague meanings, referentially indeterminate or context-dependent, produce slower priming effects than a character with precise meanings.

11. Graphic information initiates character identification, of course; however, there is evidence that graphic similarity between prime and target produces inhibition once the prime has been exposed sufficiently for phonological effects to be produced (Perfetti & Tan, 1998).

These 11 facts can be used to constrain a general model of word identification. They suggest that word identification in Chinese is a process in which graphic forms get connected to phonology as well as semantics. Furthermore, they provide no reason to think that semantic information is privileged, that is, that semantic outcomes are stronger than or earlier occurring than phonological outcomes. This last conclusion must be conditional and tentative, because it is possible that other experimental paradigms may discover some semantic privilege in the right conditions. But an important role for phonology in word identification seems to be a secure conclusion.

A THEORY OF WORD IDENTIFICATION

Given the development of models of word identification in recent years, one expects a theory of word identification to be computationally explicit or at least computationally tractable. Nevertheless, we believe it perhaps more important to be explicit about the central principles that control the details of any theory. We return to a description of a model of word identification after a discussion of the deeper principles that organize the observed facts of reading and must be represented in a theory that explains these facts.

Perfetti et al. (1992) proposed one such principle, the Universal Phonological Principle (UPP), stating that encounters with printed words activate multiple levels of phonology in all writing systems, which control only the details of activation. Thus, Chinese characters, as well as alphabetically encoded words, are read with the activation of phonology. The UPP is generally compatible with a proposal made by Frost, Katz, and Bentin (1987), the Orthographic Depth Hypothesis. However, the UPP emphasizes that phonology occurs at multiple levels, and the absence of connections at the phoneme level does not matter for phonology at other levels. The UPP assumes that all writing systems developed some degree of dependency on spoken language, even when they have a semantic–morphological component (DeFrancis, 1989; Mattingly, 1992). Reading, accordingly, must be a process in which meaning and phonology are brought together with a written word.

This description of reading suggests a further step, that orthography, phonology, and meaning are interconnected constituents of word identification. The Constituency Principle states that phonology is one of three integral constituents of word identification, and not a mere optional "code." Word identification is thus the momentary convergence of the three constituents—orthographic form, phonological form, and some nonformal (meaning) semantic substance. Defective identification occurs when one of these constituents is missing or degraded.

It is useful to emphasize that the Constituency Principle departs from accounts of reading that refer to *lexical access*, where word reading is begun by a minimal and abstract event (*access*) in which the phonological form of a word, and in some descriptions even its meaning, are irrelevant. In contrast to lexical access, word identification allows one to speak of the constituents of an identified object. (See Perfetti & Zhang [1995b] for a further discussion of the contrast between lexical access and word identification.)

Identification is the retrieval of a word-object defined as a three-way relationship among phonological word form, graphic form, and semantic possibilities. The identified object, *the word*, can be described in the language of existential propositions as the conjunction of three information sources (e.g., the lexical object such that it is spelled *dog*, has the phonological form /dawg/, and means whatever it is that *dog* means). A cognitive moment of identification suggests a short lived integration of constituents—a simultaneous availability of graphemic, phonological, and semantic information. These constituents accumulate in overlapping time courses, beginning with graphic information. Even so, a phonological word form may be available prior to the complete specification of a graphemic representation (Lukatela & Turvey, 1994; van Orden, 1987).

One consequence of the Constituency Principle is that it makes the mediation question seem odd in its usual form, that is, does phonological information bring about meaning access? In constituency, the role of phonology is not mediation in the usual sense—phonology used for something else—just as it is not some optional postlexical code transformation. Instead, constituency implies an identification moment that includes a reference to phonological form as well as meaning. Our account here is very much in tune with the resonance proposal of van Orden and Goldinger (1994). In their very apt metaphor, mediation takes on the sense of negotiations, as used in *labor mediation*, for example. A negotiation among different sources of information, in which phonology plays a major role, is carried on through recurrent feedback and feedforward, quickly settling on an identity.

In a nonalphabetic writing system, the Constituency Principle applies in the same way, although the processing details may differ. To take a Chinese case parallel to our English *dog* example, the identification of the lexical object that it is written 狗 has the syllable (word) pronunciation /gou3/ and means whatever it is that 狗 means, (which can include *dog*, *damned*, *cursed*). Chinese has graphic–phonological relationships defined over the syllable–lexical levels rather than the grapheme–phoneme level. The basic phonological constituent of word identification, accordingly, may be the syllable, which is usually also a lexical morpheme. Thus, the processes linking orthography and phonology may be different in some details from those in English. Perfetti and Tan (1998), based on time course data from primed naming, suggest that, unlike alphabetic systems, Chinese may involve a more fully specified orthographic representation prior to the activation of a phonological form. However, the Constituency Principle holds equally for Chinese and alphabetic systems.

In summary, the Constituency Principle assumes that interconnected lexical constituents are made available during word reading. Word identification is the convergence of these constituents. The word representation system itself ordinarily provides these constituents in strongly interconnected form. An acquisition corollary to this principle is that reading skill develops from poorly interconnected to highly interconnected lexical components (Ehri, 1980; Perfetti, 1992). The same acquisition corollary may hold for Chinese (Perfetti & Zhang, 1996).

THE INTERACTIVE CONSTITUENCY MODEL OF CHINESE WORD IDENTIFICATION

We developed a model of word identification over several years, first presented in Tan and Perfetti (1995), and then published in Tan and Perfetti (1997) and Perfetti and Tan (1998). In Tan and Perfetti (1997), we emphasized the parts of the model that represent the mappings among character orthographic, character phonological, and meaning lexicons, explaining the finding that the number of homophones a word has determines the extent to which phonological "mediation" is observed in the phonologically mediated priming task (Fleming, 1993; Lesch & Pollatsek, 1993). In Perfetti and Tan (1998), we emphasized aspects of the model needed to explain time course effects observed in primed naming with graphic, phonological, and semantic primes. Moreover, Tan and Perfetti (1995) elaborated parts of the relationships between character-level information and subcharacter-level information. We referred to this model as the *Interactive Constituency Model* (Perfetti & Tan, 1998; Tan & Perfetti, 1995, 1997). We present that model here as Fig. 6.1.

The model follows the basic assumptions of interactive-activation models (McClelland & Rumelhart, 1981; cf. Grainger & Ferrand, 1994). It assumes four separate constituent representation subsystems: the character orthographic subsystem, the noncharacter orthographic subsystem, the phonological subsystem, and the meaning subsystem. Each constituent subsystem consists of a set of representation units or nodes. A node whose activation value exceeds its threshold excites other nodes with which it is consistent and inhibits nodes with which it is not consistent.

Feature analysis is the first stage of visual character recognition. In Chinese, character strokes represent the basic features that characters possess; so the feature analysis is termed *stroke analysis*. As part of viewing a character, each stroke and its positional relationship with other strokes is detected. Detected features begin sending activation to the orthographic units

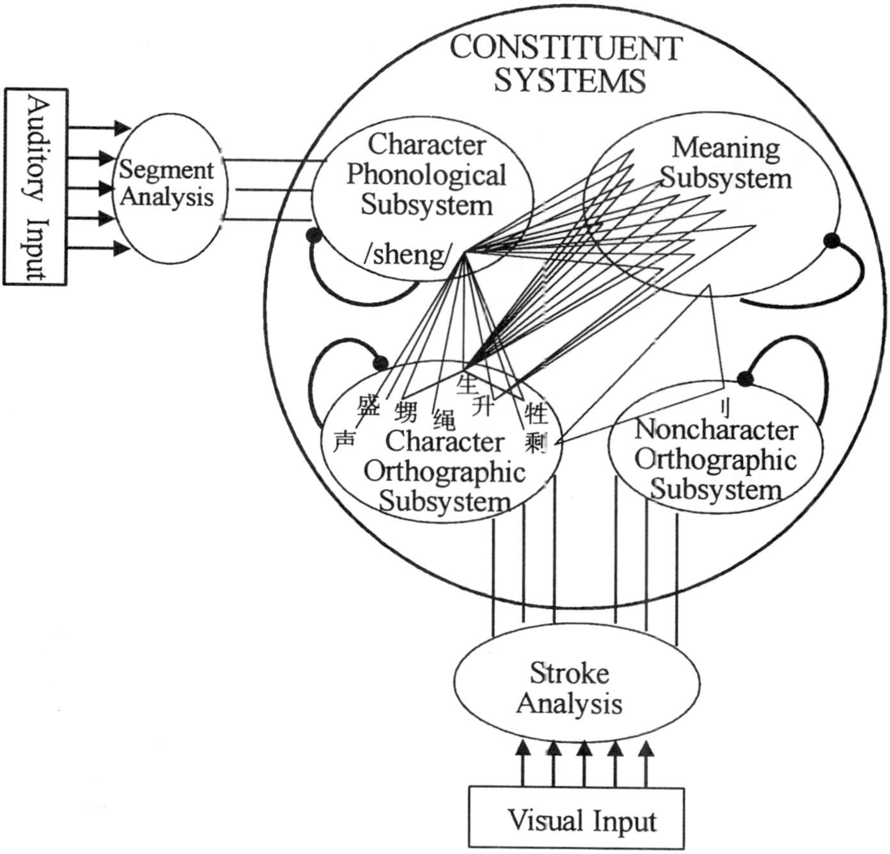

FIG. 6.1. The Interactive Constituency model of Chinese Character Identification. There are four subsystems: the noncharacter orthographic subsystem, the character orthographic subsystem, the phonological subsystem, and the meaning subsystem. Components of characters that are themselves noncharacters are represented in the noncharacter subsystem. Simple and compound characters, including components that are themselves real characters, are stored in the character orthographic subsystem. Thus, both phonetic compounds and phonetic components are organized in the same lexicon. The role of phonetics depends on the convergence and divergence of connections between orthographic units and phonological units.

in the character orthographic lexicon as well as in the noncharacter orthographic lexicon.

The character orthographic subsystem represents simple and compound characters, including those phonetics, semantic radicals, and bushou that are themselves legal characters. The noncharacter orthographic subsystem organizes those components (a few phonetics and many semantic radicals) that are not independent characters. So, in our model, both the compound character 明 (/ming2/, *bright*), and its two components 日 (/ri4/, *sun*) and 月 (/yue4/, *moon*), which are real characters, are represented in the character orthographic lexicon. The noncharacter components (e.g., 扌 艹, 犭) are stored in the noncharacter orthographic lexicon. Because both the compound character and its components are mapped to the phonological lexicon and the meaning subsystem, the model allows the possibility that compound character identification occurs (sublexically) from the activation of phonetics (character unit to phonological unit) and semantic components (character unit to meaning subsystem). The connections among 生 (/sheng1/, *give birth to, grow*, etc.), 牲 (/sheng1/, *domestic animal*), and their phonological unit /sheng1/ reflect this potential of phonetics-to-phonology mappings; and the connections among the orthographic units 剩 and 刂 and their meaning nodes illustrate the mapping of forms to meanings.

When an orthographic unit in the character orthographic lexicon exceeds threshold, it sends activation to the phonological units, the meaning system, and the corresponding units in the noncharacter orthographic lexicon simultaneously, leading to the activation of phonological units and/or relevant meaning representations and to the interaction of character orthographic units with noncharacter orthographic units. The units in the noncharacter subsystem do not connect with the phonological lexicon. However, the noncharacter orthographic units, once reaching threshold, do send activation to the orthographic units in the character lexicon and the meaning node in the meaning subsystem synchronously. The activation threshold of (character as well as noncharacter) orthographic units is determined by the frequency of occurrence of printed characters and components in daily usage. The threshold of phonological units is determined by the frequency of prior threshold level activations of the phonological form associated with the character, both through speech and print–speech experience. Activated phonological units also send activation to the meaning system and send feedback to the orthographic units.

The model assumes that meaning is represented and organized in terms of semantic attributes. Each attribute has a meaning node, and the meaning of a character consists of a set of meaning nodes. The more meanings a character has, the more attribute nodes it has, and the more complex the links among the nodes. The activation threshold of a specific meaning node relies on the frequency of encounter with this meaning. Because the activation from the orthographic or phonological units is fixed, for characters with very many meanings and no dominant one (e.g., the character 表 /biao/ with 10 distinct meanings; see Perfetti & Tan, in press, for details), each meaning node gets less activation, compared with characters having fewer meanings. Moreover, due to fewer independent occurrences of some characters in the modern-day usage (e.g., 态 /tai/), their threshold is relatively higher.

Finally, the model assumes that the activation of graphic information in the orthographic lexicon decays faster than does the activation of phonological units in the phonological lexicon or the activation of semantic nodes in the meaning subsystem. Identification depends on the *integration* of the pattern of activation over all of the nodes, reflecting the constituent integration assumption. When two units have the same excitatory strength, there is strong competition between them, leading to recognition or decision difficulty.

This model has several important features that help explain various findings from Chinese reading studies. We here enumerate a few:

1. A character orthographic unit typically connects with one phonological unit and always with more than one meaning node; a phonological unit always connects with more than one meaning node. Thus, the connections from (graphic or phonological) form units to meaning nodes are of the divergence type (Tan & Perfetti, 1997). This illustrates one of our basic assumptions that graphic form–phonological form relations are deterministic, whereas form-meaning relations are nondeterministic (Perfetti & Tan, 1998; Perfetti & Zhang, 1995b).

2. Due to pervasive homophony, the connections from orthographic units to phonological units are of the convergence type. In learning to read and in reading practice, a single phonological unit is activated by many different characters, which leads to a threshold lowering for the phonological representation. This frequency-driven phonological effect is regarded as another basis for early phonological activation (Tan et al., 1996).

3. The model provides an interpretation of the phonological mediation issue. As illustrated in Fig. 6.1, the orthographic units of "生声升牲甥绳盛剩" converge on one phonological node—/sheng/, which connects with all the meaning nodes of these characters. If a phonological node mediates access to character meanings, then the pronunciation /sheng/ will mediate activation of all the meanings of these homophones. Given the high divergence of phonological-units-to-meaning-units connections, we can expect that phonological nodes are more likely to influence access of character meanings for characters having fewer homophones (relatively low divergence) than for characters having many homophones (relatively high divergence). This is what was demonstrated by Tan and Perfetti (1997).

4. The decay hypothesis, that orthographic information decays faster than does phonological and meaning information, was used to interpret the pattern of graphic facilitation to inhibition and then to null effects found by Perfetti and Tan (1998).

5. Semantic vagueness, as reflected by high semantic uncertainty and semantic inaccessibility (Perfetti & Tan, 1998; Tan et al., 1996), influences

character recognition. This is because semantic uncertainty affects the degrees to which formal units diverge on meaning units or because of the higher threshold for semantically inaccessible characters in isolation.

6. The model predicts a complex relation between compound characters and simple characters, because most simple characters are components of compounds. They are all organized in the same character orthographic subsystem, implying the occurrence of facilitation and inhibition between compounds and their components. The pattern of facilitation and inhibition depends on connection parameters that reflect the consistency and validity of component and compound mappings to phonological and meaning subsystems.

7. Real-character and noncharacter components are represented in two separate lexicons, implying that these two kinds of subcharacters produce different roles in whole character recognition. In particular, because many significates are noncharacters and almost all phonetics are real characters, it is highly possible that significates behave differently from phonetics.

8. The model as illustrated in Fig. 6.1 emphasizes phonology that arises from a whole character, not its components. However, a kind of component phonology is possible within the model's assumptions, but because both phonetics and phonetic compounds are stored in a single-character orthographic lexicon, any sublexical phonology must arise within this single system. The effect of sublexical phonology depends on the convergence and divergence of phonological links between character orthographic units and phonological units. For example, when a compound and its phonetic link to the same phonological unit, there is convergence. Some degree of facilitation might thus occur, assuming a vital input activates both a compound and its component. When the links from the compound and from the phonetic to phonological units diverge, there is competition rather than facilitation. By requiring the orthographic subsystem to handle any component phonology, the model puts a novel perspective on sublexical phonological processes. The components are lexical in one sense (they exist separately as words) and sublexical in another sense (they can be part of a word). The same paradox applies to an attempt to define *prelexical.* The constituency idea comes in handy here: Rather than trying to salvage a definition of *sublexical* that corresponds to its use for other writing systems, we can say that phonological processes in Chinese arise through the phonological word form, which is always a constituent of identification. This phonological word form is seldom a constituent of just a single word; the occasion of component effects is set by connections among phonological and orthographic word forms, but there is no separate store of sublexical units to fire pronunciations. This process is initiated by information within the orthographic subsystem.

In connection with this issue, it is appropriate to return to the fact that Fig. 6.1 does not explicitly show some of the component links necessary to accomplish what we have discussed. This is for reasons of simplicity, to allow the parts of the system we emphasized (whole character phonology, meaning diffusion, the reliability of form-to-form mappings, etc.) to be more visible. It is a simple matter to add links between components (phonetics, especially, because they are almost always characters) in the orthographic subsystem and their corresponding phonological (and semantic) units. However, the question of how such links work in detail is complex. To take just the case of a phonological link from a phonetic (its representation as an orthographic character), its activation effect depends on the convergence or divergence of its phonological form with that of the compound. But it also depends on the connection of other compounds that also contain it. In some cases, most compounds have the same pronunciation as this component does in isolation. In other cases, relatively few do. This is the question of neighborhood effects, well known in English (Andrews, 1989) and explored in Chinese (Peng, Yang, & Chen, 1993). And whether such effects respond to type frequencies (how many different compounds have a given phonetic form) or token frequencies (summed overall types) also is a factor. We judge the evidence insufficient to postulate such connections in detail, but they can be added as evidence requires.

Actually, there is evidence that phonetic compounds facilitate identification (of low-frequency compounds) in some tasks, including naming (Fang, Horng, & Tzeng, 1986; Seidenberg, 1985) and perceptual identification when two components are viewed asynchronously (Flores d'Arcais, Saito, & Kawakami, 1995). However, in judgment tasks, Zhang (1996) found that phonetics affected judgments of pronunciation, but not judgments of meaning. The model as illustrated in Fig. 6.1 is sufficient to account for such effects. More powerful versions of the interactive model may be justified by eventual evidence, but would be extravagant at this point.

Finally, we should note that our model is different from a class of hierarchical models that have been suggested (e.g., Peng & Tan, 1987; Taft & Zhu, 1997; Zhu & Taft, 1994). Although these models differ in detail, all assume three hierarchical levels of representation from strokes through components and whole characters. Therefore, in these models, the compound character 明 (/ming2/, *bright*) is represented in the character level, whereas its components 日 (/ri4/, *sun*) and 月 (/yue4/, *moon*) are represented in the subcharacter component level. This kind of representation is adequate when the subcharacter components are themselves not real characters. However, when components are legal characters, as in the example case, the components 日 and 月 themselves need to have a character-level representation somewhere in the model. The Interactive Constituency Model does do this, introducing a kind of redundancy that gives simple characters both independent and compositional representations.

SUMMARY

Modern Chinese writing provides a system rich in interconnected morphological and syllabic components. Like all other writing systems, it rests in part on a linguistic foundation, even if its conceptual elements dominate in some respects. Indeed, Chinese appears to conform to a Universal Phonological Principle, by which reading processes make reference to speech representations. To this general observation, we add the hypothesis that word identification follows the Constituency Principle. A word identification event is the convergence of three aspects of a word's identity—orthographic form, phonological form, and nonformal (meaning) aspects. Furthermore, the deterministic mapping of form-to-form (orthographic–phonological) at the word level suggests that phonological–orthographic convergence is more reliable and more rapid than form-meaning mappings. The research we reviewed shows a pattern of robust phonological effects, even in nonphonological tasks, that is consistent with the Constituency Principle and the privilege of form-to-form mappings in bringing about constituent convergence. To account for the results of the research, we described a model based on these principles. The model, which itself is currently noncomputational, belongs to a class of models with well-known computational properties. It assumes an interactive multisystem representation, with orthographic, phonological, and meaning systems interconnected. Word identification results from patterns of activation across these subsystems. The model captures a number of important facts discovered about the graphic, phonological, and semantic processes of Chinese character identification.

REFERENCES

Andrews, S. (1989). Frequency and neighborhood effects on lexical access: Activation or search? *Journal of Experimental Psychology: Learning, Memory, and Cognition, 15*, 802–814.

Baron, J., & Strawson, C. (1976). Use of orthographic and word-specific knowledge in reading words aloud. *Journal of Experimental Psychology: Human Perception and Performance, 2*, 386–393.

Bloomfield, I. (1926). A set of postulates for the science of language. *Language, 2*, 153–164.

Chang, H.-W., Huang, J.-T., Hue, C.-W., & Tzeng, O. J. L. (Eds.). (1994). *Advances in the study of Chinese language processing.* Taiwan: National Taiwan University.

Chao, Y. R. (1968). *A grammar of spoken Chinese.* Berkeley: University of California Press.

Chen, M. J., Yung, Y. F., & Ng, T. W. (1988). The effect of context on the perception of Chinese characters. In I. M. Liu, H.-C. Chen, & M. J. Chen (Eds.), *Cognitive aspects of the Chinese language* (pp. 27–39). Hong Kong: Asian Research Service.

Chua, F. K. (1995, December). *The role of phonological information in logographic processing.* Paper presented at the International Conference on the Cognitive Processing of Chinese and Other Asian Languages, Hong Kong.

DeFrancis, J. (1989). *Visible speech: The diverse oneness of writing systems.* Honolulu: University of Hawaii.

Fang, S.-P., Horng, R.-Y., & Tzeng, O. J. L. (1986). Consistency effects in the Chinese character and pseudo-character naming tasks. In H. S. R. Kao & R. Hoosain (Eds.), *Linguistics, psychology, and the Chinese language* (pp. 11–21). Hong Kong: Centre of Asian Studies.

Feldman, L., & Siok, W. W. T. (1997). The role of component function in visual recognition of Chinese characters. *Journal of Experimental Psychology: Learning, Memory, and Cognition, 23*, 776–781.

Fleming, K. (1993). Phonologically mediated priming in spoken and printed word recognition. *Journal of Experimental Psychology: Learning, Memory, and Cognition, 19*, 272–284.

Flores d'Arcais, G. B., Saito, H., & Kawakami, M. (1995). Phonological and semantic activation in reading Kanji characters. *Journal of Experimental Psychology: Learning, Memory, and Cognition, 21*, 34–42.

Frost, R., Katz, L., & Bentin, S. (1987). Strategies for visual word recognition and orthographic depth: A multilingual comparison. *Journal of Experimental Psychology: Human Perception and Performance, 13*, 104–115.

Grainger, J., & Ferrand, L. (1994). Phonology and orthography in visual word recognition: Effects of masked homophone primes. *Journal of Memory and Language, 33*, 218–233.

Hoosain, R. (1991). *Psycholinguistic implications for linguistic relativity: A case study of Chinese*. Hillsdale, NJ: Lawrence Erlbaum Associates.

Hoosain, R. (1992). Psychological reality of the word in Chinese. In H.-C. Chen & O. J. L. Tzeng (Eds.), *Language processing in Chinese*. Amsterdam: North-Holland.

Hoosain, R., & Osgood, C. E. (1983). Information processing times for English and Chinese words. *Perception and Psychophysics, 34*, 573–577.

Kao, H. S. R., & Hoosain, R. (1986). *Linguistics, psychology, and the Chinese language*. Hong Kong: Centre of Asian Studies.

Lam, A., Perfetti, C. A., & Bell, L. (1991). Automatic phonetic transfer in bidialectal reading. *Applied Psycholinguistics, 12*, 299–311.

Leck, K. J., Weekes, B. S., & Chen, M. J. (1995). Visual and phonological pathways to the lexicon: Evidence from Chinese readers. *Memory & Cognition, 23*, 468–476.

Lesch, M. F., & Pollatsek, A. (1993). Automatic access of semantic information by phonological codes in visual word recognition. *Journal of Experimental Psychology: Learning, Memory, and Cognition, 19*, 285–294.

Lukatela, G., & Turvey, M. T. (1994). Visual lexical access is initially phonological: 1. Evidence from associative priming by words, homophones, and pseudohomophones. *Journal of Experimental Psychology: General, 123*, 107–128.

Mattingly, I. G. (1992). Linguistic awareness and orthographic form. In R. Frost & L. Katz (Eds.), *Orthography, phonology, morphology, and meaning* (pp. 11–26). Amsterdam: Elsevier Science Publishers B.V.

McClelland, J. L., & Rumelhart, D. E. (1981). An interactive activation model of context effects in letter perception: Part 1. An account of basic findings. *Psychological Review, 88*, 375–407.

Peng, D.-L., & Tan, L. H. (1987). Effects of word frequency and semantic context on Chinese two-character word recognition. *Journal of Psychology, 4*, 18–25. [In Chinese]

Peng, D.-L., Yang, H., & Chen, Y. (1993). *Consistency effect in naming Chinese phonograms*. Unpublished manuscript.

Perfetti, C. A., & Tan, L. H. (1998). The time course of graphemic, phonological, and semantic activation in visual Chinese character identification. *Journal of Experimental Psychology: Learning, Memory, and Cognition, 24*, 101–118.

Perfetti, C. A., & Tan, L. H. (1996, November). *Are there structural principles in sentence understanding across languages?* Paper presented at the Psychonomic Meeting, Chicago.

Perfetti, C. A., & Zhang, S. (1991). Phonemic processes in reading Chinese words. *Journal of Experimental Psychology: Learning, Memory, and Cognition, 17*, 633–643.

Perfetti, C. A., & Zhang, S. (1995a). Very early phonological activation in Chinese reading. *Journal of Experimental Psychology: Learning, Memory, and Cognition, 21*, 24–33.

Perfetti, C. A., & Zhang, S. (1995b). The universal word identification reflex. In D. Medin (Ed.), *The psychology of learning and motivation* (Vol. 33, pp. 159–189). New York: Academic Press.

Perfetti, C. A., & Zhang, S. (1996). What it means to learn to read. In M. F. Graves, B. M. Taylor, & P. van den Broek (Eds.), *The first R: Children's right to read* (pp. 37–60). New York: Teachers College Press.

Perfetti, C. A., Zhang, S., & Berent, I. (1992). Reading in English and Chinese: Evidence for a "universal" phonological principle. In R. Frost & L. Katz (Eds.), *Orthography, phonology, morphology, and meaning* (pp. 227–248). Amsterdam: North-Holland.

Seidenberg, M. S. (1985). The time course of phonological code activation in two writing systems. *Cognition, 19*, 1–30.

Taft, M., & Zhu, X. (1997). Sub-morphemic processing in reading Chinese. *Journal of Experimental Psychology: Learning, Memory, and Cognition, 23*, 761–775.

Tan, L. H., Hoosain, R., & Peng D.-L. (1995). Role of presemantic phonological code in Chinese character identification. *Journal of Experimental Psychology: Learning, Memory, and Cognition, 21*, 43–54.

Tan, L. H., Hoosain, R., & Siok, W. W. T. (1996). The activation of phonological codes before access to character meaning in written Chinese. *Journal of Experimental Psychology: Learning, Memory, and Cognition, 22*, 865–883.

Tan, L. H., & Perfetti, C. A. (1997). Visual Chinese character recognition: Does phonological information mediate access to meaning? *Journal of Memory and Language, 37*, 41–57.

Tan, L. H., & Perfetti, C. A. (1995). *Evidence for (restricted) phonological "mediation" in Chinese reading*. Poster presented at the Psychonomic Meeting, Los Angeles.

Tan, L. H., & Perfetti, C. A. (1998). *Phonological activation in visual identification of Chinese two-character words*. Manuscript submitted for publication.

Taylor, I., & Taylor, M. M. (1983). *The psychology of reading*. New York: Academic Press.

Tzeng, O. J. L., Hung, D. L., & Wang, W. S.-Y (1977). Speech recoding in reading Chinese characters. *Journal of Experimental Psychology: Human Memory and Learning, 3*, 621–630.

van Orden, G. C. (1987). A ROWS is a ROSE: Spelling, sound and reading. *Memory & Cognition, 15*, 181–198.

van Orden, G. C., & Goldinger, S. D. (1994). Interdependence of form and function in cognitive systems explains perception of printed words. *Journal of Experimental Psychology: Human Perception and Performance, 20*, 1269–1291.

Wang, W. S.-Y. (1981). Language structure and optimal orthography. In O. J. L. Tzeng & H. Singer (Eds.), *Perception of print: Reading research in experimental psychology* (pp. 223–236). Hillsdale, NJ: Lawrence Erlbaum Associates.

Xiandai Hanyu Duogongneng Cidian [Modern Chinese dictionary]. (1995). Beijing, China: Knowledge Press.

Xiandai Hanyu Pinlu Cidian [Modern Chinese Frequency dictionary]. (1986). Beijing, China: Beijing Language Institute Press.

Zhang, S. (1996). *The universal characteristics of phonological process in writing system comparisons*. Unpublished doctoral dissertation, University of Pittsburgh.

Zhang, S., & Perfetti, C. A. (1993). The tongue twister effect in reading Chinese. *Journal of Experimental Psychology: Learning, Memory, and Cognition, 19*, 1082–1093.

Zhang, W. T., Feng, L., & He, H. D. (1994). The activation of phonological and semantic information in Chinese character recognition. In H.-W. Chang, J.-T. Huang, C.-W. Hue, & O. J. L. Tzeng (Eds.), *Advances in the study of Chinese language processing* (pp. 185–198). Taiwan: National Taiwan University.

Zheng, X. (1983). It is easy to learn Chinese characters. *Educational Research, 4*, 56–63.

Zhou, Y. G. (1978). To what degree are the "phonetics" of present-day Chinese characters still phonetic? *Zhongguo Yuwen, 146*, 172–177.

Zhu, X., & Taft, M. (1994). The influence of perceptual experience on Chinese character processing. In H.-W. Chang, J.-T. Huang, C.-W. Hue, & O. J. L. Tzeng (Eds.), *Advances in the study of Chinese language processing*. Taiwan: National Taiwan University.

CHAPTER

7

Is There Phonologically Mediated Access to Lexical Semantics in Reading Chinese?

Xiaolin Zhou
Hua Shu
Yanchao Bi
Dongfang Shi
Beijing Normal University

Language comprehension involves retrieving from the mental lexicon the meanings associated with words a reader or listener comes across. Access to word meanings or lexical semantics is one of the major goals of the word recognition process, and it is these meanings that are the building blocks for construction of sentence and utterance level representations. For reading, it is traditionally assumed that lexical semantics can be accessed in two ways. One is by a direct visual access route, where visual features in the input are projected onto underlying orthographic representations, whose activation leads to the activation of stored lexical semantic properties. The other is by a phonologically mediated process, where the orthographic input first activates lexical phonological representations and then semantic representations.

Different theories of visual word recognition placed different emphases on these two routes, and history witnessed a pendulum-like swing between the strong phonological view and the strong visual views of access to lexical semantics (see McCusker, Hillinger, & Bias, 1981, for review). In light of recent research that provides considerable evidence for each of these positions, many researchers proposed that both visual and phonological pathways exist and operate in parallel (e.g., Besner & Smith, 1992; Coltheart, 1978; Coltheart, Curtis, Atkins, & Haller, 1993; Plaut, McClelland, Seidenberg, & Patterson, 1996; Seidenberg & McClelland, 1989). Although the dual-route theories differ in terms of the types of knowledge representations and processes involved and the relative importance of the two routes in reading alphabetic words, they share the idea that both visual and phonological routes are used in accessing lexical semantics and that each route determines semantic activation at least some of the time (Jared & Seidenberg, 1991).

On the other hand, recent years also saw the resurrection and development of the earlier position that meaning is primarily activated through phonological representations (Frost, in press; Lukatela & Turvey, 1994a, 1994b; van Orden, 1987; van Orden, Pennington, & Stone, 1990). According to this view, phonological mediation is the *predominant* process in lexical access, if not the only process.[1] In initial access, phonology is computed from the visual input and mapped, via lexical phonological representations, onto underlying semantic representations in the lexicon. Orthographic information is then used to refine the lexical semantic activation begun by phonology. Direct visual access to meaning either does not exist or plays only a minor role.

In this chapter, we question the applicability of this proposition to logographic Chinese, which evolved to represent the meaning rather than the sound of the language (e.g., Wang, 1973; Yin & Rohsenow, 1994). We argue for a view that, for written Chinese, direct visual access is the predominant way to access information in the mental lexicon, which we define as the storage of our knowledge about the phonological, orthographic, semantic, and syntactic properties of lexical items. Access to these properties must first go through orthographic representations in the lexicon. There is no "prelexical" or "nonlexical" phonological recoding in initial lexical access, and phonology becomes available only after phonological representations in the lexicon are activated by orthographic representations. More importantly, the activation of lexical semantics is normally a result of the multiple constraint-satisfaction process in which phonological mediation and direct visual access interact with each other. However, orthography and phonology in this process may not have equal status, due to the division of labor between different pathways. To the extent that orthographic and phonological effects on semantic activation can be separated out experimentally, orthographic information, for most Chinese characters, seems to play a stronger role than phonological information. Phonology by itself has limited effect on semantic activation in reading Chinese.

We begin with a model of how Chinese words are represented in the lexicon and accessed in visual word recognition (Zhou & Marslen-Wilson, 1996; Zhou, Marslen-Wilson, Shu, Bi, & Tang, 1996). This model, based on the data of previous research, makes some strong claims about the time course of semantic and phonological activation and about the roles of orthography and phonology in access to lexical semantics. Then we move to a brief review of the recent research on lexical access in reading Chinese. This

[1]We use the term *phonological mediation* here in the same sense as the term *phonologically constrained* access to lexical semantics. We do not imply that phonological mediation is a process to semantics, being independent from other processes in lexical processing.

critical review is organized according to the claims of the processing model we propose. We further report six experiments that bear on the issue of the relative time course of phonological and semantic activation in lexical access and the issue of whether phonology plays a major role in access to lexical semantics. Finally, we discuss the implications of our findings for visual word recognition in reading Chinese and for general theories of lexical processing across different orthographies.

A GENERIC MODEL OF LEXICAL REPRESENTATION AND ACCESS IN READING CHINESE

The Chinese language is perhaps best known for its logographic writing system, in which the basic meaningful units are characters. With some exceptions, each character corresponds to one morpheme and has one pronunciation, usually a syllable with a specific lexical tone. Because the number of syllables used in the language is limited to about 1,300, whereas the number of commonly used morphemes is about 5,000, Mandarin Chinese has a great many homophonic morphemes and homophonic characters in the language and in the writing system.[2]

A syllable in Mandarin Chinese has a simple phonological structure with no double consonants and with only two possible syllable-ending consonants. The tone, which is used to differentiate lexical items, is carried by the whole syllable rather than by a smaller segment. It was suggested that a syllable in Chinese can only be divided into the initial consonant and the rhyming part, with the latter being treated as one phoneme (You, Qian, Gao, 1980). A study by Read, Zhang, Nie, and Ding (1986) suggested that Chinese subjects who do not know *Pinyin*—the Chinese alphabetic system designed to specify the pronunciations of characters—cannot even divide the syllable into subparts. As we argued elsewhere (Zhou & Marslen-Wilson, 1996), the success and durability of the logographic writing system may not just be an historical accident. It may reflect the cognitive demands that a high degree of phonological ambiguity places on a writing system. A character is usually specific, so that it can be used efficiently to differentiate homophonic morphemes. It also maps directly to the syllable level, with no subcharacter and subsyllable correspondences, even for complex characters having phonetic radicals.

The Chinese language is morphologically limited, with no inflectional and derivational processes in word formation. The majority of Chinese words

[2]Syllables with the same segmental elements but different tones are treated as different syllables. Tones are used to differentiate between lexical items. If tones are not counted, there are only 415 syllables in Mandarin Chinese (e.g., Li, 1993). Throughout the chapter, the four tones were indicated by the numbers in parentheses.

are either monomorphemic words or disyllabic compounds. Compound words, which are represented by two characters in written form, are about 73.6% of the words in the language (Institute of Language Teaching and Research, 1986). Morphemes used as constituents in compounds are usually words by themselves, although there are some bound morphemes that are only used as constituent morphemes. The orthographic form of a compound word is always the concatenation of the orthographic forms (i.e., characters) of its constituent morphemes. The phonological form of a compound is usually the simple concatenation of syllables as well, although in some cases the tone of one constituent could be altered due to the influence of the neighboring tone.

In considering the nature of the logographic writing system and in considering the experimental evidence concerning lexical representation and access in visual word recognition (see the next section), we propose a generic model of how Chinese words are represented in the mental lexicon and activated in lexical processing (Fig. 7.1). We assume that each word has, at least, orthographic, phonological, and semantic representations in the lexicon. These representations are connected, so that the activation of one representation spreads to other connected representations. The solid lines between representations in Fig. 7.1 indicate the relatively efficient spread of activation between representations, whereas the broken lines between representations indicate weak effects of phonological representation on seman-

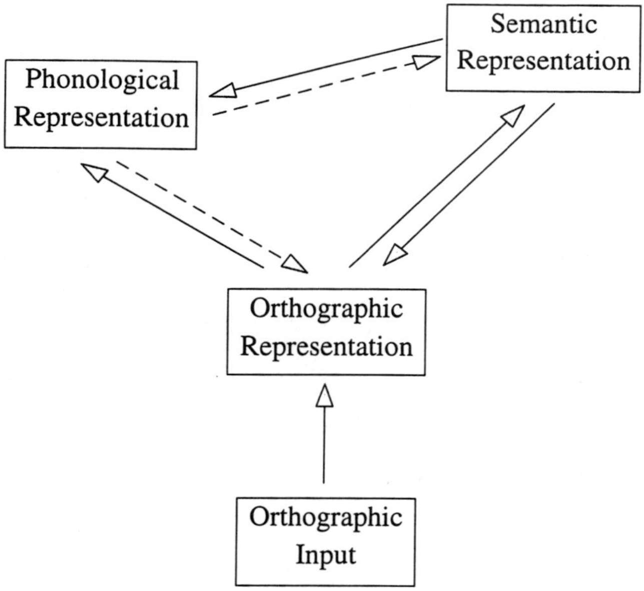

FIG. 7.1. A generic model of lexical processing in reading Chinese.

tic and orthographic activation. It should be noted that the efficiency of the flow of activation between different types of knowledge representations is functionally rather than structurally defined (see General Discussion).

The present model resembles the "lexical route" of dual-route models (e.g., Coltheart, 1978; Coltheart et al., 1993), and the connectionist structure of lexical representation and processing for alphabetic languages (e.g., Plaut et al., 1996; Seidenberg & McClelland, 1989). Here we see a triangular relation between orthographic, phonological, and semantic representations in the lexicon. However, we are neutral here about whether the lexical knowledge should be represented locally or in a distributed way. We believe that our model can be implemented using either localist or distributed terms and mechanisms. It is possible, for example, to construct a connectionist model to represent characters and syllables and their correspondences in a distributed way (Chen & Peng, 1994). However, because there is much less regularity in the construction of characters than in the construction of alphabetic words and because the orthography-sound correspondence in Chinese is rather arbitrary—even when taking into account the ability of phonetic radicals to represent the pronunciation of whole complex characters—it is not clear yet whether a connectionist model for Chinese makes functionally different predictions about lexical processing from a localist model. On the other hand, although we may not reject the idea of representing orthographic, phonological, and semantic information in a symbolic way in the lexicon, we do feel that the localist assumption of a "lexical entry" for each word is not appropriate for Chinese. This assumption, in a strict sense, requires that each distinctive character or morpheme have its own phonological representation, even though many phonological representations in different lexical entries are the same or similar (e.g., only differing on tones). If we adopt a symbolic approach to the Chinese mental lexicon, we would like to see shared representations within each of the orthographic, phonological, and semantic levels and dynamic relations between representations, perhaps based on the interactive-activation framework (e.g., McClelland & Rumelhart, 1981; Taft, 1994; Zhou & Marslen-Wilson, 1995).

We now turn to how lexical representations are activated by sensory input in real time visual word recognition. In localist terms, we may take the view that, in initial lexical access, sensory input is analyzed into orthographic units of different sizes, and these units are used in parallel to access corresponding lexical representations in the lexicon (Seidenberg, 1985). The orthographic representations in the lexicon are activated by this parallel access process, and hence the phonological and semantic representations connected to them. For a complex or compound character, for example, both the whole character and its phonetic radical are used to access their corresponding orthographic representations and the connected phonological and semantic representations (Zhou & Marslen-Wilson, chapter 3, this volume). In connectionist terms, visual word recognition involves the mapping from orthography to semantics, from orthography to phonology, and from phonology to semantics. These processes interact with each other but may contribute to different extents to the activation of a particular type of lexical representation. The efficiency of mapping depends on, among other factors, the regularity of the correspondence between different domains (e.g., Plaut et al., 1996). We return to this point in the General Discussion.

The present model makes at least four strong claims concerning lexical access, which have mostly been supported by experimental data (see the next section for a review of some of the data). First, the model assigns a predominant role to orthography in initial lexical access. Orthographic information is the primary source of constraints in initial lexical access, and the activation of an orthographic representation, which defines the nature of *lexical access* (see Frost, in press), is a prerequisite for the activation of other types of representations. The second claim of the model, related to the first claim, is that there is no "prelexical" or "nonlexical" phonological recoding in initial lexical access. Any phonological information has to be retrieved from the lexicon via direct visual access. In this way, there is no "assembled" phonology in reading Chinese, both for complex or compound characters with phonetic radicals and for "pronounceable" nonword characters. Third, the model claims that once the orthographic representation in the lexicon is sufficiently activated, the connected phonological and semantic representations are automatically activated. However, semantic and phonological activation may have different time courses. Fourth, semantic activation is predominantly orthographically constrained. The automatic phonological activation is not a prerequisite for semantic activation, although it may play some role in supporting and maintaining meaning activation begun by orthography (through the interactive-activation mechanisms). In other words, phonology by itself has limited effect on semantic activation.

These claims of our model differ in certain aspects from recent arguments stressing the importance of phonological activation in initial lexical access in Chinese and in access to meaning (e.g., Perfetti & Zhang, 1995; Tan, Hoosain, & Siok, 1996). In the next section, we critically review some of the recent studies related to the claims of our model. In particular, we argue that many studies of Chinese tended to overemphasize the role of phonological activation in reading Chinese and the similarity of phonological processes between Chinese and alphabetic languages, hence underestimating the importance of the role that orthography plays in initial lexical access and in access to meaning in reading Chinese. At least some of the arguments, furthermore, conflated the question of whether phonological information mandatorily becomes available with the question of whether phonology plays a vital role in access to meaning, and took the evidence for the first question as the evidence for the latter. Our intention in this review is to help swing the pendulum away from the strong phonological view and to highlight the necessity of conducting further research into various aspects of visual word recognition in Chinese.

PREVIOUS RESEARCH ON LEXICAL ACCESS IN READING CHINESE

The last 10 years saw an increasingly large body of research on the cognitive processing of Chinese. Although these studies cover a wide range of subjects, from semantic and syntactic cues in sentence comprehension (e.g., Li, Bates, & MacWhinney, 1993) to morphological processing of spoken compound words (e.g., Zhou & Marslen-Wilson, 1994, 1995), most of them concentrate on visual recognition of individual words. In particular, the role of phonology in lexical access attracted many valuable studies, even though there are many uncertainties concerning experimental data and theoretical interpretation.

There are at least three crucial issues concerning phonology in visual word recognition (Frost, in press; Seidenberg & McClelland, 1989). The first issue concerns access to phonology: Is phonological information necessarily and automatically activated in initial lexical processing? The second issue concerns the relation between phonological access and meaning activation: To what extent does phonology play a role in access to lexical semantics? The third issue concerns the way phonology is derived from orthography: Is phonological information a result of lexical access (addressed phonology) and/or a result of rule-based computation (assembled phonology)? Although these questions are closely related, they need to be carefully distinguished. An answer to one question (e.g., a strong view of mandatory phonological activation) does not necessarily entail a similar answer to another question (e.g., phonological mediation in access to meaning). But before we move to studies addressing the issues of phonology in visual word recognition, we first review evidence concerning the role of orthographic activation in reading Chinese.

The Predominant Role of Orthography in Initial Lexical Access

Our model assigns a predominant role to orthography in initial lexical access. It assumes that the projection of sensory input onto the lexicon is purely visual, owing nothing to phonology. Moreover, orthographic activation in the lexicon is the primary source of information to define *lexical access* and allow the processing system to locate or isolate a lexical item, in contrast with the strong phonological view of lexical access for alphabetic languages (e.g., Frost, in press). Phonological information becomes available only after the stored phonological representation is activated, as a consequence of the spread of activation from lexical orthographic representations. This view, separating orthographic and phonological representations and putting more weight on orthographic activation in initial lexical access, is supported by everyday experience. The Chinese reader often has the experience of knowing that a particular low-frequency visual pattern is a meaningful character without knowing how to pronounce it.

As pointed out by van Orden et al. (1990), unambiguous experimental evidence for direct visual access in English has been scarce or nonexistent. One reason for this is that orthography and phonology are necessarily confounded in alphabetic languages, and their effects on lexical access are difficult to separate. Many "orthographic" effects obtained in previous research can be explained away (van Orden et al., 1990) as due to the phonological properties of alphabetic words. In Chinese, however, since the relation between orthography and phonology is generally arbitrary, it is possible to manipulate orthography and phonology independently and to demonstrate directly the visual effect on lexical access.

In a stringent experiment in which only structurally simple individual characters were used, we found priming effects between words that are orthographically similar but phonologically and semantically different (e.g., 由 you(2), *because of*; and 申 shen(1), *express*) (Zhou, 1994). A significant inhibitory effect (about 20 ms) was obtained in a primed naming task, where subjects had to pronounce the second character of a pair presented consecutively on a computer screen with the stimulus onset asynchrony (SOA) of 100 ms. This effect can only be visual since there is no phonological relation between primes and targets at either the lexical or sublexical level. Specifically, the presence of a target (e.g., 申 shen(1), *express*) reinstates partially the orthographically similar prime (e.g., 由 you(2), *because of*). The reactivation of the orthographic representation of the prime, and hence its phonological representation, creates competition for the activation of the orthographic representation of the target and for the computation from orthography to phonology, leading to slower phonological activation of the target and the inhibitory effect in naming.

In another study using a primed lexical decision task, we investigated morphological, orthographic, and phonological priming effects in recognition of visually presented two-character Chinese compound words (Zhou, Taft, & Shu, 1995). A target word (e.g., 华贵 hua(2) gui(4), *luxurious*) was preceded by either a compound sharing a morpheme with it (e.g., 华丽 hua(2) li(4), *gorgeous*), or a word having a homographic and homophonic constituent (e.g., 华侨 hua(2) qiao(2), *overseas Chinese*), or a word (e.g., 滑翔 hua(2) xiang(2), *glide*) whose critical morpheme was homophonic but not homographic to the critical morpheme in the target, or a totally unrelated word (e.g., 完整 wen(2) zheng(3), *intact*). The constituent positions of related morphemes in primes and targets were also manipulated across experiments. What is relevant here is that, although there were no priming effects between words having homophonic but nonhomographic morphemes (e.g., 滑翔 hua(2) xiang(2), *glide* and 华贵 hua(2) gui(4), *luxurious*), whether primes were masked or not, we found significant facilitatory effects for words having homophonic and homographic morphemes (e.g., 华侨 hua(2) qiao(2), *overseas Chinese* and 华贵 hua(2) gui(4), *luxurious*), when the primes were masked or when the SOA between primes and targets was very short (57 ms). This effect is orthographically based, and its size was influenced by whether the homographic morphemes in primes and targets overlapped spatially in presentation. It suggested that orthography plays a much earlier and stronger role than phonology in initial lexical access.

These findings of orthographic effects in initial lexical access are consistent with the data from studies using other experimental paradigms. In a backward priming task, both Perfetti and Zhang (1991) and Tan, Hoosain, and Peng (1995) found that orthographic priming effects appeared earlier than phonological and semantic priming effects in the identification of target characters. In this experimental paradigm, a target word is presented briefly, followed immediately by a prime that is related to the target in one way or another. A pattern mask is presented following the brief exposure of the prime, and subjects are asked to write down the word they see. The primary interest in this task is the proportion of targets that subjects report in different priming conditions. The automaticity of priming is inferred from the brevity of exposure times of targets and primes (but see Verstane, Humphreys, Olson, & d'Ydewalle, 1995). Perfetti and Zhang (1991) and Tan et al. (1995) found that when targets were presented for about 50 ms and primes for 30 ms, a higher proportion of characters were reported when they were followed by orthographically similar characters than when they were followed by semantically or phonologically related or unrelated characters. In contrast, semantic and phonological primes did not have any facilitatory or inhibitory effects. Perfetti and Tan (1998) suggested further that these earlier orthographic priming effects can also be obtained in the primed naming task, although questions were raised about the reliability of the pattern of priming effects obtained in this study (Chen & Shu, 1998). In a semantic categorization task in which subjects had to judge whether an exemplar belonged to a prespecified semantic category, Chen, Cheung, and Flores d'Arcais (1995) observed more response errors and longer response times on orthographic foils than on control foils, but no effect at all for homophone foils. These data, together with our studies, demonstrated the early, independent orthographic constraints on initial lexical access in reading Chinese.

The Automatic Activation of Phonological Representation

Although our model, differing from many proposals about reading alphabetic languages or logographic Chinese (e.g., Frost, in press; Tan et al., 1996), does not assign a central role to phonology in lexical access, it does assume that the phonological representation of a word connects directly with the orthographic representation onto which the sensory input maps in initial lexical access. It assumes that the activation of orthographic representation spreads automatically to the connected phonological representation, perhaps in a cascaded way (e.g., Zhou & Marslen-Wilson, 1997). Although phonological information alone plays little role in access to meaning, as assumed in our model, it is useful in supporting semantic activation begun by orthography and, in the later stage, to maintain information in working memory and support higher level comprehension mechanisms (e.g., Tzeng & Hung, 1980; Tzeng, Hung, & Wang, 1977; Xu, 1991).

The evidence supporting phonological activation in reading Chinese comes from a number of experiments using different paradigms. Homophonic words facilitate each other in the primed naming task (e.g., Perfetti & Zhang, 1991; Zhou & Marslen-Wilson, chapter 3, this volume). Lexical decision to pseudohomophones (nonwords created by replacing the second constituents of compound words with homophonic characters) was delayed compared with matched nonwords (Zhou, Marslen-Wilson, Bi, Shi, & Shu, 1996; Zhou & Marslen-Wilson, 1998a), although this pseudohomophone effect was found mostly for subjects with overall slow response times, not for subjects with fast reaction times. At appropriate exposure times, the identification of characters was facilitated by homophone primes in the backward priming paradigm (e.g., Perfetti & Zhang, 1991; Tan et al., 1995; Tan et al., 1996). Homophones or pseudohomophones were more difficult to reject as inappropriate exemplars of predetermined semantic categories than matched words or nonwords (Leck, Weekes, & Chen, 1995; Wydell, Patterson, & Humphreys, 1993; but see Chen et al., 1995, and later discussion). In a synonym judgment task, it was more difficult for subjects to reject character pairs that were homophones but not synonyms than to reject semantically and phonologically unrelated pairs (Perfetti & Zhang, 1995; Zhou & Marslen-Wilson, 1998b).

However, a still unsettled question is to what extent the demonstrated phonological activation and its influence on experimental tasks is mandatory or automatic (Chen et al., 1995; but see Frost, in press). Although phonological effects have been found in tasks that do not require phonological information, other studies suggested that some of the evidence should be treated with caution. Recent research by Verstane et al. (1995) showed that the backward priming paradigm is not immune to strategic influences.

When subjects were discouraged from using phonological codes derived from stimuli to aid identification—for example, when targets (e.g., *piece*) have orthographically similar homophones (e.g., *peace*)—phonological priming effects disappeared. The naming task may itself have augmented phonological activation because the response requires phonology to be activated. Chen, Flores d'Arcais, and Cheung (1994) did not find a significant homophone effect in the semantic categorization task. They could not replicate the homophone priming effect in naming, either (see also Chen & Shu, 1998; Zhang, Feng, & He, 1994), even though they used essentially the same stimuli as Perfetti and Zhang (1991) or as Perfetti and Tan (1998). Zhou, Taft, and Shu (1995) did not find any significant priming effects between compound words having homophonic but nonhomographic morphemes (e.g., between 滑翔 hua(2) xiang(2), *glide*, and 华贵 hua(2) gui(4), *luxurious*) in a primed lexical decision task, although arguably, the lexical decision task may be not sensitive enough to the partial phonological activation in compounds.

The second concern about phonological activation in reading Chinese is how early it is in the time course of word recognition. Perfetti and his colleagues have strongly argued that phonological activation is "very early," earlier than semantic activation (e.g., Perfetti & Zhang, 1995; Perfetti & Tan, 1998; Tan et al., 1996). A careful analysis of the experimental evidence, however, raised strong doubts about this argument.

The Time Course of Phonological and Semantic Activation

If we accept that both semantic and phonological representations are mandatorily activated in direct visual access, an interesting question is then the relative time course of phonological and semantic activation from orthography. Perfetti and his colleagues have carried out several studies concerning this question, and their conclusion is that phonology is activated earlier than semantics. The functional reason provided is that the correspondence between orthography and phonology is more reliable and precise than the correspondence between orthography and meaning (e.g., Perfetti & Tan, 1998).

In their backward priming experiments, Perfetti and Zhang (1991) varied the exposure times of primes and targets. The authors found that the identification of target characters could be improved when they were primed by semantically related or homophonic characters. However, the semantic effect lags behind the phonological effect, as assessed against unrelated control. In Tan et al. (1995; also Tan et al., 1996), phonological effects appeared approximately 14 ms earlier than semantic effects when the meanings of target characters are "precise" and 28 ms earlier when the meanings of target characters are "vague." Further experiments using synonym and homophone judgment tasks (Perfetti & Zhang, 1995) supported this finding. The authors varied SOAs between the presence of first and second characters. The phonological interference effect between characters that were not synonyms, but happened to be homophones, appeared in a synonym judgment task when SOA was 90 ms, but the semantic interference effect between characters that were not semantically related, but happened to be synonyms, did not appear in a homophone judgment task until SOA was 140 ms. A recent study by Perfetti and Tan (1998), using a primed naming task, suggested further that although homophone priming effects can be obtained when SOA was 57 ms, semantic priming effect was not detected until SOA was about 85 ms.

Since all the preceding evidence concerning the relative time course of phonological and semantic activation is from comparisons between items and even between experimental tasks, it is legitimate to ask whether these items and tasks for phonological and semantic priming are really comparable in terms of the extent of phonological or semantic property overlap between primes and targets, in terms of the ease of carrying out an experimental task, and in terms of whether task demand may have implicitly promoted phonological effects.[3] Moreover, many of the data from these studies proved difficult to replicate. There are very large phonological and semantic priming effects (up to 103 ms) in the primed naming task, and big changes in the size of priming effects between different SOA conditions in Perfetti and Tan (1998), not observed in other studies with similar conditions (e.g., Perfetti & Zhang, 1991; Zhou & Marslen-Wilson, chapter 3, this volume). Indeed, using the same stimuli and similar experimental procedures as Perfetti and Tan (1998), Chen and Shu (1998) observed a different pattern of priming effects. Strong semantic priming effects were found at both the short and long SOA, while homophone priming effects were very weak. Zhou and Marslen-Wilson (1998b) used both lexical decision to compound words, character decision, primed naming, and semantic and phonological judgment tasks to investigate systematically the relative time course of semantic

[3]In all these studies, phonologically related pairs were either homophones or not. There were no pairs that were phonologically related but not homophones. On the other hand, even if a pair of characters are strongly semantically related or synonyms, they always have a few differences between their semantic properties. Thus, it is possible that, depending on experimental tasks, the complete overlap between phonological properties of homophones and the partial overlap between semantic properties of synonyms may have different influences on priming effects. For example, the homophone judgment task may be easier to carry out than the synonym judgment task, because there are no nonoverlapping phonological features that could delay the homophone judgment task, whereas there were always nonoverlapping semantic features that hindered the synonym judgment task. The observed differences between the time courses of phonological and semantic activation may be a consequence of the experimental designs and tasks used (see also, Perfetti & Zhang, 1995).

and phonological activation in reading Chinese. Strong semantic effects were found in all these tasks and at the very short SOA, while the appearance of phonological effects depended on whether the experimental tasks biased toward the use of phonological information. The authors concluded that semantic activation in reading Chinese is no later, and probably stronger, than phonological activation.

Sublexical "Phonological" Processing in Chinese

Because complex characters may use phonetic radicals to convey information about the pronunciation of the whole characters, the interesting issues are, first, whether Chinese readers use this information in reading, and, second, how they use this information. Experiments on regularity and consistency effects in naming complex characters (e.g., Fang, Horng, & Tzeng, 1986; Hue, 1992; Seidenberg, 1985; Peng, Yang, & Chen, 1994) suggest that Chinese readers do engage sublexical phonological processing in reading (low-frequency) complex characters. The most interesting question is where the phonological information associated with phonetic radicals comes from in processing the complex characters. Flores d'Arcais, Saito, and Kawakami (1995) suggested that, because radical-sound correspondences are overlearned, they may be used in Chinese word recognition in much the same way as grapheme–phoneme conversion rules in recognizing English words. If so, prelexical phonology is possible for Chinese, in contrast with the assumptions of the model sketched in Fig. 7.1.

Experiments reported in Zhou and Marslen-Wilson (chapter 3, this volume), however, strongly suggested that this sublexical "phonological" route is nothing more than decomposing the phonetic radical of a complex character or pseudocharacter and using it to access its corresponding representations in the lexicon, including both phonological and semantic representations. This decomposition is a general reading strategy explicitly or implicitly learned and is particularly useful in reading low-frequency characters, which tend to be regular. In real time lexical access, low-frequency complex characters are more likely to be decomposed than high-frequency characters because the frequencies of phonetic radicals are usually much higher than the frequencies of the complex characters themselves. It is easy for the phonetic radicals to be picked up in the initial analyses of input and mapped onto the corresponding lexical representations. In the cases of unfamiliar characters or pseudocharacters, the decomposition process is more likely to be under conscious control. Sublexical processing in reading Chinese, just like lexical level processing, is both a phonological and a semantic event. There are no fundamental differences between sublexical processing of phonetic radicals and lexical processing of simple and complex characters (see Zhou & Marslen-Wilson, chapter 3, this volume, for further discussion).

The Limited Effect of Phonology on Access to Lexical Semantics

According to our model, phonology by itself has limited effect on access to lexical semantics. In most cases, the direct flow of activation from orthographic representation to semantic representation is more efficient than the indirect flow of activation via phonological representations. Direct visual access is sufficient to activate the semantic representations, although it may interact with automatically activated phonology (see General Discussion). The phonologically mediated route is further hindered by the fact that the mapping from phonology to semantics is usually one-to-many, as most syllables correspond to more than one morpheme. Orthographic information has to be extensively used to discriminate lexical items even when semantic representations are activated through phonology. Therefore, access to semantic in reading Chinese is primarily orthographically constrained, rather than phonologically constrained, as in alphabetic languages (Lukatela & Turvey, 1994a, 1994b; van Orden, 1987; van Orden et al., 1990; van Orden & Goldinger, 1994).

As we pointed out earlier (see also Frost, in press), evidence for automatic or "very early" phonological activation in visual word recognition cannot be taken as evidence for a "phonologically mediated" semantic activation. For example, although Zhou, Marslen-Wilson, Bi, et al. (1996) observed that pseudohomophones based on real words were more difficult to reject than matched control nonwords in lexical decision, this pseudohomophone effect could be due to the direct influence of phonological activation on the lexical decision task (see Zhou et al., 1995; Zhou & Marslen-Wilson, 1998a), rather than due to interference from the phonologically mediated semantic activation of base words, as suggested for English (e.g., Seidenberg, Petersen, MacDonald, & Plaut, 1996).

Direct evidence concerning whether there is phonologically mediated access to semantics came from a semantic categorization task. Leck et al. (1995) asked subjects to decide whether the meaning of a presented character (e.g., 狐 hu(2), *fox*; or 湖 hu(2), *lake*) belongs to a prescribed semantic category (e.g., animal). The empirical question here was whether those characters that were homophonic to and/or shared orthographic components with the correct categorical examples were rejected more slowly than unrelated control characters. What they found was that although characters sharing orthographic components with the base characters (e.g., 弧 hu(2), *curve*; or 呱 gua(1), *making noise*), whether they are homophones or not, were difficult to reject, the latencies for rejecting homophonic characters (e.g., 湖 hu(2), *lake*) that had no orthographic similarity with the base characters were *not* significantly different from the baseline. Using a similar design, Chen et al. (1995) did not observe any interference from homophone foils when these homophones were orthographically dissimilar to real exemplars. Thus, these studies showed the predominant role of orthography in access to meaning, but revealed no phonological effect of its own in lexical processing. In another study, Wydell et al. (1993) investigated whether there was a phonologically mediated access to meaning in Japanese Kanji compound words. The homophone foils were either orthographically similar to the category exemplars, sharing one character with them, or orthographically dissimilar. While orthographically similar words, whether they were homophonic to category exemplars or not, were more difficult to reject than matched control words, the visually dissimilar homophones were *no more* difficult to reject than control words. Moreover, the interference effect was larger for orthographically similar homophones than for orthographically similar nonhomophones—a finding different from Leck et al. (1995). The authors suggested that reading Kanji words is characterized by parallel access to semantics from orthographic and phonological representations.

We believe the discrepancy between Leck et al. (1995) and Wydell et al. (1993) concerning whether orthographically similar homophones were more difficult to reject than orthographically similar nonhomophones in the semantic categorization task is due to the fact that Wydell et al. (1993) used semantic categories with few members (e.g., a kind of debt—*overdraft*; a chance—*opportunity*), whereas Leck et al. (1995) used semantic categories with many members (e.g., animal). (See Jared & Seidenberg, 1991, for discussion about the effect of category size.) It is likely that the extra interference effect for homophones compared with nonhomophones in Wydell et al. (1993) was due to the semantic priming from category names (e.g., *debt*) to exemplars (e.g., *overdraft*) whose phonological activation interfered with the decision of rejecting their homophones (e.g., *red cloth*, in English translation). This account is consistent with the failure of finding a homophone interference effect for orthographically dissimilar homophones (Chen et al., 1995; Zhou & Marslen-Wilson, 1998a). Therefore, data from the semantic categorization task demonstrated both the inefficiency of "pure" phonology and the major effect of orthography in constraining access to lexical semantics.

PHONOLOGY IN ACCESS TO LEXICAL SEMANTICS: TO WHAT EXTENT DOES IT PLAY A ROLE?

The main purpose of the experiments reported next is to provide convergent evidence concerning the limited effect of phonology in access to lexical semantics. We used a phonologically mediated semantic priming technique (e.g., Fleming, 1993; Lesch & Pollatsek, 1993; Lukatela & Turvey, 1994a) that has better control over the time course of phonological activation than the semantic categorization task. Before we move to the rationale behind this technique and our experiments, however, we first report one experiment using the direct phonological priming technique. Homophonic compound words that shared both segmental and suprasegmental (i.e., tonal) elements (e.g., 洁净 (jie(2) jing(4), *clean*; 捷径, jie(2) jing(4), *shortcut*) were used as primes and targets in Experiment 1. The purpose of this experiment was to investigate, first, whether phonological activation is mandatory and how early it is, and second, whether semantic effects on visual word recognition can be detected earlier than phonological effects.

Experiment 1: Homophone Priming

The phonological priming technique, in which a target word is preceded by a phonologically related word or nonword, has been extensively used in research into phonological processing in visual word recognition. One of the general findings in the research on alphabetic languages is that targets (e.g., *toad*) are responded to faster or identified more accurately when they are preceded by homophones (e.g., *towed*) than when they are preceded by unrelated words (e.g., Grainger & Ferrand, 1994; Humphreys, Evett, & Taylor, 1982; Lukatela & Turvey, 1994b). This homophone priming effect appears to be automatic because it can be found with very short SOAs between primes and targets or when primes are masked and not identifiable. It can be interpreted as evidence for phonologically constrained initial lexical access (e.g., Lukatela & Turvey, 1994b).

The homophone priming effects have also been obtained in Chinese for single-character words, although, as we pointed out earlier, it is not clear to what extent these effects reflected the mandatory component of phonological activation and to what extent they reflected the demands of the experimental tasks being used. Perfetti and Zhang (1991) and Tan et al. (1995) showed in the backward priming paradigm that, at certain exposure times, single-character words or morphemes were more likely to be identified when they were followed by homophone characters than when followed by unrelated characters. They were also named faster when they were preceded by homophonic characters than when they were preceded by unrelated control characters (e.g., Perfetti & Tan, 1998; Perfetti & Zhang, 1991; Zhou & Marslen-Wilson, chapter 3, this volume; Zhou & Marslen-Wilson, 1998b).

In the present experiment, we compared the pattern of priming effects in different tasks. Using a within-item design, we compared directly the homophone priming effect with the semantic priming effect to track the relative time course of semantic and phonological activation. Each target word,

TABLE 7.1
Experiment 1: Experimental Design, Sample Stimuli, and Average Frequencies of Primes and Targets

	Prime Type			
	Semantic	*Homophone*	*Control*	*Target*
Written Form	卫生	捷径	流放	洁净
Pronunciation	wei(4) sheng(1)	jie(2) jing(4)	liu(2) fang(4)	jie(2) jing(4)
Meaning	*hygiene*	*shortcut*	*banish*	*clean*
Frequency	852	186	177	345

therefore, had three types of prime (Table 7.1). A target word (e.g., 洁净, jie(2) jing(4), *clean*) was preceded by its semantic associate (e.g., 卫生 wei(4) sheng(1), *hygiene*), or by a word homophonic to the target (e.g., 捷径, jie(2) jing(4), *shortcut*), or by an unrelated word (e.g., 流放, liu(2) fang(4), *banish*). There were no orthographic similarities between primes and targets. Three experimental tasks were used: visual–visual priming, masked priming lexical decision, and visual–visual primed naming. Since there were no word–nonword pairs in the experiment that were homophones, the use of the masked priming technique should be able to provide additional information concerning subjects' possible response strategies.

Forty-eight targets were chosen from a large database. Because the main interest here was whether there was a phonological priming effect in lexical decision or naming, control primes were chosen to match the frequency of homophone primes rather than of semantic primes. The frequency information came from the Institute of Language Teaching and Research (1986), with about 1.8 million characters (and about 1.3 million words) in the corpus.[4] A Latin square design was used to assign the critical primes and their targets to three test versions in a counterbalanced way. There were 153 word–word and word–nonword pairs used as fillers in all three test versions. The filler prime–target pairs had no semantic, phonological, or orthographic similarities between them. In the naming version of the experiment, nonwords used in the lexical decision task were deleted from the test and their primes were rearranged to created unrelated word–word pairs.

All stimuli were generated using a computer word processing program and captured as pictures on the screen by a snapshot program. Each word or nonword was excised and stored as an individual image file on a computer hard disk. Because there is no case variation in Chinese, the primes

[4]Throughout the chapter, the frequency counts, whether of compound words or single characters and whether from the Institute of Language Teaching and Research (1986) or from Liu, Liang, Wang, and Zhang (1990), were all normalized to a rate per million.

and targets were created in *kaiti* and *songti* fonts respectively to maximize visual dissimilarity between primes and targets. In visual–visual priming lexical decision or naming, each trail began with an eye fixation signal ("+") presented at the center of the screen for 300 ms, followed by a 300 ms blank interval. A prime was then presented for 100 ms, followed immediately by the corresponding target, which was presented for 400 ms. The prime and target were also presented at the center of the screen, with the target overwriting the prime. A target word was about 3.5×5.5 cm in size and the prime was slightly smaller. In masked priming, a pattern mask composed of lines was first presented for 300 ms, followed immediately by the prime. The prime was presented for 57 ms, and the target overwrote the prime immediately after its presentation. There was a 3-second interval between the disappearance of the last target and the appearance of the next eye fixation point or the pattern mask. The presentation of stimuli and the recording of reaction times were accurate to 1 ms. Subjects, all undergraduate students at Beijing Normal University and native speakers of Mandarin Chinese, were tested in groups of three or less for lexical decision, and one by one for naming. Each test version had at least 10 subjects.

Mean reaction times and response error percentages for each task are reported in Table 7.2. In the analyses of reaction times, there were highly significant main effects of prime type in all three tasks. The semantic priming effects were significant in both visual–visual and masked priming lexical decision tasks ($p < .01$ or $p < .05$), but were not in the naming task ($p > .1$). The homophone priming effects were not significant in either the visual–visual priming ($p > .1$) or the masked priming lexical decision task, but were significant in the naming task ($p < .01$).

These patterns of priming effects across the three tasks are not consistent with the strong view of phonology in visual word recognition. Although Zhou et al. (1995, December) already demonstrated that compound words sharing one syllable did not prime each other, the absence of homophone priming in visual–visual priming lexical decision is still striking. This is not

TABLE 7.2
Experiment 1: Mean Response Latencies (ms) and Error Percentages

	Prime Type		
	Semantic	*Homophone*	*Control*
Visual–Visual	604	634	645
Lexical Decision	(3.6)	(7.6)	(7.8)
Masked	575	597	597
Lexical Decision	(3.2)	(5.7)	(4.2)
Visual–Visual	611	599	621
Naming	(2.1)	(2.4)	(2.4)

only because such a priming effect was observed in studies on alphabetic languages, but also because subjects could use response strategies to speed up their responses in the priming condition, given that there were no pseudohomophones in the stimuli set and a target must be a real word if it sounds the same as its prime. The appearance of the semantic priming effects in lexical decision, especially in the masked priming where the primes were presented for only 57 ms, demonstrated that semantic activation in reading Chinese is a very early process, possibly even earlier than phonological activation, in contrast with claims made by Perfetti and his colleagues (e.g., Perfetti & Zhang, 1995; Perfetti & Tan, 1998).

A proponent of the strong phonological view could argue that phonology *is* automatically activated and could be detected by a task focusing on phonology (i.e., naming), but not by a task focusing on semantics (i.e., lexical decision). However, if we accept that the lexical decision task taps mainly into the semantic rather than the phonological activation of compound words, the absence of a significant homophone priming effect in the present experiment could be interpreted as demonstrating that, although phonological representations are automatically activated, they do not play an active role in access to semantics. If there exists phonological mediation in access to meaning in reading Chinese, the activation of the phonological representations of homophone primes, which were shared with target words, should spread to all the corresponding semantic representations, facilitating lexical decisions to targets.

The weakness of semantic priming in the naming task indicates that the naming of compound words begins as soon as the phonological representations of initial morphemes are activated during lexical access (Zhang & Peng, 1992; Zhou & Marslen-Wilson, 1994, 1995). The input of a compound is morphologically decomposed, and the two characters, although they are used to access the semantic representations of the whole word, are at the same time used to access their corresponding individual semantic and phonological representations in the lexicon (Zhou, Taft, & Shu, 1995). The naming of a compound word could benefit from the early activation of its constituent morphemes, with naming being initiated as soon as the phonological representation of the initial morpheme is activated. This morpheme-based procedure may have helped the naming of target words when they were preceded by unrelated words, and hence reduced the size of semantic priming effects (see also Experiment 2).

Experiment 2: Phonologically Mediated Semantic Priming: Compound Words

Experiment 1 demonstrated that, when an experimental task focused on phonology (and possibly when phonological activation of primes was allowed to accumulate over time), the processing of target words was influenced by the phonological activation of primes. This finding indicated that phonological activation of Chinese compounds accrues over time and can be later than semantic activation, but they can be accelerated or tapped more efficiently by the demand of experimental tasks. This experiment, however, did not tell us whether this phonological activation plays an important role in access to lexical semantics. The following experiments, on both two-character compound words and single-character words, used the phonologically mediated semantic priming technique to address this question.

This technique works in the following way: If semantic representations are activated through the activation of phonological representations, a facilitatory priming effect should be observed not only between semantic primes and targets, but also between primes that are homophonic to the semantic primes and targets. When a prime is presented, whether it is the semantic prime or the homophone prime, the shared phonological representation is activated and used to access to lexical semantic representations corresponding to the phonological representation. The activation of semantic representations spreads to other words, including the target word, that share semantic properties with one of the activated semantic representations. The target word is thus preactivated, and the response to the subsequently presented target is facilitated.

This technique has been successfully used in studies investigating the role of phonology in reading English. Lukatela and Turvey (1994a), for example, found that at a brief SOA of 50 or 100 ms, the naming of target words (e.g., *frog*) was facilitated to a similar extent when they were preceded by their semantic associates (e.g., *toad*), by words homophonic to the semantic primes (e.g., *towed*), or by pseudohomophones (e.g., *tode*). The occurrence of priming effects between homophone and pseudohomophone primes and targets was interpreted as showing that, in reading alphabetic words, phonology plays a predominant role in initial lexical access and in access to meaning. If phonology also plays such a role in reading logographic Chinese, we should obtain similar patterns of priming effects in similar designs for Chinese.

Except for three new triplets, the same 48 pairs of homophone and semantic primes and their corresponding targets used in Experiment 1 were used in this experiment, but with the positions of semantic primes and targets reversed. The experimental design is presented in Table 7.3. A target word (e.g., 卫生 wei(4) sheng(1), *hygiene*) was preceded by its semantic associate (e.g., 洁净, jie(2) jing(4), *clean*), or by a word homophonic to the associate (e.g., 捷径, jie(2) jing(4), *shortcut*), or by an unrelated word (e.g., 流放, liu(2) fang(4), *banish*). There were no orthographic or phonological similarities between primes and targets and no orthographic similarities among the three primes. If phonology was used to mediate the access to semantics in Chinese, we should find not only a facilitatory priming effect

TABLE 7.3
Experiment 2: Experimental Design, Sample Stimuli, and Average Frequencies of Primes and Targets

	Prime Type			
	Semantic	*Mediated*	*Control*	*Target*
Written Form	洁净	捷径	流放	卫生
Pronunciation	jie(2) jing(4)	jie(2) jing(4)	liu(2) fang(4)	wei(4) sheng(1)
Meaning	*clean*	*shortcut*	*banish*	*hygiene*
Frequency	325	181	170	869

between 洁净 jie(2) jing(4), *clean* and 卫生 wei(4) sheng(1), *hygiene*, but also an effect between 捷径 jie(2) jing(4), *shortcut*) and 卫生. Both lexical decision and naming tasks were used, with SOAs between primes and targets set at 100 ms.

A Latin square design was used to assign the critical primes and their targets to three test versions in a counterbalanced way. The same word–word and word–nonword fillers (153 pairs in total) used in Experiment 1 were used here. In the naming version of the experiment, nonwords used in the lexical decision task were deleted from the test, and their primes were rearranged to created unrelated word–word pairs.

The mean reaction times and error percentages are summarized and presented in Table 7.4. Statistical analyses of the data revealed a significant semantic priming effect in both lexical decision and naming latencies, although this effect, consistent with Experiment 1, was weaker in the naming task ($p < 05$). In contrast, there was no significant phonologically mediated semantic priming in either task ($p > .1$). The presence of the mediated homophone primes did not facilitate the processing semantic targets.

The dissociation between semantic priming and mediated priming in the present experiment, in contrast with the findings in alphabetic languages (e.g., Lukatela & Turvey, 1994a), has two important implications for the processing of Chinese words. First, the direct semantic priming effect (i.e.,

TABLE 7.4
Experiment 2: Mean Response Latencies (ms) and Error Percentages

	Prime Type		
	Semantic	*Mediated*	*Control*
Lexical Decision	581	620	623
	(2.7)	(6.6)	(4.9)
Naming	609	621	626
	(2.1)	(1.8)	(2.7)

between 洁净 jie(2) jing(4), *clean*; and 卫生, wei(4) sheng(1), *hygiene*) in the naming task suggests that the preactivation of the semantic representation of the target, due to the presence of the semantic prime, spreads immediately to its phonological representation and facilitates the naming process (see Zhou & Marslen-Wilson, 1997, for a discussion of this process). Second, the absence of mediated priming effects suggests that there is little mediated access from phonology to lexical semantics. Either the presence of a homophone prime 捷径 jie(2) jing(4), *shortcut* does not lead to automatic activation of the phonological representation shared with the semantic prime 洁净 jie(2) jing(4), *clean* (but see earlier discussion), or the automatic phonological activation does not lead to the activation of semantic representations of 洁净. Either way, phonology alone has no or limited effect on access to meaning in visual recognition of Chinese compounds.

Experiment 3: Pseudohomophones

Experiment 2 demonstrated that there was little phonologically mediated semantic activation in processing two-character compound words. However, one might argue that the absence of the mediated priming effect could be due to competition between semantic representations corresponding to the semantic and homophone primes, which are activated through phonology. If this competition was strong, the semantic representations corresponding to semantic primes could have been suppressed by the semantic activation of homophone primes by the time the targets were presented. Consequently, no mediating priming could be observed.

We exploit here the fact that pseudohomophones (e.g., 安权, an(1) quan(2)) that sound like real words do not have semantic representations of their own, even though their constituents, like their counterparts in real compounds, do have their own meanings. Compared with real homophones, pseudohomophones could not compete efficiently with the semantic representations of their base words (e.g., 安全 an(1) quan(2), *safe*). If there is phonologically mediated semantic activation in reading Chinese, the presentation of pseudohomophones should be most likely to activate the semantic representations of their base words, and this activation should be easily detected in subjects' responses to target words (e.g., 危险, wei(1) xian(3), *dangerous*) that have strong semantic relations with the base words.

We chose 50 priming words (e.g., 安全) that had clearly semantically related words (e.g., 危险), which were used as targets in lexical decision. For each semantic priming word, a pseudohomophone (e.g., 安权) was created by replacing the second constituent of the semantic prime with another homophonic character. Rather than replacing the initial characters of the semantic prime with another homophonic character, we kept the original initial character to maximize the chance of semantic activation of the base word when the pseudohomophone was presented. For each semantic prime, an unrelated control priming word (e.g., 连续 lian(2) xu(4), *continuous*) was chosen, matched on frequency with the semantic prime. For each pseudohomophone prime, two unrelated nonword control primes were created, one an unrelated pseudohomophone itself and the other the random combination of two characters. The purpose of having both pseudohomophone and nonword control primes in the design was to check whether pseudohomophony of primes alone affected the processing of subsequently presented target words. Table 7.5 presents the design and sample stimuli.

Critical stimuli were split into five test versions, and 110 pairs of unrelated word–word, word–nonword, nonword–word, and nonword–nonword fillers were added into each version. The preparation of stimuli and test of subjects were conducted in the same way as Experiment 1. The SOAs between primes and targets were set at 100 ms, and subjects were asked to make lexical decisions to targets as quickly and as accurately as possible. The mean reaction times are shown in Fig. 7.2. Very few response errors were committed by subjects.

Statistic analyses revealed a highly significant main effect of prime type ($p < .001$). The Newman–Keuls post hoc tests showed a significant effect of semantic priming, against semantic controls and against other priming conditions ($p < .01$). However, there was no significant effect of pseudohomophone priming, either against pseudohomophone controls or against nonword controls ($p > .1$). This null effect of pseudohomophone priming in Chinese contrasts sharply with the robust effects in alphabetic languages (e.g., Lukatela & Turvey, 1991, 1994a; van Orden, Johnston, & Hale, 1988), but is consistent with the findings of Experiment 2. It suggests that the null effect of phonological mediation in semantic activation in Experiment 2 was not due to competition between homophones at the semantic level, but due to the limited effect of phonology on semantic activation, as assumed in the model sketched in Fig. 7.1.

Experiment 4: Single-Character Words

Experiment 4 applied the same mediated priming technique to single-character words. We also manipulated the relative frequency of semantic primes and the mediated primes, even though Lukatela and Turvey (1994a) did not find an interaction between this frequency manipulation and the size of mediated priming in English. For some targets (e.g., 舞, wu(3), *dance*), the frequencies of the semantic primes (e.g., 歌 ge(1), *song*) were higher than the frequencies of the mediated primes (e.g., 鸽 ge(1), *pigeon*). For other targets (e.g., 罐 guan(4), *jar*), the frequencies of the mediated primes (e.g., 谈 tan(2), *talk*) were higher than the frequencies of the semantic primes (e.g.,

TABLE 7.5
Experiment 3: Experimental Design and Sample Stimuli

	Prime Type					
	Semantic	*Semantic Control*	*Pseudohomophone*	*Pseudohomophone Control*	*Nonword Control*	*Target*
Written Form	安全	连续	安权	生误	改桌	危险
Pronunciation	an(1)	lian(2)	an(1)	sheng(1)	gai(3)	wei(1)
	quan(2)	xu(4)	quan(2)	wu(4)	zhuo(1)	xian(3)
Meaning	*safe*	*continuous*				*dangerous*

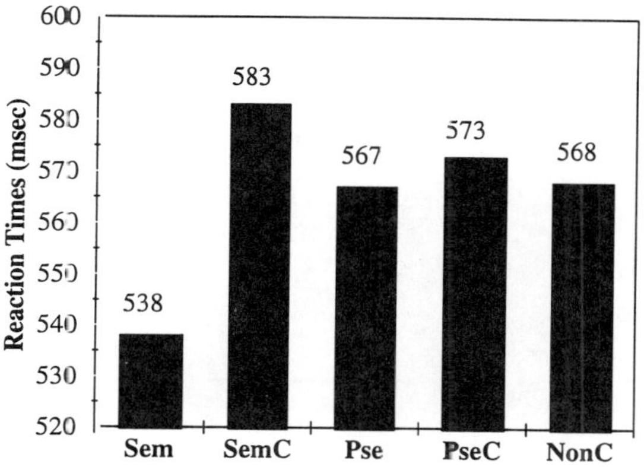

FIG. 7.2. Experiment 3: Mean Reaction Times (ms).
Sem = semantic primes, SemC = control primes matched with semantic primes, Pse = pseudohomophone primes, PseC = pseudohomophone control primes, NonC = nonword control primes.

TABLE 7.6
Experiment 4: Experimental Design and Sample Stimuli

		Prime Type				
		Semantic	*SemCon*	*Mediated*	*MedCon*	*Target*
High–Low	Written Form	歌	派	鸽	锤	舞
	Pronunciation	ge(1)	pai(4)	ge(1)	chui(2)	wu(3)
	Meaning	*song*	*send*	*pigeon*	*hammer*	*dance*
	Frequency	858	849	24	25	790
	Stroke	9.0	8.5	9.6	9.8	8.8
Low–High	Written Form	坛	浩	谈	村	罐
	Pronunciation	tan(2)	hao(4)	tan(2)	cun(1)	guan(4)
	Meaning	*pot*	*vast*	*talk*	*village*	*jar*
	Frequency	27	27	642	642	253
	Stroke	9.3	9.0	8.6	8.3	9.4

坛 tan(2), *pot*; see Table 7.6). If there exists strong competition between semantic representations of semantic and homophone primes, and/or if mediated phonological access plays an important role in semantic activation, one might predict some sort of interaction between the relative frequencies of semantic and homophone primes and the size of the phonologically mediated priming effects.

There were 48 critical target characters in total, 24 for the High–Low condition and 24 for the Low–High condition. Each target was paired with a semantically related prime and a mediated prime that was homophonic, but not orthographically similar, to the semantic prime. Because frequencies

TABLE 7.7
Experiment 4: Mean Naming Latencies (ms) and Error Percentages

	Prime Type			
	Semantic	*SemCon*	*Mediated*	*MedCon*
High–Low	543	553	569	562
	(0.8)	(2.1)	(1.7)	(2.5)
Low–High	560	576	569	574
	(2.1)	(3.3)	(1.3)	(2.5)
Mean	551	564	569	568

of semantic primes and mediated primes varied rather drastically, we chose different control primes for them, matched with them respectively on frequency and visual complexity (in terms of the number of strokes). Only the naming task was used, with the SOA between primes and target set at 43 ms. One purpose of using a shorter SOA here than the SOAs used in Experiments 2 and 3 was to check whether the null effect of phonologically mediated priming in the previous experiments was due to the relatively long SOA.[5]

Mean naming latencies and response error rates are reported in Table 7.7. Statistical analyses showed a significant main effect of prime type ($p < .01$) and no interaction between prime type and frequency manipulation ($p > .1$). Post hoc tests revealed a significant effect of semantic priming, but no effect of mediated priming.

These results replicate the findings for compound words. The direct semantic priming effect showed that the preactivation of semantic representations of single-character targets or compound targets—due to the presence of semantic primes—spreads to their phonological representations and facilitates the naming of the subsequently presented targets. The null effect of mediated priming in both High–Low and Low–High conditions was consistent with the view that there was little spread of activation from phonological representations to semantic representations.

Experiment 5: Homophone Density

Although Experiment 4, consistent with previous experiments, did not find significant effects of phonology on semantic activation, one might argue that the appearance of mediated priming has more to do with the homophone density than with the relative frequency of homophones. *Homophone density*

[5]The present experiment was also run with the SOA of 100 ms and 200 ms. Essentially the same pattern of priming effects was found in these levels as in the SOA of 47 ms.

is defined as a measure of how many characters have exactly the same pronunciations (at both the segmental and suprasegmental levels). Mediated primes having many homophonic competitors are less likely to prime targets than mediated primes having few homophonic competitors. This is because the mapping from phonology to semantics is modulated by the density of homophony, where semantic representations are less likely to be activated through a phonological representation and the competition between the activated semantic representations is strong if a phonological representation corresponds to many semantic representations. On this account, Experiment 4 did not observe a phonologically mediated semantic activation because the phonological representation of a mediated prime corresponded to, on average, 8.6 homophonic morphemes. On the other hand, homophone density may not be a crucial factor in determining whether a phonological mediated semantic priming appears. Both English studies (e.g., Lukatela & Turvey, 1994a) and our Experiment 3 used words with very low homophone density (i.e., usually with just two words corresponding to the phonological form), but the mediated priming appeared in English, not in Chinese. Pseudohomophones in Experiment 3, which did not have whole word semantic representations of their own and whose phonology corresponded to no words other than their base words, did not show priming, either.

In this experiment, we examined these arguments by explicitly manipulating homophone density. We chose a set of semantic and mediated primes whose phonological representations correspond to 5 or less homophonic morphemes and another set of semantic and mediated primes whose phonological representations correspond to 10 or more homophonic morphemes.[6] By using a design similar to Experiment 5, we would be able to

TABLE 7.8
Experiment 5: Experimental Design and Sample Stimuli

		Prime Type				
		Semantic	*SemCon*	*Mediated*	*MedCon*	*Target*
Low	Written Form	甜	稳	填	扰	苦
Homophone	Pronunciation	tian(2)	wen(3)	tian(2)	rao(3)	wu(3)
Density	Meaning	*sweet*	*steady*	*fill*	*disturb*	*bitter*
	Frequency	211	210	396	373	
	Stroke	9.3	9.3	9.2	9.1	
High	Written Form	叔	射	舒	援	爷
Homophone	Pronunciation	shu(1)	she(4)	shu(1)	yuan(2)	ye(2)
Density	Meaning	*uncle*	*shoot*	*stretch*	*aid*	*jar*
	Frequency	189	189	186	186	
	Stroke	9.8	9.7	8.9	8.9	

TABLE 7.9
Experiment 5: Mean Naming Latencies (ms)

	Prime Type			
	Semantic	*SemCon*	*Mediated*	*MedCon*
Low Density	598	618	614	607
High Density	596	617	607	611
Mean	597	618	611	609

examine whether homophone density played a role in the absence or appearance of phonologically mediated semantic priming.

Forty semantic and mediated primes with low density of homophone neighbors (from 1 to 5, 4 on average) and 40 semantic and mediated primes with high density of homophone neighbors (above 10, 14 on average) were chosen (Table 7.9). Control primes were selected to match respectively with semantic primes and mediated primes on frequency and visual complexity (in terms of number of strokes). Table 7.8 presents the design and sample stimuli. The SOA between primes and targets was set at 57 ms. The mean naming latencies are presented in Table 7.9. Very few errors were made in naming, and they were not analyzed further.

The analyses of naming latencies reveal a significant main effect of prime type ($p < .01$) and no interaction between prime type and density of homophony ($p > .1$). Post hoc tests showed a significant semantic priming effect against the semantic control ($p < .01$) and no effect of mediated priming against its control ($p > .1$).Therefore, the density of homophony has no obvious effect on the absence of phonologically mediated semantic activation.[7]

We note that a recent study by Tan and Perfetti (1997) used essentially the same technique and manipulation as this experiment, but found a dif-

[6]Here the number of characters included into a homophone neighborhood was counted according to the Institute of Language Teaching and Research (1986), which listed 4,574 different characters based on a corpus of 1.8 million characters. Although there are more modern Chinese characters (about 7,000; see Li, 1993; Yin & Rohsenow, 1994), most of them are unlikely to be known to university students.

[7]A reviewer of this chapter observed that although the mediated primes in the high density condition produced a nonsignificant 4 ms facilitatory effect compared with their matched control primes, they produced a much bigger 10 ms facilitatory effect compared with the semantic control primes, which were matched with both mediated primes and mediated control primes on frequency (see Table 7.8). However, semantic control primes were visually more complex than mediated primes and mediated control primes. This difference in visual complexity between semantic control primes and mediated control primes may be responsible for the difference between the mean naming latencies in the two conditions. The mediated priming effect could be overestimated if it comes from the comparison between mediated priming and semantic control priming.

ferent pattern of priming effects. They obtained not only semantic priming, but also reduced but significant mediated priming for the low homophone density condition. Moreover, they observed bigger semantic priming effects in the low density condition than in the high density condition. Although it is possible that this discrepancy with our results was caused by the differences in defining homophone density,[8] and in experimental procedures, a replication of this study using the same stimuli and procedures as Tan and Perfetti (1997), however, found no mediated priming effects at all for either the high or low homophone density condition (Zhou & Marslen-Wilson, 1998c).

GENERAL DISCUSSION

The main results of the five experiments can be summarized as follows. For phonological priming, the processing of target words, as measured in the lexical decision task, was not influenced by the preceding homophone primes at short SOAs (Experiment 1). Semantic priming effects appeared much earlier than homophone priming in lexical decision, although the phonological priming effect can be detected in the naming task (Experiment 1). Semantic priming effects were also strong in the naming task for single character words (Experiments 4 and 5), although not strong for compound words (Experiments 1 and 2). For phonologically mediated semantic priming, there were no effects for compound words in either naming or lexical decision tasks (Experiment 2), no effect for pseudohomophones in lexical decision (Experiment 2), and no effect for single-character words in naming (Experiments 4 and 5). The homophone density and the relative frequency between semantic and mediated homophone primes had no apparent influence on phonologically mediated semantic priming, which had been observed consistently in alphabetic languages (e.g., Lesch & Pollatsek, 1993; Lukatela & Turvey, 1994a).

We see, therefore, a dissociation between mediated semantic activation and mediated phonological activation. Although the presence of semantically related words can lead to the activation of phonological representation, the presence of phonologically related words does not lead to strong activation of semantic representations. The presence of semantic priming in naming demonstrates the efficient flow of activation from semantics to phonology. The absence of phonologically mediated semantic priming and the absence of homophone priming in lexical decision imply two possibilities. Either phonology is not automatically activated when a mediated or homophone prime is presented, or the activation of phonological

[8] Tan and Perfetti disregarded lexical tone when counting homophonic characters.

representations of primes does not play an important role in access to lexical semantics.

Automatic Phonological Activation and Limited Effect of Phonology on Access to Lexical Semantics

Despite uncertainties about some of the evidence concerning the automaticity of phonological activation in reading Chinese, we believe that the homophone priming effect in naming in Experiment 1 reflected, at least partly, the automatic component of phonological activation in processing compound words. However, the data from the experiment do not support the argument that phonological activation from orthography *must* be earlier than semantic activation from orthography (e.g., Perfetti & Tan, 1998). For compound words, we actually observed earlier semantic activation than phonological activation. We believe further stringent research is needed to address the relative time course of semantic and phonological activation (see Zhou & Marslen-Wilson, 1998b, for such a study). Rather than assuming that there exists an across-the-board answer to this issue for all types of words and under any experimental conditions, this research should consider at least the possibility of different time courses for different types of words and the potential influence of experimental tasks and designs on the pattern of data.

The fundamental reason for the automaticity of semantic and phonological activation is that logographic Chinese, like any other writing system, was created to represent spoken language and to convey lexical information, especially meanings of words. Although the correspondence between a character and its morphemic meaning and between a character and its pronunciation may be rather arbitrary, such correspondences are overlearned in individual development. Phonology is also required to keep lexical information in working memory and for higher level language comprehension. The efficient flow of activation from semantics to phonology conforms to this requirement (see Zhou & Marslen-Wilson, 1997).

Our central argument here is that although phonology may be automatically activated in reading Chinese, it does not play a major role in access in lexical semantics. In order to retrieve word meanings from the lexicon in real time language comprehension, the visual word recognition uses the most efficient way to access lexical semantics, although other less efficient routes, given time, may contribute to this access. In other words, different routes to semantics interact with each other in real time lexical processing. The optimal route may be different in different orthographies and may vary according to the relations between different components or domains of the lexicon. These relations or the spread of activation between different types of lexical representations can be better understood in terms of the mapping between different domains.

For Chinese, the mapping from orthography to semantics is rather arbitrary, although less so than in alphabetic languages. For simple characters, there is no systematic correspondence between orthography and semantics. Orthographically similar characters (e.g., 由 you(2), *because of*; 甲 jia(3), *first*; 田 tian(2), *field*) usually do not overlap semantically, whereas orthographically different characters (e.g., 水 shui(3), *water*; 干 gan(1), *dry*) may share semantic properties. For complex characters, although semantic radicals have the function of indicating the semantic category of the morphemes, in many cases this function has been obscured in the development of the writing system. For example, although the semantic reference of 犭, which indicates the category *mammal*, is relatively stable and obvious across different characters containing it (e.g., 狗 (*dog*), 猫 (*cat*), 猴 (*monkey*), 猿 (*ape*), and 猪 (*pig*)), it has exceptions (e.g., in 猜, *guess*). Moreover, not all characters in the same semantic category have this semantic radical (e.g., in 牛 (*ox*), 马 (*horse*) 驴 (*donkey*)) and not all semantic radicals refer to synchronically obvious semantic categories.

The mapping from orthography to phonology is also rather arbitrary, more so than orthographically "deep" alphabetic languages. Orthographically similar simple characters are mostly pronounced in different ways, whereas orthographically different simple characters may have similar pronunciations. For complex characters, although about one third of them have the same pronunciations as their phonetic radicals, this is far from consistent, even for most radicals involved in "regular" complex characters. For example, all the following complex characters are pronounced in the same way as their phonetic radical: 平 (ping(2)), 评 (*comment*), 坪 (*level ground*), 苹 (*apple*), 萍 (*duckweed*). However, there exists an exceptional or irregular character 秤 (cheng(4), *scale*), which is pronounced differently from the phonetic radical. In some extreme cases, characters having the same component (也 ye(3), *also*; 他 ta(1), *he*; 地 di(4), *earth*; 池 chi(2), *pool*) may have little similarity in pronunciation. Moreover, orthographically different characters (e.g., 鸽 ge(1), *pigeon*; 歌 ge(1), *song*; 割 ge(1), *cut*) may have the same pronunciations.

These relatively arbitrary and unpredictable relations between orthography and semantics and between orthography and phonology create difficulties for the mapping between them. Nonetheless, these mappings are easier to accomplish than the mapping between phonology and semantics, which is not just arbitrary but also plagued by ambiguity. For single-character or monosyllabic words, the same phonological form maps onto the different semantic properties of its homophonic morphemes. For compound words, although different words may share common morphemes, these morphemes may have different contributions to the meanings of compounds, which often have their own semantic properties that cannot be computed from morphemic meanings.

It is along this line that we can explain why phonology does not play a major role in access to lexical semantics. In the time course of lexical processing, the mapping from orthography to semantics is far more efficient than the mapping from orthography to phonology to semantics. Although the mapping from orthography to semantics and the mapping from orthography to phonology are both rather arbitrary—in some cases the former could be slower than the latter—the extra step from phonology to semantics makes the phonologically mediated route to semantics, which takes time to accomplish, comparatively underused. Thus, the inefficiency of spread of activation from phonology to semantics, as indicated by the broken line between phonology and semantics in Fig. 7.1, is not structurally but functionally defined. In the time course of visual word recognition, the semantic representations are optimally activated through the direct visual route, leaving the indirect route little time to manifest its effect. The fact that lexical semantics are accessed from phonological representations in spoken language comprehension does not contradict the present view. It is simply that this route is the more efficient way of access to lexical semantics in spoken word recognition. (This is also the reason for the broken line from phonology to orthography in Fig. 7.1.)

The preceding argument does not imply a simple race between two independent, autonomous routes to lexical semantics. The generic model sketched in Fig. 7.1 can easily be implemented as an interactive-activation based localist model, or more importantly, as a distributed connectionist model, in the framework of Seidenberg and McClelland (1989). In such a model, the knowledge of a word's orthographic form, meaning, and pronunciation is represented as patterns of activation distributed over sets of processing units connected in different weights. In visual word recognition, the presentation of visual stimuli initiates the computation from orthographic units to semantics and to phonology. It also starts the computation from phonological units to semantics when these phonological units are sufficiently activated. In fact, we prefer a parallel constraint–satisfaction approach in which semantic representations can be partially activated from different sources and there is continuous interaction between different types of information in real time (Zhou & Marslen-Wilson, 1998a, 1998b, 1998c). The meaning of a word becomes available when its semantic activation accrues over time from the spread of activation from both orthography and phonology, which interact with each other in real time. However, the relative contribution for the semantic activation from the two sources is determined by the weights between orthographic and semantic units, between orthographic and phonological units, and between phonological and semantic units. These weights, which determine the time course of activation, are largely determined by the systematicity of the correspondences between orthography, phonology, and semantics.

Clearly, given the characteristics of these correspondences in the logographic system, the build-up of semantic activation in reading Chinese is predominantly from orthography. Jared and Seidenberg (1991) argued for a division of labor for words of different frequencies. According to them, semantic representations for high-frequency words are principally activated directly from orthography before activation has spread from orthography to phonology to meaning. For low-frequency words, semantic activation from orthography is sufficiently slow to allow more activation of meaning from phonology. Such an argument does not readily apply to Chinese, however. Although the computation from orthography to semantics is slowed down for low-frequency words, the computation from orthography to phonology, unlike in English, is also slowed down due to the relatively arbitrary relations between orthography and phonology. Although low-frequency complex characters with regular phonetic radicals may have more phonological activation, the stronger competition in the semantic system, due to the activation of semantics of radicals, could obscure the semantic activation of complex characters (Zhou & Marslen-Wilson, chapter 3, this volume). Thus, while we do not deny that, under certain circumstances, the phonological activation may influence semantic activation and, more importantly, phonology interacts with orthography in constraining semantic activation (e.g., Zhou & Marslen-Wilson, 1998c), in order to account for our data and to swing the pendulum away from the strong phonological view, we assign a predominant role to orthography in access to meaning in reading Chinese.

Access to Semantics Across Different Orthographies

Our demonstration of the limited role of phonology in access to word meaning in reading Chinese is complementary rather than contradictory to the argument of a predominant role of phonology in reading English (Lukatela & Turvey, 1994a, 1994b; van Orden, 1987; van Orden et al., 1990; but see Jared & Seidenberg, 1991). In Chinese, because the mapping from orthography to phonology is roughly as arbitrary as the mapping from orthography to semantics, phonology does not have many opportunities to influence semantic activation in the time course of processing. In English, however, the mapping from orthography to phonology is much more systematic than the mapping from orthography to semantics. Phonology could be activated much earlier than semantics. Because the mapping from phonology to semantics is also highly overlearned (van Orden et al., 1990), phonology has more chances to constrain semantic activation or to influence experimental tasks involving semantics. Nevertheless, we believe that the fundamental architecture of lexical processing is the same across different orthographies, perhaps in the connectionist framework (e.g., Plaut et al.,

1996), and different routes to semantics interact with each other in lexical processing. What differs for different orthographies are the language-specific global parameters and word-specific local environmental factors that determine the weights between different processing units and the time course of orthographic, phonological, and semantic activation. For Chinese, global parameters are perhaps more important than local factors, and the direct visual constraints on lexical semantic activation dominate over the indirect phonological constraints.

ACKNOWLEDGMENTS

The research reported here was supported in part by a grant from UK Economic and Social Research Council to William Marslen-Wilson and Xiaolin Zhou and in part by a grant from China National Science Foundation to Xiaolin Zhou. We thank Yinghong Tang, Yaxu Zhang, Yan Sun, and Ningning Wu for helping us test subjects in Beijing, and William Marslen-Wilson, Gareth Gaskell, and several anonymous reviewers for their comments on earlier versions of this chapter.

REFERENCES

Becker, C. A. (1979). Semantic context and word frequency effects in visual word recognition. *Journal of Experimental Psychology: Human Perception and Performance, 5*, 252–259.

Besner, D., & Smith, M. C. (1992). Models of visual word recognition: When obscuring the stimulus yields a clearer view. *Journal of Experimental Psychology: Learning, Memory, and Cognition, 18*, 468–482.

Chen, H.-C., Cheung, S. L., & Flores d'Arcais, G. B. (1995). Orthographic and phonological activation in recognizing Chinese characters. *Psychological Research, 58*, 144–153.

Chen, H.-C., Flores d'Arcais, G. B., & Cheung, S. L. (1994). *Is phonological activation automatic in character recognition in Chinese?* Paper presented at the Inaugural Asian–Australian Workshop on Cognitive Processing of Asian Languages, Sydney.

Chen, H.-C., & Shu, H. (1997). *Lexical activation during the recognition of Chinese characters.* Unpublished manuscript, The Chinese University of Hong Kong.

Chen, Y., & Peng, D.-L. (1994). A connectionist model of Chinese character recognition and pronunciation. In H.-W. Chang, J.-T. Huang, C.-W. Hue, & O. J. L. Tzeng (Eds.), *Advances in the study of Chinese language processing* (Vol. 1, pp. 211–240). Taipei: National Taiwan University.

Coltheart, M. (1978). Lexical access in simple reading tasks. In G. Underwood (Ed.), *Strategies of information processing* (pp. 151–216). San Diego, CA: Academic Press.

Coltheart, M., Curtis, B., Atkins, P., & Haller, M. (1993). Models of reading aloud: Dual-route and parallel-distributed-processing approaches. *Psychological Review, 100*, 589–608.

Fang, S.-P., Horng, R.-Y., & Tzeng, O. (1986). Consistency effects in the Chinese character and pseudo-character naming tasks. In H. S. R. Kao & R. Hoosain (Eds.), *Linguistics, psychology, and the Chinese language* (pp. 11–21). Hong Kong: Centre of Asian Studies, University of Hong Kong.

Fleming, K. K. (1993). Phonologically mediated priming in spoken and printed word recognition. *Journal of Experimental Psychology: Learning, Memory, and Cognition, 19*, 272–284.

Flores d'Arcais, G. B., Saito, H., & Kawakami, M. (1995). Phonological and semantic activation in reading Kanji characters. *Journal of Experimental Psychology: Learning, Memory, and Cognition, 21*, 34–42.

Frost, R. (in press). Towards a strong phonological theory of visual word recognition: True issues and false trails. *Psychological Bulletin*.

Grainger, J., & Ferrand, L. (1994). Phonology and orthography in visual word recognition: Effects of masked homophone primes. *Journal of Memory and Language, 33*, 218–233.

Hue, C.-W. (1992). Recognition processes in character naming. In H.-C. Chen & O. J. L. Tzeng (Eds.), *Language processing in Chinese* (pp. 93–107). Amsterdam: North-Holland.

Humphreys, G. W., Evett, L. J., & Taylor, D. E. (1982). Automatic phonological priming in visual word recognition. *Memory & Cognition, 10*, 576–590.

Institute of Language Teaching and Research. (1986). *A frequency dictionary of modern Chinese*. Beijing: Beijing Language Institute Press.

Jared, D., & Seidenberg, M. (1991). Does word identification proceed from spelling to sound to meaning? *Journal of Experimental Psychology: General, 120*, 358–394.

Leck, K. J., Weekes, B. S., & Chen, M. J. (1995). Visual and phonological pathways to the lexicon: Evidence from Chinese readers. *Memory & Cognition, 23*, 468–476.

Lesch, M. F., & Pollatsek, A. (1993). Automatic access of semantic information by phonological codes in visual word recognition. *Journal of Experimental Psychology: Learning, Memory, and Cognition, 19*, 285–294.

Li, D. (1993). *A study of Chinese characters*. Beijing: Peking University Press.

Li, P., Bates, E., & MacWhinney, B. (1993). Processing a language without inflections: A reaction time study of sentence interpretation in Chinese. *Journal of Memory and Language, 32*, 169–192.

Liu, Y., Liang, C., Wang, D., & Zhang, X. (1990). *Dictionary of usage frequency of modern Chinese words*. Beijing: Beijing University of Aeronautics and Astronautics Press.

Lukatela, G., & Turvey, M. T. (1991). Phonological access of the lexicon: Evidence from associative priming with pseudohomophones. *Journal of Experimental Psychology: Human Perception and Performance, 19*, 166–178.

Lukatela, G., & Turvey, M. T. (1994a). Visual lexical access is initially phonological: 1. Evidence from associative priming by words, homophones, and pseudohomophones. *Journal of Experimental Psychology: General, 123*, 107–128.

Lukatela, G., & Turvey, M. T. (1994b). Visual lexical access is initially phonological: 2. Evidence from phonological priming by homophones and pseudohomophones. *Journal of Experimental Psychology: General, 123*, 331–353.

McClelland, J. L., & Rumelhart, D. E. (1981). An interactive activation model of context effects in letter perception: Part 1. An account of basic findings. *Psychological Review, 88*, 375–407.

McCusker, L. X., Hillinger, M. L., & Bias, R. C. (1981). Phonological recoding and reading. *Psychological Bulletin, 88*, 217–245.

Peng, D., Yang, H., & Chen, Y. (1994). Consistency and phonetic-independence effects in naming tasks of Chinese phonograms. In Q. Jing, H. Zhang, & D. Peng (Eds)., *Information processing of the Chinese language* (pp. 26–41). Beijing: Beijing Normal University Press.

Perfetti, C. A., & Tan, L. H. (1998). The time course of graphic, phonological, and semantic activation in Chinese character identification. *Journal of Experimental Psychology: Learning, Memory, and Cognition, 24*, 101–118.

Perfetti, C. A., & Zhang, S. (1991). Phonological processes in reading Chinese words. *Journal of Experimental Psychology: Learning, Memory, and Cognition, 17*, 633–643.

Perfetti, C. A., & Zhang, S. (1995). Very early phonological activation in Chinese reading. *Journal of Experimental Psychology: Learning, Memory, and Cognition, 21*, 24–33.

Plaut, D. C., McClelland, J. L., Seidenberg, M. S., & Patterson, K. E. (1996). Understanding normal and impaired word reading: Computational principles in quasi-regular domains. *Psychological Review, 103*, 56–115.

Read, C. A., Zhang, Y., Nie, H., & Ding, B. (1986). The ability to manipulate speech sound depends on knowing alphabetic reading. *Cognition, 24*, 471–485.

Seidenberg, M. S. (1985). The time course of phonological code activation in two writing systems. *Cognition, 19*, 1–30.

Seidenberg, M. S., & McClelland, J. L. (1989). A distributed, developmental model of visual word recognition and naming. *Psychological Review, 96*, 523–568.

Seidenberg, M. S., Petersen, A., MacDonald, M. C., & Plaut, D. C. (1996). Pseudohomophone effects and models of word recognition. *Journal of Experimental Psychology: Learning, Memory, and Cognition, 22*, 48–62.

Taft, M. (1994). Interactive-activation as a framework for understanding morphological processing. *Language and Cognitive Processes, 9*, 271–294.

Tan, L. H., Hoosain, R., & Peng, D.-L. (1995). Role of early presemantic phonological code in Chinese character identification. *Journal of Experimental Psychology: Learning, Memory, and Cognition, 21*, 43–54.

Tan, L H., Hoosain, R., & Siok, W. W. T. (1996). Activation of phonological codes before access to character meaning in written Chinese. *Journal of Experimental Psychology: Learning, Memory, and Cognition, 22*, 865–882.

Tan, L. H., & Perfetti, C. A. (1997). Visual Chinese character recognition: Does phonological information mediate access to meaning? *Journal of Memory and Language, 37*, 41–57.

Tzeng, O. J. L., & Hung, D. L. (1980). Reading in a nonalphabetic writing system: Some experimental studies. In J. F. Kavanagh & R. L. Venezky (Eds.), *Orthography, reading and dyslexia*. Baltimore: University Park Press.

Tzeng, O. J. L., Hung, D. L., & Wang, W. S.-Y. (1977). Speech recoding in reading Chinese characters. *Journal of Experimental Psychology: Learning, Memory, and Cognition, 3*, 621–630.

van Orden, G. C. (1987). A ROWS is a ROSE: Spelling, sound, and reading. *Memory & Cognition, 15*, 181–198.

van Orden, G. C., & Goldinger, S. D. (1994). Interdependence of form and function in cognitive systems explains perception of printed words. *Journal of Experimental Psychology: Human Perception and Performance, 20*, 1269–1291.

van Orden, G. C., Johnston, J. C., & Hale, B. L. (1988). Word identification in reading proceeds from spelling to sound to meaning. *Journal of Experimental Psychology: Learning, Memory, and Cognition, 14*, 371–385.

van Orden, G. C., Pennington, B. F., & Stone, G. O. (1990). Word identification in reading and the promise of subsymbolic psycholinguistics. *Psychological Review, 97*, 488–522.

Verstane, A., Humphreys, G. W., Olson, A., & d'Ydewalle, G. (1995). Are phonemic effects in backward masking evidence for automatic prelexical phonemic activation in visual word recognition? *Journal of Memory and Language, 34*, 335–356.

Wang, W. S.-Y. (1973). The Chinese language. *Scientific American, 228*, 50–56.

Wydell, T. N., Patterson, K. E., & Humphreys, G. W. (1993). Phonologically mediated access to meaning for Kanji: Is a ROWS still a ROSE in Japanese Kanji? *Journal of Experimental Psychology: Learning, Memory, and Cognition, 19*, 491–514.

Xu, Y. (1991). Depth of phonological recoding in short-term memory. *Memory & Cognition, 19*, 263–273.

Yin, B., & Rohsenow, J. S. (1994). *Modern Chinese characters*. Beijing: Sinolingua.

You, R., Qian, N., & Gao, Z. (1980). On the phonological system in Mandarin Chinese. *Zhongguo Yuwen, 148*, 328–334.

Zhang, B. Y., & Peng, D.-L. (1992). Decomposed storage in the Chinese mental lexicon. In H.-C. Chen & O. J. L. Tzeng (Eds.), *Language processing in Chinese* (pp. 131–149). Amsterdam: North-Holland.

Zhang, W., Feng, L., & He, H. (1994). The activation of phonology and semantics in Chinese character recognition. In H.-W. Chang, J.-T. Huang, C.-W. Hue, & O. J. L. Tzeng (Eds.), *Advances in the study of Chinese language processing* (Vol. 1, pp. 185–198). Taipei: National Taiwan University.

Zhou, X. (1994). *Sub-lexical processing in reading Chinese.* Paper presented at the Inaugural Asian–Australian Workshop on Cognitive Processing of Asian Languages, Sydney.

Zhou, X., & Marslen-Wilson, W. (1994). Words, morphemes and syllables in the Chinese mental lexicon. *Language and Cognitive Processes, 9*, 393–423.

Zhou, X., & Marslen-Wilson, W. (1995). Morphological structure in the Chinese mental lexicon. *Language and Cognitive Processes, 10*, 545–601.

Zhou, X., & Marslen-Wilson, W. (1996). Direct visual access is the only way to access the Chinese mental lexicon. In G. W. Cottrel (Ed.), *Proceedings of the Eighteenth Annual Conference of the Cognitive Science Society* (pp. 714–719). Mahwah, NJ: Lawrence Erlbaum Associates.

Zhou, X., & Marslen-Wilson, W. (1997). Spread of activation in the mental lexicon. In M. G. Shafto & P. Langley (Eds.), *Proceedings of the Nineteenth Annual Conference of the Cognitive Science Society* (pp. 838–843). Mahwah, NJ: Lawrence Erlbaum Associates.

Zhou, X., & Marslen-Wilson, W. (1998a). *Pseudohomophone effects and lexical processing in reading Chinese.* Manuscript submitted for publication.

Zhou, X., & Marslen-Wilson, W. (1998b). *The relative time course of semantic and phonological activation in reading Chinese.* Manuscript submitted for publication.

Zhou, X., & Marslen-Wilson, W. (1998c). *Orthography, phonology, and lexical semantic activation in reading Chinese.* Manuscript submitted for publication.

Zhou, X., Marslen-Wilson, W., Bi, Y., Shi, D., & Shu, H. (1996). *Pseudohomophone effects in reading Chinese.* Manuscript, Birkbeck College, University of London.

Zhou, X., Marslen-Wilson, W., Shu, H., Bi, Y., & Tang, Y. (1996). *Phonology in reading Chinese.* Manuscript, Birkbeck College, University of London.

Zhou, X., Taft, M., & Shu, H. (1995, December). *Morphological, orthographic, and phonological structure in the Chinese mental lexicon.* Paper presented at the Seventh International Conference on the Cognitive Processing of Chinese and Other Asian Languages, Hong Kong.

CHAPTER

8

Character and Word Recognition in Chinese

In-mao Liu
National Chung-Cheng University
and
National Taiwan University

This chapter presents a model of character and word recognition in Chinese. It becomes apparent that the present model also applies to word recognition in other languages. Before presenting a detailed model of character and word recognition, it is necessary to clarify some definitions.

DEFINITIONS

Characters are visual units of the Chinese writing system. Each character appears as a compact, square pattern of strokes. Characters that consist of two subpatterns of strokes are very common (e.g., 詩). These characters, called phonetic compounds, consist of two parts (subpatterns of strokes): (a) the left part (言), called a *semantic radical*, that contributes some aspect of its meaning to the character, such as to which category the referred object belongs, and (b) the right part (寺), called a *phonetic*, that cues the sound of the character. There are, of course, deviations from these norms; for example, the left part may be a phonetic component and the right part a semantic radical, or the two parts may be arranged vertically, etc.

Although characters are generally considered as morphemes, there is a dispute as to whether characters are words. DeFrancis (1989) obtained a random sample of 200 characters from the Chao and Yang dictionary (1947) and found the following distribution: 44% free, 45% semibound, and 11% completely bound. DeFrancis compared Chinese "yuan2" (員) to English *er*, both of which are semibound morphemes. However, there are many differences. First, "yuan2" can be the first character of a two-character coordinate compound, such as "yuan2 e2" (員額) that means something like *quota*. Second, "yuan2" can be combined with a numeral to denote a certain number of persons (e.g., 三員一工). In the latter case, "yuan2" is used nearly as a free morpheme. A similar argument applies to the case of completely bound morphemes in Chinese. It may be concluded that nearly all Chinese characters may be used as words. In the sequel, therefore, a character will be used interchangeably with a single-character word, and a word used to refer to a two-character word. That is because most multicharacter words are two characters long, and because the process of recognizing single-character words differs significantly from the process of recognizing two-character words.

There is quite an agreement on the usage of the term *recognition*. In a standard recognition experiment consisting of a study and a test session, a fixed number of items are presented during the study session. In the test session, the studied items are intermixed with a set of new items and presented one at a time. If subjects identify an item as being presented in the study session, they are said to *recognize* the item. In other words, recognition refers to the process of finding a match between a stimulus and a stored representation of that stimulus.

In the *lexical decision task*, the subject is presented a letter string (e.g., "tiger" or "siger") and asked to decide as quickly as possible whether or not it comprises a word. The time taken by the subject to arrive at this decision is called the *lexical decision time*. The lexical decision task is a sort of recognition experiment, because the experimenter attempts to measure the time taken by subjects to recognize a letter string as a word that has appeared in any "list" encountered in the past. In this sense, the lexical decision task is a word classification task.

MODELS OF WORD RECOGNITION

In the Western literature, several influential models of word recognition have been proposed: the logogen (a word representation in the lexicon) model (e.g., Morton, 1969), the search model (e.g., Forster, 1976), the interactive-activation model (e.g., McClelland & Rumelhart, 1981), the verification model (e.g., Becker, 1976), and the parallel distributed processing (PDP) model (e.g., Seidenberg & McClelland, 1989). In a visual recognition task, there are two features that are common to all models. First, a visually presented stimulus is encoded for extracting information that gives rise to a set of candidates. Second, a word representation is selected from the candidate set. The first and second steps are combined in the search model. The second step is somewhat different for the PDP model from the rest of the models because the PDP model does not assume the existence of the lexicon.

There are several shortcomings with existing models of word recognition. First, the interplay between word recognition and phonological activation (or semantic activation) is either vaguely specified or unspecified. Second, all models were proposed for explaining word recognition in English. A model described next is proposed to overcome these shortcomings.

AN ENCODING-ASSOCIATIVE MODEL OF CHARACTER RECOGNITION

The major assumption of the encoding-associative model of character recognition is that the context in which a character has been experienced determines how the character is recognized in a later occasion. The two concepts in this assumption, context and experience, need to be clarified.

Encoding or *experiencing* a character represents a procedure of registering a visual input. The same character may be printed in different fonts on different occasions. It may appear in different handwritings. Therefore, each encoding may represent a different procedure of registering the same character. However, there is some common denominator that is shared by all the encoding procedures of the same character. A similar procedural view was proposed earlier (e.g., Kolers & Roediger, 1984).

Experiencing a character always takes place in some *context*. Characters are encountered in the contexts of various reading materials. In initial encounters, beginning readers are taught characters' pronunciation as well as their meaning. In later encounters, characters continue to arouse their meaning or to activate their phonology. This is a process of associating characters with their meaning and pronunciation. This kind of associative learning continues so long as meaning and pronunciation are necessary parts of reading. As a consequence of associative learning, when characters are individually presented, they tend to give rise associatively to their meaning and pronunciation.

There are thus two stages in the encoding-associative model: (a) the stage of encoding in which a character is experienced; and (b) the stage of association. In the first stage, a character is encoded in its characteristic way to give its identification. In the stage of association, the common denominator of the contexts in which a character is experienced comes to be associatively linked to the encoding of the character. Let us consider these two stages separately.

The Nature of Character Encoding

The basic nature of character encoding is that the encoding of a certain fraction tends to facilitate the encoding of the whole. The assumption of this basic nature of character encoding is generally made in all existing models of word recognition. Moreover, subjects come to recognize an item by encoding its smaller fraction as a function of frequency of experiencing the item in the past. If two parts co-occur frequently, then the encoding of one part tends to facilitate the encoding of the other part.

As for empirical evidence from studies using English words or nonwords, it is known that initial fragments serve as the best cues for recognizing the items, final fragments serve as the next best cues, and middle fragments serve as the poorest cues (e.g., Horowitz, White, & Atwood, 1968). By manipulating fragment size and number of training trials, Liu (1969) further showed that subjects come to recognize the items by encoding smaller fragments as the number of trials increases. These experiments were conducted using "episodic" memory tasks. However, as the following studies show, their conclusions should apply to a "semantic" memory task with appropriate modifications.

With respect to recognition of Chinese characters using a semantic memory task, Liu (1983) deleted an upper left quarter, upper right quarter, lower left quarter, or lower right quarter. It was found that deleting an upper quarter of each character resulted in more reading errors than deleting a lower quarter. In other words, an upper quarter serves as a better cue than does a lower quarter for recognizing the character. Therefore, the cueing function of Chinese character fragments is more or less similar to that of English words. In a further study (Liu, 1984), Chinese characters were classified on the basis of how two separable groups of strokes are arranged to form a character: (a) horizontal (e.g., 技), (b) vertical (e.g., 音), (c) inside–outside (e.g., 困). The results support the hypothesis that beginning strokes are more important than ending strokes in serving as the cue for recognizing characters. The cueing power of a fragment-deleted character of a certain composition, therefore, depends on how beginning strokes are concentrated in that fragment. This finding also supports the view that subjects encode a character in the way it is written.

The evidence that encoding one part facilitates the encoding of another part when the two parts co-occur frequently is also available. Sue and Liu (1996) used characters of the type consisting of three groups of strokes in their study of the character superiority effect (e.g., 詩). Let us use A (言), B (土), and C (寸) to stand for the three groups of strokes. Then, A occupies the left part, and B and C the right part of a character, with B on the top of C. For the sake of exposition, let us consider only some part of this study. The subjects' task in this part of the study was to identify B or C in characters or noncharacters or in isolation. For noncharacters, B–C combinations (寺) that were identical to those in characters were paired with different A's (斤), the results of which did not form characters.

The crucial finding was that C was identified better in characters and noncharacters than in isolation. An interpretation is as follows. We first note that B and C co-occur regularly as a phonetic and combine with a different radical (A) to form a legal character. In the case of a noncharacter, both B and C still co-occur regularly. When C is to be identified in the context of characters or noncharacters, there are two cases. In one case, subjects may not identify B. In the other case, subjects have identified B at the time of trying to identify C. In the latter case, suppose that subjects identify C better in the context of characters and noncharacters than in isolation. If this result obtains, then it provides the evidence that identification of B aids in the identification of C, because B and C co-occur regularly. This is what was obtained in the Sue and Liu (1996) study. They called this phenomenon the *partial-character superiority effect*.

The Nature of Associative Processes

People encounter characters usually in the context of sentences for the purpose of gaining the meanings expressed by the sentences. After encountering some characters more frequently than others, some characters tend to give a sense of familiarity more readily than do others. This familiarity refers to the familiarity of a visual item as it is used as a character, that is, the property of "characterness" or "wordness" (Liu, Wu, & Chou, 1996). The experience of character familiarity is a consequence of association, because characters are experienced always in "reading" contexts.

In gaining meanings from characters in reading sentences, people tend to recode spelling into phonological representations because the latter can be processed effectively in working memory for comprehending sentences. Therefore, characters become associated not only with their meaning, but also with their phonological representation.

In conclusion, because characters are experienced in the context of reading, their encoding will associatively activate "characterness" experience in the next encounter. Because characters have both phonological and semantic functions, their encoding also tends to associatively activate phonological and semantic representations in the next reading. By making use of characterness experience in recognizing characters, the lexical decision task attempts to measure lexical access time. Since both characterness experience and phonology are associatively activated by visual presentation of words, there is a dispute as to whether phonology is activated prior to recognizing words.

By emphasizing the encoding and associative aspects involved in character recognition, the present model can explain (a) how phonological and semantic codes become activated as a consequence of recognizing characters, (b) why phonological activation differs from one language to another, and in particular, (c) why there is no prelexical phonology in reading Chinese characters. Because of space limitation, however, each topic is discussed only briefly.

BASIC ELEMENTS OF THE MODEL

The encoding-associative model has four basic elements: one encoding and three associated responses. The first associated response is wordness or characterness experience. Because every time subjects encounter a character in the context of reading materials, it is experienced as a script unit, the association between a character and its familiarity of being used for reading quickly develops. The second associated response is phonological activation, the result of vocal or subvocal responses accompanying reading. The third associated response is an arousal of semantic meaning that is extracted from sentences in reading. Because the three types of associated responses usually co-occur in the same context, it is possible for one type of response to be associated with another. For the sake of simplicity, this complication is not considered in the sequel.

Encoding

In this model, a characteristic way in which a character is encoded stands for how the character is recognized. A characteristic encoding of a character is naturally accompanied by some familiarity experience ("characterness experience") through association. It is not necessary to assume hierarchical levels of representations that reside somewhere in the brain. In this sense, the present model is closer to a PDP model than to the interactive-activation model of word recognition. As a consequence of a characteristic encoding of a character, the encoding of some parts of a character may facilitate the encoding of the remaining parts simply because the character has been experienced or encoded in its characteristic way many times.

Wu, Chou, and Liu (1994) selected two sets of high-frequency and low-frequency characters. All the characters were phonetic compounds such that phonetic components were also characters. These investigators also controlled for frequency of phonetics. The distributions of semantic radicals were about the same for high-frequency and low-frequency characters. Therefore, so far as frequencies of character components were concerned, they were approximately equated for the two sets of characters. The time of encoding phonetics was further controlled by presenting phonetics one second prior to presentation of the entire characters. It was found that the lexical access time of high-frequency characters was still shorter than that of low-frequency characters. It may be concluded that the encoding of one component facilitates the encoding of another component when the two components co-occur frequently with the component frequencies being controlled.

The encoding of characters results in activating graphic codes that are generally short-lived. Perfetti and Tan (in press) conducted primed-naming experiments by varying prime type and prime–target stimulus onset asynchrony (SOA). Their results showed that graphic prime facilitation at 43 ms SOA gave way to inhibition at longer SOAs. The onset of graphic inhibition coincided with the onset of phonological facilitation.

Characterness Experience

The cue that produces characterness experience from a particular character may be identified as a set of common procedures involved in encoding the character. By identifying the cue as a set of common procedures involved in encoding the character, it naturally follows that the more specific procedures are involved, the longer it takes for its recognition. This explains what is called the visual degradation effect for English words (e.g., Meyer, Schvaneveldt, & Ruddy, 1974). This is because the procedures involved in identifying degraded items should include some specific or uncommon procedures.

Characterness experience is better studied in the lexical decision task. This is a task in which subjects are timed for classifying a presented letter string as a word or nonword. Since wordness experience is not all-or-none, a decision component necessarily enters to complicate the response measure. Liu et al. (1996) used a lexical naming task as well as a regular naming task and a lexical decision task for locating the frequency effects in lexical decision and naming.

A task is called a lexical naming task because it is an analogue of the lexical decision task. It differs from a regular naming task in that pseudocharacters are included as well as characters. When an item (a character or a pseudocharacter) is presented in the lexical naming task, the subject is to name it if the item is a character. If the presented item is a pseudocharacter, the subject is to say aloud "jia3" (假), meaning a pseudocharacter, because it has no pronunciation. Thus, the naming of Chinese characters in the lexical naming task involves decision processes absent in the regular naming task. Because naming a Chinese character necessarily involves lexical access, a decision component of the frequency effect in lexical decision can be isolated. This procedure does not work for alphabetic writing systems because sublexical processes underestimate the frequency effect in the regular naming task. As a consequence, Liu et al. (1996) were able to estimate a prelexical component of the frequency effect in lexical decision. Their results confirm the early finding (e.g., Forster & Chambers, 1973) that the frequency effect is prelexical and substantial.

Furthermore, because word recognition is based on wordness experience, nonwords whose spellings are similar to words may slow responses and produce recognition errors. This results in what are called the *nonword legality effect* and the *word similarity effect* (see Taft, 1991). These effects are readily accounted for by noting that some common procedures are involved in encoding nonwords whose spellings are similar to words, etc.

Phonological Activation

Although words/characters differ for different writing systems, the main principles of encoding words/characters for arriving at wordness/characterness experience are the same. The processes of phonological activation, however, differ markedly from alphabetic to logographic writing systems. The principle of association that governs the development of characterness/wordness experience is, however, the same in the development of phonological activation. When characters are presented visually, the spelling–sound correspondence (stimulus–response compatibility) is now an important factor that affects the development of phonological activation.

Spelling–sound correspondences are generally understood in terms of cue properties for activating sounds. What is the cue for activating phonology in each language? In English, many words are regular. Therefore, it is possible to rely on word components (letters) to *start* pronouncing a word correctly. In other words, the cue for activating phonology is not necessarily the entire word. As a consequence, phonology may be activated without necessarily encoding the entire word. This provides a case for phonological activation without recognizing a word, that is, prelexical phonology. As an example, let us consider how *prime* is pronounced. In this case, by encoding the initial letters *pr* alone, the subject may be able to prepare pronouncing the initial *p* sound correctly.

The picture is different in the case of reading Chinese characters. Although most Chinese characters are phonetic compounds, estimates on the validity of phonetics are not high, ranging from 27% (Fan, Gao, & Ao, 1984) to 48% (Zhou, 1978). If the validity of phonetics for character sounds is less than 50%, it is doubtful that the phonology will be activated by encoding phonetics alone. In other words, phonological activation follows only after encoding both phonetic and semantic radical.

Experimental findings generally support the notion that there is no prelexical phonology in reading Chinese characters. Paap, Chen, and Noel (1987) presented 50 characters of very low frequency to 12 skilled Chinese readers. When they were unable to recognize the characters, they were asked to guess a pronunciation based on whatever subcharacter cues they could find. Paap et al. (1987) found that only 13% of the pronunciations were correct. Recently, Wu and Liu (1997) obtained direct evidence in support of the conclusion that phonological activation follows only after encoding both phonetic and semantic radical.

Liu (1988, 1995) proposed a way through which it is possible to compute how many letters (or components) one has to encode in order to arrive at a correct pronunciation for each letter (component). The computational procedure is as follows. First, subjects start to extract any detectable pronunciation cues from some smallest component. Second, if the pronunciation of such component is uniquely determined as it is contained in all words in the subjects' vocabulary, then this unique pronunciation is a correct pronunciation for that component. Third, if the pronunciation of such component is not uniquely determined because this component is pronounced differently when it is contained in different words, then this component is combined with an adjacent component to obtain a larger component. Subjects make the same test with this larger component until a unique pronunciation is found. Fourth, the same testing procedure is applied to all components, and the pronunciation of a word is obtained by blending the component pronunciations. This algorithm applies to any writing system and has been tested for the pronunciation of English words (So, 1990).

As an example, let us compute how the first letter *t* is pronounced in *tape*. First, we note that *t* is not uniquely pronounced when it is used as a beginning letter of those words starting with *t* (consider, e.g., *thin*). Second, *t* is uniquely pronounced as /t/ for all those words beginning with *ta*. Therefore, two tests are required to determine how *t* is pronounced in *tape*. It was found that the number of tests required for arriving at a correct pronunciation for a constituent letter is a good measure for accounting for a difference in the pronunciation latencies between exception and regular words (Liu, 1995). The number of tests required for arriving at a correct pronunciation for each constituent can also be obtained for Chinese characters. It was found that the number of tests required for Chinese characters was much greater than the number of tests required for comparable English words (i.e., consisting of "onset" and "body").

It is known that there is a problem with any lexical analogy model because a person is assumed to store pronunciations of all words in an analogy set for computing a word's pronunciation. Instead of computing a word's pronunciation from an analogy set, then, isn't it easier to retrieve it from the analogy set? The same problem seems to apply to the present conceptualization. It should be noted, however, that a low-frequency word is still encoded letter by letter. How many tests are required to arrive at a correct pronunciation for a particular letter may form internalized knowledge, which is then used to compute the pronunciation of a low-frequency word.

Semantic Activation

Recent findings (e.g., Tan, Hoosain, & Siok, 1996) using Chinese characters as stimuli suggested that phonological activation precedes semantic activation. In the Tan et al. study, when the target was exposed at threshold + 14 ms in a backward masking paradigm, the effect of homophonic masks was significant, whereas semantic masks in the absence of associative relatedness to the targets yielded no significant effect on target identification. However, as these investigators are also aware, this is entirely a measurement problem. Semantic activation may precede phonological activation. The problem is that it is difficult to find appropriate semantic masks to capture activated semantic representations.

The process of semantic activation should not differ between writing systems because according to the encoding-associative model of word recognition, semantic activation merely represents responses associated to the encoding of words. Aaronson and Ferres (1986), among others, hypothesized that Chinese characters are different from alphabetic words, and are directly encoded as abstract images that are themselves representations of meanings. Liu, Zhu, and Wu (1992) tested this hypothesis by asking subjects to judge, as fast as possible, whether a Chinese character or English word or picture contained a meaning component specified in advance. Chinese characters were classified into two types: phonograms, which contain a phonetic component as a clue to the sound of characters; and pictograms, which are pictographic in origin. The results showed that (a) Chinese characters were not responded to faster than English words, (b) reaction times were not faster to pictograms than to phonograms, and (c) reaction times were faster to pictures than to Chinese characters.

In conclusion, as in the case of phonological activation, the development of semantic activation follows the same principle of association. When characters are visually presented, a pronunciation becomes directly associated with a character. There is often a one-to-one association between script units and sounds. The development of a semantic code is, however, quite different. This is because various meanings become associated with many sentences or phrases, each of which includes a character as a constituent. There is controversy as to whether phonological activation is earlier than semantic activation. Nevertheless, this controversy becomes transparent in the encoding-associative framework of character recognition.

RECOGNITION OF TWO-CHARACTER WORDS

As noted, Chinese characters may be generally used as words. However, in modern times many two-character words are used to represent single ideas, such as "dian4 hua4" (電話 for *telephone*), "si1 xiang3" (思想 for *thought*), etc. It is estimated that more than 60% of words in Chinese consist of two characters. Therefore, a separate section is used for discussing recognition of those words constructed from two characters. Although there are a number of three-character words, the same model for two-character words naturally applies to three-character words. Four-character or more-than-four-character words are generally considered as phrases.

Frequency Effects in Lexical Decision and Naming

The frequency effects in lexical decision and naming were obtained for two-character words as in single-character words (Wu et al., 1994). The size of the frequency effect in lexical decision tends to be larger than the size of the frequency effect in naming for two-character words as in English. This contrasts markedly with the case of characters (e.g., Liu et al., 1996). In the case of characters, the frequency effect in naming is two to three times larger than the frequency effect in lexical decision. The reasons for a reverse in the relative size of the frequency effects for two-character words may be as follows. First, it must be more difficult to judge whether a combination of two characters is a two-character word than to judge whether a combination of a radical and a phonetic component forms a character. Second, it is rarely the case that both characters of a low-frequency two-character word are all of low frequency. Suppose that the first character of a low-frequency two-character word is of high frequency, and that the second character is of low frequency. In this case, subjects may start pronouncing the first character while they attempt to process the second character in the meantime. This is apparently the case of prelexical phonology. The lexicon here refers to that of two-character words.

Facilitation of Encoding in the Lexical Decision Task

Although the same encoding-associative model of character recognition applies to the case of two-character words, there are some characteristic features. Wu et al. (1994) attempted to study the strength of intraword connections as a function of word frequency by using the lexical decision task. It was hypothesized that the lexical-access frequency effect should be localized largely at the assembling of subword units (characters). In other words, although high-frequency words and low-frequency words generally involve different subword units, it takes longer for the latter than for the former to assemble subword units in lexical access.

The lexical access time may consist of (a) first-character encoding time, (b) second-character encoding time, and (c) assembling time, although some part of the encoding time may overlap with the assembling time. The strategy of Wu et al. (1994) was to equate the encoding time of first characters and to control for the encoding time of second characters in measuring the assembling time as a function of word frequency.

To equate the encoding time of first characters, for every high-frequency word a low-frequency word was selected to start with the same character. For instance, there are characters standing for *sea* ("hai3", 海), *army* ("jun1", 軍), and *strip* ("dai4", 帶). Then, a high-frequency and low-frequency word can be constructed by using "hai3 jun1" (海軍, meaning *navy*) and "hai3 dai4"

(海帶, meaning *seaweed*), respectively. To control for the encoding time of second characters, in the successive condition the second character of each word was presented 500 ms before presenting the first character for subjects to make lexical decision. In the simultaneous condition, both characters were presented for lexical decision. By comparing the frequency effects obtained in the two conditions, it was observed that the assembling (integration) of two characters to form a meaningful word comprises the major component of the obtained frequency effect. This is because the mean frequency effect of 70 ms in the successive condition, which is attributable to the assembling of word components, is approximately 85% of the mean frequency effect of 82 ms that is attributable to both processing and assembling of word components in the simultaneous condition.

Intraword Connections

The concept of unit formation or response integration was early proposed in the context of paired-associate learning. This concept of intraword connections becomes explicit in connectionist models, although such intraword connections are dealt with at quite a low level of analysis.

Mandarin Chinese, the standard language of China, has rather simple syllable structures (Ho, 1987). For instance, consonant clusters are not allowed, and final consonants are reserved for nasals. Thus, syllables always have clear boundaries, and ambisyllabic consonants that cross syllabic boundaries hardly exist. The lexical tone also has a bearing on syllabic units, characterizing each syllable by a fixed tone pattern that is lexically contrastive. Based on these observations, Liu and Wu (1997) undertook to measure not only the latency of pronouncing the first character, but also the latency of pronouncing the second character, in naming a Chinese two-character word.

Liu and Wu (1997) found that word frequency affected not only the latency of pronouncing a word (first-character latency), but also the latency of pronouncing its second character (intercharacter latency). Repetition priming also obtained the same result. Their control experiment ruled out a possibility that the latency of second syllables is nothing but the duration of first syllables. Therefore, they provided evidence that intraword connections are strengthened as a function of frequency.

Postlexical Phonology in Naming Two-Character Words

In their fourth experiment, Liu and Wu (1997) presented (a) two-character words, (b) only the first characters of two-character words, and (c) two characters that were not lexicalized, for naming. By measuring not only the first-character RT (reaction time), but also the second-character RT, they found, first, that subjects encode the first character as an entry point to a two-character word. Consequently, frequency of first characters has an effect on first-character RTs in naming two-character words. Second, subjects process the second character as part of the two-character word, that is, the second character is processed in the context of the first character. Since the second character is not independently processed, frequency of second characters has no effect on first-character RTs nor on second-character RTs. Third, naming two-character words is essentially postlexical, that is, subjects fully encode both constituent characters before naming a two-character word. As in the study of Baluch and Besner (1991), however, it should be noted that the naming of two-character words may no longer be postlexical if two-character nonwords are included in the experiment.

CONCLUSION

In visual recognition of a word, subjects perform a characteristic encoding, which gives rise to graphic codes. Because a word is generally experienced in the context of reading, visual recognition of a word is not only associated with wordness experience, but also associatively activates phonological codes and semantic codes. The way in which phonological codes are activated depends on alphabetic or logographic writing systems, because the cues for activating phonological codes differ markedly among writing systems. By their basic nature, all the other elements of the encoding-associative model do not differ among writing systems and are universal.

By noting the encoding and associative aspects involved in word recognition, not only can the process of word recognition be clearly described, but also the developments of phonological and semantic codes become transparent.

ACKNOWLEDGMENT

This study was supported by National Science Council Grant NSC 85-2413-H-194-004.

REFERENCES

Aaronson, E., & Ferres, S. (1986). Sentence processing in Chinese-American bilinguals. *Journal of Memory and Language*, *25*, 136–162.

Baluch, B., & Besner, D. (1991). Visual word recognition: Evidence for strategic control of lexical and nonlexical routines in oral reading. *Journal of Experimental Psychology: Learning, Memory, and Cognition*, *17*, 644–652.

Becker, C. A. (1976). Allocation of attention during visual word recognition. *Journal of Experimental Psychology: Human Perception and Performance, 2*, 556–566.

Chao, Y. R., & Yang, L. S. (1947). *Concise dictionary of spoken Chinese*. Cambridge, MA: Harvard University Press.

DeFrancis, J. (1989). *Visible Speech: The diverse oneness of writing systems*. Honolulu: University of Hawaii Press.

Fan, K. Y., Gao, J. Y., & Ao, X. P. (1984). Pronunciation principles of Chinese characters and alphabetic writing scripts. *Chinese Character Reform, 3*, 23–27.

Forster, K. I. (1976). Accessing the mental lexicon. In R. J. Wales & E. Walker (Eds.), *New approaches to language mechanisms* (pp. 257–287). Amsterdam: North-Holland.

Forster, K. I., & Chambers, S. M. (1973). Lexical access and naming time. *Journal of Verbal Learning and Verbal Behavior, 12*, 627–635.

Ho, D. A. (1987). *Sheng Yun Xue Zong De Gai Nian He Fang Fa* [Concepts and methods in phonology]. Taipei: Da-An.

Horowitz, L. M., White, M. A., & Atwood, D. A. (1968). Word fragments as aids to recall: The organization of a word. *Journal of Experimental Psychology, 76*, 219–226.

Kolers, P. A., & Roediger, H. L., III (1984). Procedures of mind. *Journal of Verbal Learning and Verbal Behavior, 23*, 425–449.

Liu, I. M. (1969). Cuing function of fragments of verbal items. *Journal of Experimental Psychology, 82*, 107–114.

Liu, I. M. (1983). Cuing function of fragments of Chinese characters in reading. *Acta Psychologica Taiwanica, 25*, 85–90.

Liu, I. M. (1984). Recognition of fragment-deleted characters and words. *Computer Processing of Chinese & Oriental Languages, 1*, 276–287.

Liu, I. M. (1988). Context effects on word/character naming: Alphabetic versus logographic languages. In I. M. Liu, H. C. Chen, & M. J. Chen (Eds.), *Cognitive aspects of the Chinese language* (Vol. 1, pp. 81–92). Hong Kong: Asian Research Service.

Liu, I. M. (1995). Script factors that affect literacy: Alphabetic vs. logographic languages. In I. Taylor & D. R. Olson (Eds.), *Script and literacy* (pp. 145–162). Netherlands: Kluwer Academic Publishers.

Liu, I. M., & Wu, J. T. (1997). *Latencies of both syllables in naming bisyllabic words*. Manuscript submitted for publication.

Liu, I. M., Wu, J. T., & Chou, T. L. (1996). Encoding operation and transcoding as the major loci of the frequency effect. *Cognition, 59*, 149–168.

Liu, I. M., Zhu, Y., & Wu, J. T. (1992). The long-term modality effect: In search of differences in processing logographs and alphabetic words. *Cognition, 43*, 31–66.

McClelland, J. L., & Rumelhart, D. E. (1981). An interactive activation model of context effects in letter perception: Part 1. An account of basic findings. *Psychological Review, 88*, 375–407.

Meyer, E. C., Schvaneveldt, R. W., & Ruddy, M. G. (1974). Functions of graphemic and phonemic codes in visual word recognition. *Memory & Cognition, 2*, 309–321.

Morton, J. (1969). The interaction of information in word recognition. *Psychological Review, 76*, 165–178.

Paap, K. R., Chen, L., & Noel, R. W. (1987). *Word recognition: Is the sky falling on top-down processing?* A paper presented at the annual meeting of the Psychonomic Society, Seattle, WA.

Perfetti, C. A., & Tan, L. H. (in press). The time course of graphic, phonological, and semantic activation in Chinese character identification. *Journal of Experimental Psychology: Learning, Memory, and Cognition*.

Seidenberg, M. S., & McClelland, J. L. (1989). A distributed, developmental model of word recognition and naming. *Psychological Review, 96*, 523–568.

So, F. M. (1990). *An analogy model of reading aloud for English words*. Unpublished bachelor's thesis, Chinese University of Hong Kong, Hong Kong.

Sue, I. R., & Liu, I. M. (1996). Word and character superiority effects in Chinese. *Chinese Journal of Psychology, 38*, 11–30.

Taft, M. (1991). *Reading and the mental lexicon*. Hillsdale, NJ: Lawrence Erlbaum Associates.

Tan, L. H., Hoosain, R., & Siok, W. W. T. (1996). Activation of phonological codes before access to character meaning in written Chinese. *Journal of Experimental Psychology: Learning, Memory, and Cognition, 22*, 865–882.

Wu, J. T., Chou, T. L., & Liu, I. M. (1994). The locus of the character/word frequency effect. In H.-W. Chang, J.-T. Huang, C.-W. Hue, & O. J. L. Tzeng (Eds.), *Advances in the study of Chinese language processing* (pp. 31–58). Taipei: National Taiwan University.

Wu, J. T., & Liu, I. M. (1997). Phonological activation in pronouncing characters. In H.-C. Chen (Ed.), *The cognitive processing of Chinese and related Asian languages* (pp. 47–64). Hong Kong: Chinese University Press.

Zhou, Y. (1978). To what degree are the "phonetics" of present-day Chinese characters still phonetic? *Zhongguo Yuwen, 146*, 172–177.

CHAPTER

9

Eye Movements in Reading Chinese and English Text

Fuchuan Sun
Shanghai Institute of Physiology, Chinese Academy of Sciences

Dagan Feng
Department of Computer Science, University of Sydney

Eye movements have been extensively investigated as an indicator of visual information processing and cognition (Inhoff, Pollatsek, Posner, & Rayner, 1989; Kowler, 1990; Noton & Stark, 1971; O'Regan, 1981, 1990; Rayner, 1978, 1983, 1992, 1995; Rayner & Pollatsek, 1987; Stark & Ellis, 1981; Yarbus, 1967), since Javal (1878) first observed *par saccades* in reading. It was discovered that, during the reading process, the eyes do not scan smoothly along text lines. There are two types of eye behavior in reading: *saccade* and *fixation*. In saccades, eyes change gaze directions (or visual axes) rapidly from one position to another, the main function of which is to bring a new part of text into the field of foveal vision. On the other hand, the fixation pauses are necessary for extracting the visual information from the text. It was found from many vision studies that the visual threshold for object detecting is increased during saccadic eye movement. Therefore, the vision system is nearly "turned off" and no visual input signal is sent to the brain during this rapid eye movement. This phenomenon is called *saccadic suppression* (Hubel, 1988; Latour, 1962; Stevenson, Volkmann, Kelly, & Riggs, 1986; Volkmann, 1962, 1976, 1986). Experimental data from reading investigations also demonstrated that the acquisition of visual information from text exists only in fixation pauses (Wolverton & Zola, 1979, 1983). There are microsaccades, slow drifts, and high-frequency tremors in a fixation pause; however, their individual function in reading is still controversial (Ditchburn, 1980; Kowler & Steinman, 1980). The main eye-movement parameters, employed to represent the temporal and spatial reading behavior, are the fixation duration and saccade length (including both forward and regressive saccade) or recognition span (Adler-Grinberg & Stark, 1978; McConkie, 1983; McConkie & Rayner, 1975; Morris, Rayner, & Pollatsek, 1990; O'Regan, 1990; Rayner, 1983, 1995; Rayner & Inhoff, 1981; Rayner & Pollatsek, 1987; Sun, Morita, & Stark, 1985). In this chapter, we often refer to saccade length as the *reading recognition span* because this term can be used to delineate spatial processing even when no eye movements are recorded, as occurs in the Rapid Serial Visual Presentation (RSVP) technique (Adler-Grinberg & Stark, 1978; Fang & Sun, 1993; Sun, Morita, & Stark, 1985). For the sake of practical experimental data processing and recording, the fundamental parameters of reading eye movements could be summed up as the fixation position and the fixation duration because the saccade length or recognition span can be easily deduced from fixation positions (Sun & Chen, 1995).

Chinese written language differs greatly from English written language in both the linguistic concept and the visual formation aspects. The logographic Chinese writing system was developed from earlier pictographs, therefore a Chinese character has a more direct connection between its shape and its meaning. Chinese written language consists mainly of invariable monosyllabic morphemes (meaningful linguistic units) called *characters* or *Hanzi*. These give the form and meaning to morphemes. All the characters are constructed from two types of smaller elements called *radicals* and *strokes*. There are about 20 different strokes and 200 radicals (Huang, 1983; Wang, 1973). A stroke is a line or a curve that is ended when the writing pen leaves the paper, and a radical is built with one or several strokes. In some cases a radical itself is a character, but generally a character contains one radical and an additional component called a remainder.

All Chinese characters can be divided into six categories according to their formation. The main category is the *pictophonetic*, which contains about 80% of all the character population. In a pictophonetic character, the ideographic meaning is usually decided by its radical part, whereas the pronunciation is mainly determined by its remainder. Two other categories are also used frequently. One is called the *associative compounds* character (*Huiyi*), in which a character is formed by a radical in conjunction with one or more other elements. Each of these elements has an independent meaning. The overall meaning of the character can then be derived from their combination. The other is called the *self-explanatory* character (*Zhishi*), in which the meaning of the character is taken directly from its form (Huang, 1983). Examples of these three categories are shown in Fig. 9.1. Because there is a close connection between meaning and form in Chinese characters, a skilled native reader can often guess roughly the meaning and the pronunciation of a character that he has never seen before (Wang, 1973).

Although most Chinese Hanzi are monocharacter words, there is also quite a number of double-character and triple-character words. According

A: 米 粮 糕
(rice) (grain) (cake)

B: 木 林 森
(wood) (trees) (forest)

C: 一 二 三
(three) (two) (three)
十 廿 卅
(ten) (twenty) (thirty)

FIG. 9.1. Examples of Chinese characters from three categories with their English translations given in parentheses: (A) the first pictophonetic character means "rice" and the next two are related to food; (B) associative compound characters related to wood; (C) self-explanatory characters related to numbers.

to statistics, the information quantity of 1.5 Chinese characters is equivalent to one English word, which has been called "1.5 factor" (Sun, 1993; Sun, Morita, & Stark, 1985; Sun & Stark, 1988). From the visual formation aspect, Woodworth suggested very early in his book *Experimental Psychology* (Woodworth, 1938) that Chinese characters with their box-like shapes might be better adapted to exploit the small disc fovea in the retina than alphabetic English words with their rectangular outlines. In addition to the traditional character Hanzi, a new system of Chinese written language called *Pinyin* (means *spell sound*) was developed in 1956. Pinyin is an alphabetic phonetic language, in which the Chinese words are built phonetically by using the 26 Latin letters from A to Z.

It is generally agreed that eye movements reflect the human perceptual and cognitive processes in reading (Inhoff, 1987, 1989; Just & Carpenter, 1980; Kowler, 1990; Rayner, 1992, 1995; Rayner & Pollatsek, 1987, 1989; Vitu &

O'Regan, 1995), hence eye-movement data for reading nonalphabetic Chinese could provide a fresh insight into the understanding of the reading process (Holender, 1987; Osaka, 1989; Osaka & Oda, 1991; Sun, 1993; Sun et al., 1985).

COMPARATIVE STUDY OF EYE MOVEMENTS IN READING CHINESE AND ENGLISH

During each fixation in reading, written language is converted into nerve signals by the retina. The nerve signals are then decoded by the brain. Meanwhile the visuomotor center sends oculomotor signals to feedback-control the eye movements for extracting additional visual information. The structure of to-be-encoded text and its spatial layout are, therefore, likely to affect the planning of eye movements. The comparisons of eye movements during the reading of two completely different languages, such as Chinese and English, should thus provide insight into the types of information that are used to control eye movements (Sun, 1993; Sun et al., 1985; Sun & Stark, 1988). In the following, we report recent work from our laboratory that compared eye movements during the reading of Chinese and semantically equivalent English text, and that examined the functional role of eye movements during the reading of Chinese text.

Experimental subjects consisted of one group of 13 native Chinese graduate students and one group of 13 native English-speaking graduate students. All subjects reported normal or corrected-to-normal vision. To reduce the effects of cultural differences, reading materials were short paragraphs of popular science articles selected from *Scientific American* and the counterpart articles in published Chinese translations, which were fairly simple for all subjects to understand. Ten passages in both languages were selected for the experiments, and each passage has about 110 to 150 Chinese characters or 70 to 100 English words. The reading materials were presented on a computer screen. The display area was about $20° \times 20°$ in visual angle, and the distance was about 57 cm, about 1° of visual angle for a single Chinese character. Eye movements were measured by an infrared TV eye tracker and infrared LED-photocell eyeglasses (Fang & Sun, 1995; Sun, Dai, Zhao, & Qin, 1979; Zhang & Sun, 1989). The data was recorded via the A/D converter, and the results were processed by using on-line software or the neural network (Sun & Chen, 1995; Sun et al., 1985). At the beginning of the experiment, each subject was first asked to read three sample paragraphs to practice the experimental procedure. During the experiment, the subject was instructed to do his or her best to read the text and to report the gist of the paragraph immediately after reading. Before and after the experiment, eye-movement calibrations were taken to minimize error in the eye-movement recording.

TABLE 9.1
Comparison of Eye Movements in Reading Chinese and English

	Chinese Subjects Reading Chinese	*English Subjects Reading English*
Fixation Duration		
(ms)	257	265
Recognition Span		
(characters/fixation)	2.57	
(words/fixation)	(1.71)	1.75
Reading Rate		
(characters/min)	580	
(words/min)	(386)	382

The results are summarized in Table 9.1. The fixation duration of Chinese subjects who read Chinese characters amounted to 257 ± 63 ms (milliseconds). The recognition span (saccade size) is 2.57 ± 0.5 character spaces per fixation, equivalent to 1.7 ± 0.3 words per fixation. The reading rate was 580 characters per minute, the equivalent of 386 words/min. The fixation duration of native English speakers who read English text with same content amounted to 265 ± 72 ms; the recognition span was 1.75 ± 0.6 words per fixation, and the reading rate was 382 words per minute. None of the effects of text type was significant: fixation duration, $t(26) = 0.365$, $p > .05$; recognition span, $t(26) = 0.229$, $p > .05$; and reading rate, $t(26) = 0.087$, $p > .05$.

In these experiments, although the visual forms of the two languages are strikingly different their reading eye-movement patterns and parameters are very similar because their contents are the same. The similarities of eye-movement patterns support the evidence that eye movements during reading are controlled mainly by the linguistic content, rather than by the visual form of the particular language. In other words, the results support the High-Level Control Model, instead of the Low-Level Control Model (Rayner & Inhoff, 1981). Furthermore, even if the shape of the Chinese character fits the fovea of retina better than the shape of the English word, resemblance of reading rate and recognition span between reading Chinese and English suggested that the bottleneck of decoding speed in reading is at the higher level central nervous system, instead of the peripheral nervous system of vision.

In spite of the great significance of these experiments, very few comparative studies of Chinese–English reading eye movements have been undertaken because of the limited number of eligible subjects and the complexity of the experimental setup. The earliest article of comparative studies of eye movements in reading Chinese and English was written by Miles and Shen (1925). They reported, "Chinese is apparently read more rapidly than English" (4.7 words per second for native speakers of English reading English text; 7.1 words/sec for Chinese reading Chinese text). However, they equated Chinese "characters" with Chinese "words." If we normalize Shen's data with the word factor of 1.5 (Sun et al., 1985), his results could be converted into 282 words/min for English and 286 equivalent words/min for Chinese. Consequently, the maximum speeds for reading Chinese and English texts are similar and appear to be determined by limitations of the linguistic processor. When saccade size is measured in word units, rather than visual angle, a typical saccade length of 1.4 to 1.8 words emerges for the reading of English text (Rayner, 1995; Wang, 1973), which is pretty similar to our data that revealed a recognition span of 1.75 words for English readers and of 1.71 equivalent words for Chinese readers.

EYE MOVEMENTS OF BILINGUALS IN READING CHINESE AND ENGLISH

Experimental data were obtained from 10 bilingual subjects who were educated Chinese Americans and who had learned Chinese for more than 15 years at an early age and English for more than 10 years later on. The reading paragraphs, experimental apparatus, and procedures were the same as those in the previous section. Six of the bilinguals could read Chinese much faster than English: 570 characters/min (or 380 equivalent words/min) for Chinese and 255 words/min for English. These six subjects' reading performances were similar to the native Chinese readers in reading Chinese (386 equivalent words/min), but were much slower than the native English subjects in reading English (382 words/min). However, in contrast, the other four subjects read Chinese much slower than they read English. Their performances in reading English were identical to those of the English readers (385 words/min for English and 233 equivalent words/min for Chinese). One can conclude from these results that each of the 10 subjects has a distinctive bias in reading skill and none has roughly equal reading skill in both languages. These results suggested that during a certain period for a subject, only one language is dominant. In other words, there exist no "real" Chinese–English bilinguals (see Sun, 1993, for quantitative data; see Sun & Stark, 1988, for qualitative results).

READING CHINESE WITHOUT SACCADE EYE MOVEMENT—RSVP ON CHINESE READING

To investigate the influences of eye movements on the reading process, the rapid serial visual presentation (RSVP) method (Forster, 1970) allows us to display successive words of text at the same physical location on a display screen in a temporally ordered sequence (Chen, 1985; Potter, 1983; Potter,

Kroll, & Harris, 1980; Kennedy & Murray, 1984). In one recent study (Kong & Sun, 1992), we used RSVP to eliminate eye movements during the reading of Chinese text and to examine its effects on reading. Forty Chinese graduate students read 44 short texts selected from popular Chinese science magazines with easy words and contents. Each text contained 110 to 250 characters. Four display modes were used, consisting of an RSVP mode, a whole page mode, a moving-window mode, and a cumulative moving-window mode. In the RSVP mode, the text was presented successively at the center of the screen. The display window was chosen to be the variable size of two or three Chinese characters in order to keep the lexical completeness (Chan & Chen, 1991; Cocklin, Ward, Chen, & Juola, 1984) and to match the recognition span, which is about 2.6 characters per fixation in reading Chinese (Sun et al., 1985). The presentation speed was set by the subject during the warm-up practice before the formal experiment. In the whole page mode, the text was simply displayed on the screen as in normal reading. The eye movement was recorded and analyzed for monitoring the saccade. The average reading rate was 853 ± 196 characters/min for the fixed window RSVP mode and 640 ± 211 characters/min for the whole page mode. The reading rates here are a little bit faster than other reading experiments because of the difference in text difficulties.

All the results indicate that the ability to read Chinese could be improved by eliminating eye movements such as that in the RSVP method. Most studies of the RSVP reading in English also demonstrated that the reading performance without eye movements is usually better, or at least not worse, than normal reading (Chen, 1986; Juola, Haugh, Trast, Ferraro, & Liebhaber, 1987; Juola, Ward, & McNamara, 1982; Potter, 1983; Potter, Kroll, & Harris, 1980; Ward & Juola, 1982). In addition to the saccadic suppression effect, eye-movement control and saccade length determination are also involved in the normal reading process. They might contribute to the slow-down of information decoding in the cognition center.

EYE MOVEMENTS IN READING MOVING CHINESE TEXT—OKN READING WITHOUT EYE FIXATION

Eye movements elicited by moving visual patterns that are larger than a single point target are usually called *Optokinetic Responses* or *Optokinetic Nystagmus* (OKN), in which the major component is a smooth tracking of the visual patterns, defined as the *slow phase* of the OKN. The smooth tracking movement during the slow phase is interrupted irregularly by fast movements towards the opposite direction, called the *fast phase* of the OKN. When moving patterns are presented to a subject, a sequence of slow phase and fast phase of involuntary eye movements, Optokinetic Nystagmus, is

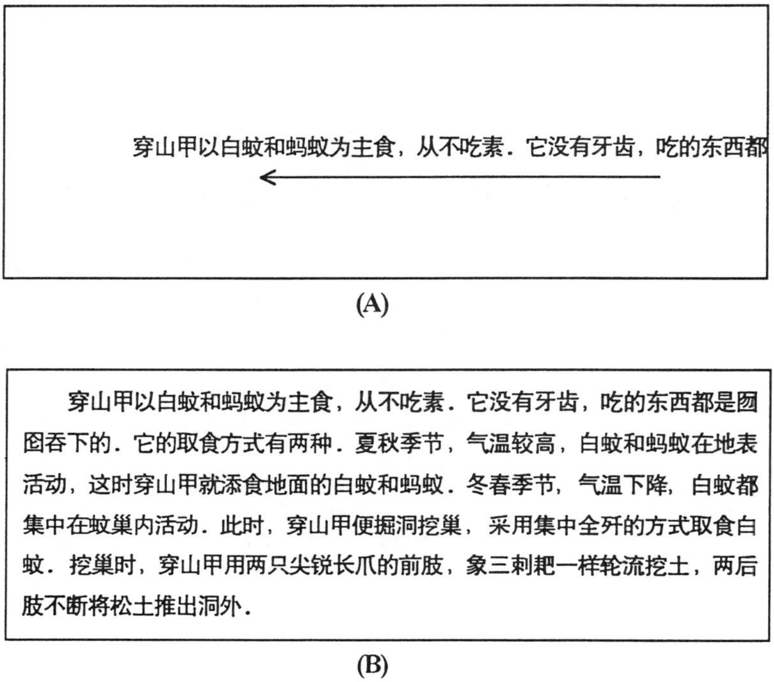

FIG. 9.2. The presentation modes in reading moving text: (A) the moving text mode; (B) the whole page mode.

evoked (Collewijn, 1981; Miles, 1995). The eye-movement experiment for sliding presentation of Chinese text was designed by Fang and Sun (1993) to answer the following questions:

1. Can we extract information from the moving text without fixation pauses? In other words, are the eye-movement fixation pauses indispensable in the reading process?
2. What is the relationship between the reading processing and OKN eye movement during moving text reading?

In this sliding text experiment, 44 Chinese passages were selected from popular scientific articles. Each passage contained 100 to 250 Chinese characters. The reading material was presented on the center line of a computer CRT, moving leftward (see Fig. 9.2). During warm-up practice, the gliding speed of each passage was stepwise decreased until it reached the maximum reading speed of the subject. The visual distance was 38 cm and the visual angle of the moving line, which contained 33 characters (about 0.84

degrees per character), was about 28°. The moving gratings and the whole page presentation are also provided for comparison.

Experiment results showed that:

1. The eye-movement pattern in reading moving text is very similar to the OKN elicited by moving grating. It is a sequence of smooth tracking eye movement (the slow phase) and quick jumping back to the opposite direction (the fast phase). The gain of slow phase (the velocity of tracking eye movement divided by the velocity of moving stimulus) for both stimuli is close to 1. The nystagmus frequency is higher for moving text than for moving grating, as shown in Fig. 9.3.

2. In the case of the moving Chinese text, although there is no fixation pause, subjects can extract the contents from the moving text and report the text gist after reading. The average maximum acceptable speed of gliding text for the 34 subjects was about 12 characters/second. At this maximum average speed, the average duration of the slow phase of eye movements is about 317 ms, the recognition span in each slow phase is approximately 3.67 characters, and the reading rate is about 723 characters/min. Because the smooth tracking of OKN-like eye movements stabilized the image of characters on the retina, it is possible to read the moving text without fixations. Thus the slow phase eye movement corresponds to the fixation pauses in normal reading. The eye-movement parameters from these two reading methods are summarized in Table 9.2.

3. Under the experimental condition in which the visual angle is 28° for 33 characters, the average maximum speed for reading moving text is ap-

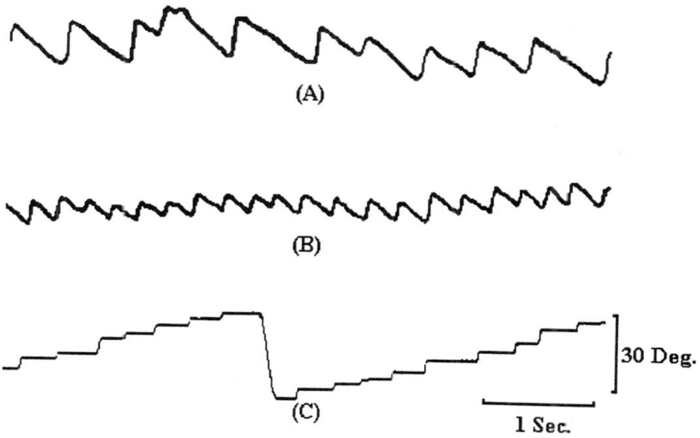

FIG. 9.3. Eye movement responses to the three different stimuli in which the rising edges correspond to the eye rightward movements: (A) the OKN responses to moving gratings; (B) the eye movements in reading moving text without fixation pauses; (C) the eye movements in normal reading.

TABLE 9.2
Comparison of Eye Movements in Reading Moving Text and Whole Page

	Moving Text Reading	*Whole Page Normal Reading*
Eye movement pattern	OKN	Saccades and fixations
Duration (ms)	Slow phase 317	Fixation 284
Recognition Span		
(characters/slow phase)	3.67	
(characters/fixation)		2.65
Reading Rate (characters/min)	723	576

proximately 10.2°/second (or 12 characters/second). When the gliding speed increases from 15° to 80°/second, the OKN eye movements can still be evoked, but subjects are unable to extract any meaning from the moving text. When the size of characters is increased to 1.6°/character or 3.2°/character, or decreased to 0.55°/character, the speed limit in terms of degree/second for reading moving text would vary accordingly. However, the speed limitation in terms of characters/second remained unchanged. These results provided additional evidence that the limitation of reading processing is determined by linguistic information decoding in the higher level brain, rather than the low level visual performance.

The rate of reading moving text with the OKN eye movements appears to be higher than that of normal reading. Perhaps the slow, smooth tracking and fast jumps in the OKN are involuntary eye movements, which are mainly controlled by the low-level oculomotor reflex. Contrarily, the foveating saccades in normal reading are voluntary eye movements, and its control procedure would intrude into the linguistic processing. Of course, this conjecture needs to be verified carefully by further experiments.

EYE MOVEMENTS IN RECOGNIZING A SINGLE CHINESE CHARACTER

The eye-movement patterns, during recognition of a single Chinese character, depend on the subjects' knowledge and familiarity with the character (Sun, 1993; Sun, Zhao, & Gu, 1994). During these experiments, the subjects were instructed to close their eyes immediately after recognizing the character. When the single Chinese character is well known to the subjects, the eye movements made only one fixation for cognition of the character. This

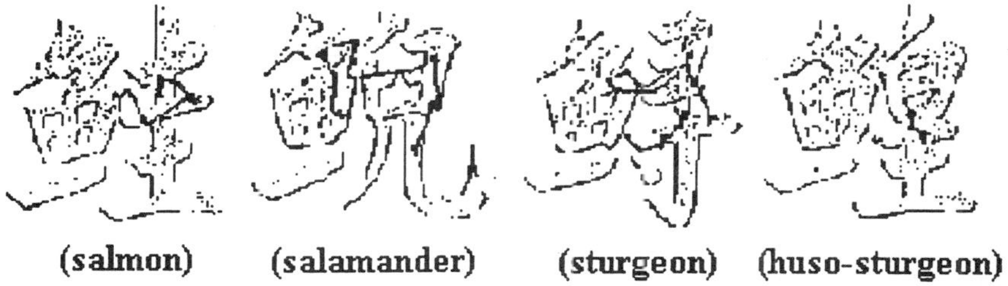

FIG. 9.4. Eye movement patterns for recognition of a single pictophonetic character. Because the left radicals ("fish") of these four characters are well known to the subject, most fixations are concentrated on the remainder phonetic parts that define the specific type of fish.

phenomenon indicated that the subjects already have an efficient recognition model in their brains from their previous experience. Thus one fixation pause is enough for the fovea, parafovea, and peripheral visual regions to extract sufficient information for recognition. A well-known character has only a little information entropy to the subject, and it is unnecessary to make extra fixations. When subjects read an unknown pictophonetic character, the fixations were mainly focused on the remainder part, that is, the non-radical part (Fig. 9.4). According to the formation of Chinese characters, the phonetic remainder contains more detailed linguistic information for those characters formed with the same radical. In reading complex Chinese characters, the difficult words could often be divided into several simple well-known components. The eye-movement pattern in recognizing such a character demonstrates a "scanpaths"-like trajectory to guess the meaning of the character, as shown in Fig. 9.5.

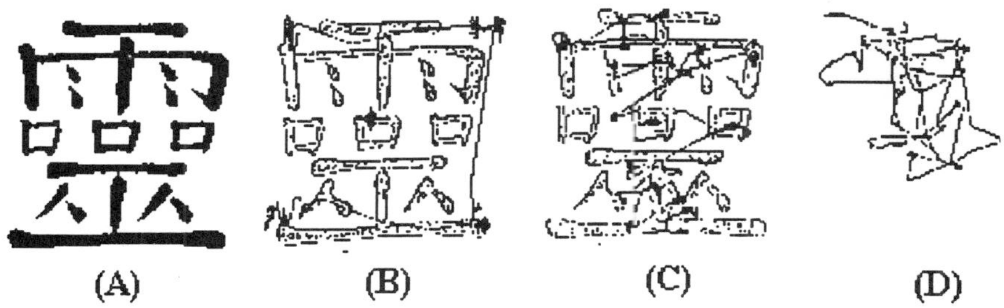

FIG. 9.5. Eye movement patterns from three different subjects to recognize a complex character: (A) the form of the character (means "soul"); (B) the eye movement is only a single fixation from the subject who knew the character very well. The four fixations on the corners are calibration eye movements; (C) the eye movement pattern is scanning the components of the character when the subject is familiar with components only; (D) the eye movement pattern from a non-Chinese subject who is not familliar with the Chinese language. These data come from Sun (1993).

Quantitative eye-movement data in recognition of a single Chinese character exhibited that the number of fixations and recognition duration are correlated with the "character frequency," rather than with the number of the character strokes (Sun, Chen, & Zhao, 1996). Eye-movement research in reading English also demonstrated shorter fixation on high-frequency than on low-frequency target words (Inhoff & Rayner, 1986). Although there was some difference between the procedures in these two experiments, both concluded that the recognition time is controlled by the information entropy of the word or character.

COMPARISON OF READING EYE MOVEMENTS BETWEEN CHINESE HANZI AND PINYIN

Three groups of subjects—Chinese primary-school students, high-school students, and graduate students—have attended these comparative experiments. Reading materials are simple paragraphs selected from the textbooks of the Chinese primary schools. Each paragraph was presented in the Hanzi version and the Pinyin version respectively. Experimental results are briefly summarized in Table 9.3. The analyses of variance for these data indicate that for all subjects, regardless of their educational difference, the average fixation duration for reading Hanzi is much shorter than that for reading Pinyin. Similarly, the average recognition span is larger and the reading rate is faster for Hanzi (Sun, 1993).

Two-factor analyses of variances for these data were used, one describing script difference between Chinese Hanzi and Pinyin, and the other expressing the educational levels. The difference between the fixation duration for

TABLE 9.3
Comparison of Eye Movements in Reading Chinese Hanzi and Pinyin for Three Groups of Subjects

	Subject Groups		
	Primary-School Students	*High-School Students*	*Graduate Students*
Fixation Duration			
Hanzi (ms)	268 ± 50	266 ± 61	240 ± 54N
Pinyin (ms)	326 ± 65	326 ± 70	353 ± 77
Recognition Span			
Hanzi (characters/fix)	2.08 ± 0.62	2.11 ± 0.59	2.61 ± 0.56
Pinyin (characters/fix)	0.89 ± 0.22	0.79 ± 0.26	0.77 ± 0.21
Reading Rate			
Hanzi (characters/min)	459 ± 168	491 ± 190	598 ± 197
Pinyin (characters/min)	171 ± 59	145 ± 60	137 ± 55

reading Hanzi and Pinyin was highly significant $F(1, 66) = 26.47$, $p < .001$. However there was no significant difference for the fixation duration among these three education levels $F(2, 66) = .015$, $p > 0.5$, nor was there a significant interaction $F(2, 66) = 1.45$, $p > 0.2$. Similarly, for recognition span and reading rate, the difference between reading Hanzi and Pinyin was highly significant $F(1, 66) = 188.0$, $p < .001$ and $F(1, 66) = 127.0$, $p < .001$ for recognition span and reading rate respectively. The difference between educational levels was not significant for either the recognition span $F(2, 66) = 2.01$, $p > 0.2$ or the reading rate $F(2, 66) = 1.10$, $p > 0.5$. Also, no interaction effects were found.

The eye-movement performances in reading Hanzi are obviously superior over Pinyin. The large number of homophone characters in Chinese may make the phonetic Pinyin more difficult to identify by meaning. The other possible reason is that the written Pinyin is less frequently used in Chinese daily life. Thus, it is too early to conclude that Pinyin script is less suitable to Chinese language.

The educational level failed to exhibit its influence with our reading performance. The reason was probably due to the ease of reading materials, which were primary-school texts and easy for all of the groups. The easy texts need to be selected to fit all groups with different educational levels.

SUMMARY

The similarities of eye movements in reading two strikingly different languages, logographic Chinese and alphabetic English, suggest that: (a) reading eye-movement patterns, fixation duration, and recognition span are controlled by the linguistic information rather than by the geometric forms of the text, and (b) the bottleneck of decoding speed in reading, or the capacity limitation of the reading information channel, is at the high-level recognition center of the brain instead of at the peripheral nervous system of vision.

The eye-movement data obtained from bilinguals in reading demonstrates that there probably are no "real Chinese–English bilinguals." No one really possesses equal ability for reading Chinese and English. During a certain period for a subject, only one language is dominant.

Experiments in reading Chinese moving text provided a novel approach to investigate reading processes without fixation pauses and to explore the role of smooth tracking in reading.

Three patterns of eye movement in recognition of single Chinese characters have been found. They are (a) a single fixation for a well-known character, (b) more fixations concentrated on the no-radical part for a stranger pictophonetic character, and (c) fixations scanned the components of the character for an unknown complex character. Three patterns are determined by the degree of familiarity of the subject to the character.

The primary eye-movement data obtained from the comparative studies showed the result of reading Chinese Hanzi is superior to that of reading Pinyin.

ACKNOWLEDGMENTS

This work is partially supported by the Australia–China Exchange Programs, ARC grant, National Science Foundation of China, and Visual Information Processing Lab of Biophysics Institute, Chinese Academy of Sciences. The authors also gratefully acknowledge the help of Prof. Lawrence Stark of the University of California at Berkeley.

REFERENCES

Adler-Grindberg, D., & Stark, L. (1978). Eye movements, scanpaths, and dyslexia. *American Journal of Optometry and Physiological Optics*, *55*, 557–570.

Chan, K. T., & Chen, H.-C. (1991). Reading sequentially-presented Chinese text: Effects of display format. *Ergonomics*, *34*, 1083–1093.

Chen, H.-C. (1985). Reading Chinese text in sequential display format: Effects of display size. *Perceptual and Motor Skills*, *61*, 595–598.

Chen, H.-C. (1986). Effects of reading span and textual coherence on rapid-sequential reading. *Memory & Cognition*, *14*, 202–208.

Cocklin, L. G., Ward, N. J., Chen, H.-C., & Juola, J. F. (1984). Factors influencing readability of rapidly presented text segments. *Memory & Cognition*, *12*, 431–442.

Collewijn, H. (1981). The optokinetic system. In B. L. Zuber (Ed.), *Models of oculomotor behavior and control* (pp. 111–138). Boca Raton, FL: CRC Press.

Ditchburn, R. W. (1980). The function of small saccades. *Vision Research*, *20*, 271–272.

Fang, L., & Sun, F. (1993). OKN eye movements during reading moving text. *Acta Biophysica Sinica*, *9*(2), 247–252.

Fang, L., & Sun, F. (1995). Dual mode control of head movements during eye-head coordination. *Acta Physiologica Sinica*, *47*(1), 45–53.

Forster, K. I. (1970). Visual perception of rapidly presentation word sequences of varying complexity. *Perception & Psychophysics*, *8*, 215–221.

Holender, D. (1987). Synchronic description of present-day writing systems: Some implications for reading research. In J. K. O'Regan, & A. Lévy-Schoen (Eds.), *Eye movements: From physiology to cognition* (pp. 397–420). North-Holland: Elsevier Science B. V.

Huang, B. (1983). *Modern Hanyu*. China Gansu: Gansu People's Publisher Company.

Hubel, D. H. (1988). *Eye, brain and vision*. New York: Scientific American Library.

Inhoff, A. W. (1987). Lexical access during eye fixations in sentence reading: Effects of word structure. In M. Coltheart (Ed.), *Attention and performance 12* (pp. 403–418). Hillsdale, NJ: Lawrence Erlbaum Associates.

Inhoff, A. W. (1989). Lexical access during eye fixation in sentence reading: Are word codes used to integrate lexical information across interword fixations? *Journal of Memory and Language*, *28*, 444–461.

Inhoff, A. W., & Rayner, K. (1986). Parafoveal word processing during eye fixations in reading: Effects of word frequency. *Perception & Psychophysics*, *40*, 431–439.

Inhoff, A. W., Pollatsek, A., Posner, M. I., & Rayner, K. (1989). Covert attention and eye movements during reading. *The Quarterly Journal of Experimental Psychology*, *41A*, 63–89.

Javal, K. E. (1878). Essai sur la physiologie de la lecture. *Annales d'Oculistique*, *79*, 98–117.

Juola, J. F., Haugh, D., Trast, S., Ferraro, F. R., & Liebhaber, M. (1987). Reading with and without eye movements. In J. K. O'Regan & A. Lévy-Schoen (Eds.), *Eye movements: From physiology to cognition* (pp. 499–508). North-Holland: Elsevier Science, B.V.

Juola, J. F., Ward, N. J., & McNamara, T. (1982). Visual research and reading of rapid, serial presentations of letters, strings, words and text. *Journal of Experimental Psychology: General*, *111*, 208–227.

Just, M. A., & Carpenter, P. A. (1980). A theory of reading: From eye fixations to comprehension. *Psychological Review*, *87*, 329–354.

Kennedy, A., & Murray, W. S. (1984). Reading without eye movements. *Theoretical and applied aspects of eye movement research*, The Second European Conference on Eye Movements, 187–191.

Kong, Y., & Sun, F. (1992). Reading Chinese texts without eye movements—A rapid serial visual presentation study. *Acta Biophysica Sinica*, *8*(3), 521–526.

Kowler, E. (1990). The role of visual and cognitive processes in the control of eye movement. In E. Kowler (Ed.), *Eye movements and their role in visual and cognitive processes* (pp. 1–70). North-Holland: Elsevier Science, B.V.

Kowler, E., & Steinman, R. M. (1980). Small saccades serve no useful purpose. *Vision Research*, *20*, 273–276.

Latour, P. (1962). Visual threshold during eye movements. *Vision Research*, *2*, 261–262.

McConkie, G. W. (1983). Eye movements and perception during reading. In K. Rayner (Ed.), *Eye movements in reading* (pp. 65–96). London: Academic Press.

McConkie, G. W., & Rayner, K. (1975). The span of the effective stimulus during a fixation in reading. *Perception & Psychophysics*, *17*, 578–586.

Miles, F. A. (1995). The sensing of optic flow by the primate optokinetic system. In J. M. Findlay, R. Walker, & R. W. Kentridge (Eds.), *Eye movement research: Mechanisms, processes and applications* (pp. 47–62). North-Holland: Elsevier Science, B.V.

Miles, W. R., & Shen, E. (1925). Photographic recording of eye movements in the reading of Chinese in vertical and horizontal axis: Methods and preliminary results. *Journal of Experimental Psychology*, *8*(5), 344–362.

Morris, R. K., Rayner, K., & Pollatsek, A. (1990). Eye movement guidance in reading: The role of parafoveal letter and space information. *Journal of Experimental Psychology: Human Perception and Performance*, *16*(2), 268–281.

Noton, D., & Stark, L. (1971). Scanpaths in eye movements during pattern perception. *Science*, *171*, 308–311.

O'Regan, J. K. (1981). The "convenient viewing position" hypothesis. In D. F. Fisher, R. A. Monty, & J. W. Senders (Eds.), *Eye movements: Cognition and perception* (pp. 289–298). Hillsdale, NJ: Lawrence Erlbaum Associates.

O'Regan, J. K. (1990). Eye movements and reading. In E. Kowler (Ed.), *Eye movements and their role in visual and cognitive processes* (pp. 395–453). North-Holland: Elsevier Science.

O'Regan, J. K., Levy-Schoén, A., & Jacobs, A. M. (1983). The effect of visibility on eye movement parameters in reading. *Perception & Psychophysics*, *34*, 457–464.

O'Regan, K. (1979). Saccade size control in reading: Evidence for the linguistic control hypothesis. *Perception & Psychophysics*, *25*(6), 501–509.

Osaka, N. (1989). Eye fixation and saccade during kana and kanji text reading: Comparison of English and Japanese text processing. *Bulletin of the Psychonomic Society*, *27*, 548–550.

Osaka, N., & Oda, K. (1991). Effective visual field size necessary for vertical reading during Japanese text processing. *Bulletin of the Psychonomic Society*, *29*, 345–347.

Potter, M. C. (1983). Rapid serial visual presentation (RSVP): A method for studying language processing. In D. Kieras & M. Lust (Eds.), *New methods in reading comprehension research*. Hillsdale, NJ: Lawrence Erlbaum Associates.

Potter, M. C., Kroll, J. F., & Harris, C. (1980). Comprehension and memory in rapid-sequential reading. In R. Nickerson (Ed.), *Attention and performance 8* (pp. 395–418). Hillsdale, NJ: Lawrence Erlbaum Associates.

Rayner, K. (1978). Eye movements in reading and information processing. *Psychological Bulletin, 85*(3), 618–660.

Rayner, K. (1983). *Eye movements in reading: Perceptual and language process*. New York: Academic Press.

Rayner, K. (Ed.). (1992). *Eye movements and visual cognition*. New York: Springer-Verlag.

Rayner, K. (1995). Eye movements and cognitive processes in reading, visual search, and scene perception. In J. M. Findlay, R. Walker, & R. W. Kentridge (Eds.), *Eye movement research: Mechanism, processes and applications* (pp. 3–22). North-Holland: Elsevier Science, B.V.

Rayner, K., & Inhoff, W. (1981). Control of eye movements during reading. In B. L. Zuber (Ed.), *Models of oculomotor behavior and control* (pp. 209–231). Boca Raton, FL: CRC Press.

Rayner, K, & Pollatsek, A. (1987). Eye movements in reading: A tutorial review. In M. Coltheart (Ed.), *Attention and Performance 12* (pp. 327–362). Hillsdale, NJ: Lawrence Erlbaum Associates.

Rayner, K, & Pollatsek, A. (1989). *The psychology of reading*. Englewood Cliffs, NJ: Prentice-Hall.

Stark, L., & Ellis, S. (1981). Scanpaths revisited: Cognitive models of direct active looking. In D. F. Fisher, R. A. Monty, & J. W. Senders (Eds.), *Eye movements: Cognition and visual perception* (pp. 183–226). Hillsdale, NJ: Lawrence Erlbaum Associates.

Stevenson, S. B., Volkmann, F. C., Kelly, T. P., & Riggs, L. A. (1986). Dependence of visual suppression on the amplitude of saccades and blinks. *Vision Research, 26*, 1815–1824.

Sun, F. (1993). Eye movements in reading Chinese: Paragraphs, single characters and Pinyin. In S. F. Wright & R. Groner (Eds.), *Facets of dyslexia and its remediation. Studies in visual information processing, 3*, (pp. 245–255). North-Holland: Elsevier Science Publishers, B.V.

Sun, F., & Chen, L. (1995). Identification of fixations in reading eye movements by a multi-layer neural network. *1995 IEEE International Conference on Neural Networks*, 2309–2313.

Sun, F., Chen, L., & Zhao, X. (1996). Eye movements during recognition of the single Chinese characters for different word frequency. *Acta Biophysica Sinica*. (in submission).

Sun, F., Dai, S., Zhao, X., & Qin, Y. (1979). An inferred instrument television for measurement of dynamic pupil area. *Acta Electronica Sinica, 3*, 61–67.

Sun, F., Morita, M., & Stark, L. (1985). Comparative patterns of reading eye movement in Chinese and English. *Perception and Psychophysics, 37*, 502–507.

Sun, F., & Stark, L. (1988). Visual information processing: Eye movements during reading Chinese and English, *Acta Biophysica Sinica, 4*(1), 1–6.

Sun, F., Zhao, X., & Gu, F. (1994). Top-down information processing in cognition eye movements. *Acta Biophysica Sinica, 10*(3), 431–438.

Vitu, F., & O'Regan, J. K. (1995). A challenge to current theories of eye movements in reading. In J. M. Findlay, R. Walker, & R. W. Kentridge (Eds.), *Eye movement research: Mechanism, processes and applications* (pp. 381–392). North-Holland: Elsevier Science.

Volkmann, F. C. (1962). Vision during voluntary saccadic eye movements. *Journal of the Optical Society of America, 52*, 571–578.

Volkmann, F. C. (1976). Saccadic suppression: A brief review. In R. A. Monty & J. W. Senders (Eds.), *Eye movements and psychological processes* (pp. 73–83). Hillsdale, NJ: Lawrence Erlbaum Associates.

Volkmann, F. C. (1986). Human visual suppression, *Vision Research, 26*, 1401–1416.

Wang, W. S.-Y. (1973). The Chinese language. *Scientific American, 228*, 50–60.

Ward, N. J., & Juola, J. F. (1982). Reading with and without eye movements: A reply to Just, Carpenter and Woolley. *Journal of Experimental Psychology: General, 111*, 239–241.

Wolverton, G. W., & Zola, D. (1979). *The acquisition of visual information during fixations and saccades in reading*. ERIC Document Service No. ED 178 861.

Wolverton, G. W., & Zola, D. (1983). The temporal characteristics of visual information extraction during reading. In K. Rayner (Ed.), *Eye movements in reading* (pp. 41–51). New York: Academic Press.

Woodworth, R. S. (1938). *Experimental Psychology* (chap. 28). New York: Holt.

Yarbus, A. (1967). *Eye movement and vision*. New York: Plenum Press.

Zhang, M., & Sun, F. (1989). Eye-movement measurement by the low-noise inferred photo-electronic method. *Journal of Biomedical Engineering (Chinese)*, *6*(4), 282–286.

CHAPTER

10

Reading Chinese: Some Basic Eye-Movement Characteristics

Hsien-Ming Yang
National Tainan Teachers College

George W. McConkie
University of Illinois at Urbana–Champaign

People read by making a series of saccadic eye movements that bring different parts of the text into the region of clearest vision. Both the lengths of these saccades and the time between them, called *fixation durations*, are variable, reflecting to some degree the processing that is taking place. A number of textual and processing variables were found to affect the eye movements (McConkie, 1983; Rayner & Pollatsek, 1989); where the eyes land and the durations of the fixations also affect the reading (O'Regan, 1990; O'Regan & Jacobs, 1992; O'Regan & Lévy-Schoen, 1987). Thus, eye behavior both reflects and affects the processing taking place. The recording of eye movements provided a primary source of dependent variables for studying the processing taking place during reading because it produces a rich, continuous record of reading behavior, yet is minimally intrusive of the reading act itself (Just & Carpenter, 1987; Rayner & Pollatsek, 1989).

It is only natural, then, that investigators who seek to understand the nature of the processing taking place during the reading of Chinese text should turn to the use of research methods involving the recording of eye movements. Several such studies have already been conducted (Peng, Orchard, & Stern, 1983; Sun, Morita, & Stark, 1985; Zhang & Feng, 1992) and others are currently being pursued.

In order to help lay the groundwork for future eye-movement research, this chapter presents a general summary of eye-movement characteristics of a group of Taiwanese college students studying in the United States who participated in a study designed to investigate several issues related to the control of eye movements when reading Chinese text. The more detailed analyses related to the specific issues that originally motivated the experiment are reported elsewhere. This chapter provides a more global overview of the readers' eye behavior.

SOME FEATURES OF CHINESE CHARACTERS AND WORDS

Chinese Characters

Written Chinese is generally regarded as a logographic language, with a graphic symbol, or character, representing a spoken unit, or syllable. A character is made up of a configuration of eight types of strokes. In alphabetic languages, words vary in length, or number of constituent letters; Chinese characters differ in number of strokes, varying from 1 to more than 20 strokes. However, because each character is completely written within a standard character space, this leads to large differences among characters in their density or visual complexity. More complex characters are, in general, more difficult to perceive (Huang, Wang, & Liou, 1991; Yeh & Liu, 1972), particularly in peripheral vision (Yang, 1994).

Although there is great variation in stroke numbers, Chinese characters are not a chaotic array of strokes. Rather, they are constructed in a systematic way, so the structure of characters can be captured easily. First, Chinese characters often contain multistroke components that are generally assumed to be the basic units of Chinese characters. These components are combined in different ways to form a character. Most Chinese characters are compound characters with one component that is called a *radical* (or *signific*), often carrying information associated with the meaning of the character, and the other (or the rest of the character) called a *phonetic*, often carrying information associated with the pronunciation of a character. Signific radicals are often located at the left or the top of characters, and phonetics at the right or the bottom (Hoosain, 1991).

Two components may be aligned either horizontally or vertically to form a character, or a part of a more complex character. Likewise, three components can be combined as a character with two components horizontally aligned, and the third component placed either above or below to form the whole. Some characters have four components, with two aligned horizontally on top and two on the bottom.

Three other types of structure are also quite common in Chinese characters: an enclosed structure in which a component is encompassed within another component; a multiparallel structure in which three components are horizontally aligned; and an independent structure in which one component makes up the whole of a character, creating a character that cannot be divided into independent components more complex than the strokes themselves. These types of structures account for most of the variation in orthographic configuration of Chinese characters, and are widely employed in the control of experimental variables in studies of Chinese word perception (Cheng & Wu, 1993; Huang & Wang, 1992; Yu, Hu, & Pan, 1993). They also lead to characters that differ greatly in overall complexity, as indexed by the number of strokes.

Chinese Words

The concept of *word* is less central in written Chinese than it is in European alphabetic languages, where writers must continually make decisions about where words begin and end. In fact, there is considerable disagreement among educated Chinese as to the locations of word boundaries in text. Still, it is clear that most lexical units in written Chinese have more than one character. Although most words in the language are represented by two characters (and, correspondingly, spoken with two syllables), some words are written with a single character, and others with three or more characters (Hoosain, 1991). Words written with more than one character are normally called *compound words*.

There are four features of Chinese words that may affect eye behavior during reading. First is their variation and configuration in visual complexity. Words vary greatly in the number of strokes in their component characters. In addition, compound words can have high and low complexity at different character positions within the word. For example, in a two-character word, a high complexity character can reside at the first, second, both or neither position in the word.

Second, compound words differ in their frequency of use in the language. Just as frequency of use affects word processing in English, we would expect it to affect word processing in Chinese, with this effect being reflected in eye behavior (Inhoff & Rayner, 1986). However, differences in usage frequency also exist among characters, and among character components. How these combine in their influences on processing is not currently known.

Third, compound words differ in their distribution of meaning. Since characters themselves carry meaning, it is possible to consider the question of the degree to which the individual characters contribute to the meaning of the whole word. In some cases, both characters contribute about equally to the meaning of the word, whereas in other cases one of the characters conveys more of the meaning than does the other. Of course, it is possible to make more fine-grained distinctions concerning the nature of the contribution being made by each character, as well.

Fourth, an important characteristic of written Chinese is that, unlike European alphabetic languages, there is no visual indicator of where words begin and end. The width of the spacing between words is identical to that between characters within a word. Consequently, there is no basis for low-level perceptual processes to parse the text into word-units, as is possible in English. This raises an interesting set of issues regarding how the Chinese reader parses the text into supersyllabic lexical units during reading.

EXPLORING READING EYE MOVEMENTS IN CHINESE

Our exploration of the characteristics of eye behavior of Chinese readers was carried out in three stages. The first stage examined statistics and frequency distributions regarding global aspects of eye movements. The second stage investigated whether the decision of where to send the eyes is word-based, as it appears to be in English (McConkie, Kerr, Reddix, & Zola, 1988). The third stage looked specifically at the effect of three variables on eye behavior: character complexity, word complexity configuration, and word frequency.

In the reading of English, there is evidence that, when the eyes are moved, they are being sent to a selected word. For example, there tend to be more fixations near (usually slightly to the left of) the center of a word, referred to as the *preferred landing position*, than at its beginning or ending. In Chinese, a similar phenomenon would require that words be located in peripheral vision before moving the eyes to them, without the assistance of low-level perceptual cues such as word spacing in English. Thus, it is of some interest to know whether a word-based preferred landing position exists for Chinese words.

Alternatively, because the most obvious perceptual unit in Chinese text is the character, it may be that the eyes are being sent to a selected character when they are being moved. If this were the case, we would expect to see a character-based preferred landing position phenomenon, with the eyes going most often to the centers of characters and less often to the spaces between characters.

We would expect characteristics of words and characters to influence eye behavior in several ways. First, since higher frequency and less visually complex words are easier to process, fixations on these words should have shorter mean durations. Second, there is evidence that the landing position of a saccade is influenced by characteristics of the stimulus pattern, with objects of higher intensity, density, or contrast or larger size tending to draw the eyes (Deubel & Hauske, 1988; Deubel, Wolf, & Hauske, 1984; Findlay, 1982, 1992; Menz & Groner, 1987). Landing positions within words can be influenced by the luminance pattern, a phenomenon referred to as the *center of gravity effect* (Coeffe & O'Regan, 1987; Coren & Hoenig, 1972; Findlay, 1981,

1982). These results suggest that the reader's eyes should be drawn toward more complex characters when reading Chinese text. Alternatively, because more complex characters can be identified less far into the visual periphery (Yang, 1994), if eye guidance is related to character or word identification, it is likely that complex characters are less likely to be skipped.

The analyses presented next, that begin the exploration of the issues just discussed, provide initial steps in investigating the basis for eye-movement guidance in the reading of Chinese text.

Method

Subjects

Thirteen Taiwanese students at the University of Illinois at Urbana–Champaign participated in the experiment. All are native speakers of Chinese and have normal, uncorrected vision.

Materials and Procedure

In order to observe the effects of word complexity and frequency, 20 sets of five two-character words were identified. A word set consists of a group of five words that could each fit into the same word position in sentences, being both syntactically and semantically appropriate. The words are distinguished by their word frequency (high vs. low) and the complexity (high vs. low) of each of the two characters of the word. The five word types are:

1. H–LL. The H indicates a high-frequency word (occurrence greater than 300 per 0.88 million words). The LL indicates that both the first and second characters were low-complexity words, each having less than 10 strokes.
2. L–LL. The initial L indicates a low-frequency word (occurrence less than 10 per 0.88 million words).
3. L–HH. A low-frequency word, with both characters being of high-complexity, each having more than 13 strokes.
4. L–HL. A low-frequency word that has a high-complexity initial character and a low-complexity final character.
5. L–LH. A low-frequency word that has a low-complexity initial character and a high-complexity final character.

For each word set, five sentences were written, each containing a word location at which any of the words in that set could appropriately reside. Thus, from each word–sentence set, five versions of each of five sentences could be generated With 20 word–sentence sets and 100 basic sentences (five per word–sentence set), 500 sentence versions could be generated. All sentence versions were reviewed by five native Chinese speakers, including one Chinese linguist, and unanimously found to be acceptable.

Each subject read all 100 basic sentences, but with different words occupying the target locations, a factor that was varied from subject to subject using a Latin-square design. With 15 subjects, this design would be completely counterbalanced; with 13 subjects the counterbalancing was not quite complete.

In addition to the 100 sentences, subjects also read 25 test sentences, each being a paraphrase or inference of the meaning of the immediately preceding nontest sentence. On each test sentence, subjects indicated, by pressing one of two buttons, whether its meaning was or was not consistent with that of the immediately preceding sentence. All subjects performed very well on the test sentences, indicating that they were reading for comprehension.

Subjects began by reading 15 warm-up sentences presented one at a time, five of which were test sentences. They pressed a button to bring the next sentence on the screen. After any questions they had were resolved, they then read the experimental and test sentences in five blocks of 25 sentences each, with rest breaks between the blocks.

Each sentence was displayed in the center of the monitor with characters arranged horizontally. Characters were generated from the Big Five font, with characters being 16 pixels across with an 8-pixel (3 mm) space between characters. A line of 22 characters subtended 18° of visual angle, indicating that each character, together with an adjacent space, occupied slightly more than ¾°.

During reading, subjects' eye movements were recorded using a Dual Purkinje Eye Tracker, sampling eye position at 1000 Hz. Head movement was minimized by the use of a bite bar. Subjects completed a calibration task prior to reading each block of sentences.

Results

Global Statistics

Saccade Length. For the measurement of distance, an 8-pixel unit was defined, which is equivalent to half the width of a Chinese character, as displayed in this study, and also to the width of the space between adjacent characters. For convenience of comparison, this is also the width of a letter as displayed on the screen in our English reading studies.

An analysis of the data indicates that 81.6% of all saccades are progressive in direction, with 18.4% being regressive. The mean lengths of these saccades are 6.02 units versus 4.42 units respectively, a difference that is sta-

TABLE 10.1
Frequency Distribution of Progressive Saccade Lengths Across Subjects
(1 Unit = Half the Width of a Character)

Length units	1	2	3	4	5	6	7
Percentage %	1.5	2.6	6.2	11.1	15.4	17.1	15.0
Length units	8	9	10	11	12	13	14
Percentage %	12.2	7.6	5.1	3.1	1.6	1.0	0.5

tistically significant $F(1, 12) = 18.46$, $p < .001$. Thus, as in English, progressive saccades tend to be substantially longer than regressive saccades.

Table 10.1 presents the frequency distribution of progressive saccade length across subjects, as measured in these units. As the table shows, most saccades were 4 to 8 units in length. The mode of the distribution is 6 units, which is close to the mean. Thus, the average saccade length is equivalent to the most common word length in the language; it is equal to the distance from one location in a two-character word to the corresponding location in an immediately following two-character word.

We wondered whether the progressive saccade length distribution might be multimodal, showing modes corresponding the distances from one character to another, either the immediately following character, or skipping one or more characters. If this were the case, local maxima would be seen in the frequency distribution for saccades of lengths that are multiples of three. However, Table 10.1 provides no evidence for such a pattern; the distribution is monomodal.

Table 10.2 shows the mean and standard deviation of saccade lengths for each subject. These means range from 4.1 to 8.4 units, with standard deviations that range from 1.5 to 2.7 units. Thus, there are substantial individual differences in saccade length among subjects who are excellent readers of Chinese.

Fixation Duration. The data in Table 10.3 show individual subject means and standard deviations for the durations of fixations following regressive and progressive saccades. An analysis of variance showed no reliable dif-

TABLE 10.2
Mean Progressive Saccade Length and Standard Deviation for Each Subject (1 Unit = Half the Width of a Character)

Subject No.	1	2	3	4	5	6	7
Mean	5.5	6.2	6.7	6.9	4.1	5.6	5.4
SD	2.0	1.8	2.4	2.7	1.5	2.5	1.8
Subject No.	8	9	10	11	12	13	
Mean	4.7	8.4	6.3	8.0	4.5	6.0	
SD	2.0	2.6	1.8	2.3	1.6	2.5	

TABLE 10.3
Mean Fixation Durations(in Milliseconds) and Standard Deviations Following Regressive and Progressive Saccades for Each Subject

Subject No.	1	2	3	4	5	6	7
Reg. Fix.							
Mean	243	234	226	227	317	276	265
SD	117	98	84	146	148	143	130
Prog. Fix.							
Mean	251	232	214	248	278	300	248
SD	105	82	96	113	103	134	85
Subject No.	8	9	10	11	12	13	
Reg. Fix.							
Mean	225	219	235	259	215	305	
SD	88	87	92	119	94	135	
Prog. Fix.							
Mean	209	251	239	251	224	277	
SD	73	110	82	92	70	114	

ference between these two kinds of fixation durations, $F(1, 12) = 0.15$ $p > .05$. The result indicates that the durations of fixations following regressive and progressive saccades do not differ.

Preferred Viewing Position

In order to determine whether the eyes tend to go to centers of words when reading Chinese text, as is the case in reading English, we calculated the frequency with which the initial fixation on two-character words was located at different positions in the word. This was done separately for the target words and for all two-character words in the sentences. The data are presented in Table 10.4. In that table, position 0 is the space to the left of the word, position 1 is the left half of the first character, position 2 is the right half of that character, position 3 is the space between the characters, and so on.

The data in Table 10.4 do not show a noticeable maximum around the centers of words, as occurs in English. Analyses of variance of the fixation proportions also provide no evidence for differences for different locations within the word $F(5, 60) = 1.93$, $p > .10$ for target words only, and $F(5, 60) =$

TABLE 10.4
The Probability of Initial Fixation at Different Word Locations in Two-Character Words

	Space	Character		Space	Character	
Fixation location	0	1	2	3	4	5
Target words only (%)	17.7	16.9	18.0	17.4	15.8	14.2
All two-char. words (%)	17.7	16.4	17.3	16.5	15.6	15.3

0.61, $p > .69$ for all 2-character words. Thus, there appears to be no preferred landing position in words when reading Chinese text. Table 10.4 also shows no tendency for the eyes to go selectively to characters rather than to the spaces between characters: frequency of landing on the spaces is as great as that of landing on the characters themselves. Thus, these analyses provide no evidence for either a word- or character-based eye guidance process during the reading of Chinese text.

Effect of Word Variables on Eye Behavior

The final set of analyses investigated the effect of word and character variables on the eye behavior of Chinese readers.

Frequency of Skipping Characters. The fact that the average saccade length for these readers was the size of two characters indicates that there are many characters on which the eyes do not directly land; in this sense, the characters are "skipped." A plausible hypothesis is that simpler (i.e., less complex) and more frequent words and characters are likely to be identified without the need for direct fixation, and thus are skipped more often. As an initial test of this hypothesis, the likelihood of skipping at least one, two, and three characters at the locations of the target words, for each of the five types of target words, was calculated. The results are displayed in Table 10.5, which shows the same pattern for word skips of each length: highest for the H–LL (high-frequency, low-complexity) target words, and decreasing in the following order: L–LL, L–LH, L–HL, L–HH. This effect was statistically significant for each skip length: $F(4, 48) = 7.6$, $p < .001$, $F(4, 48) = 12.1$, $p < .001$, and $F(4, 48) = 5.9$, $p < .001$ for one-, two- and three-letter skips.

To test for word frequency effects, the difference between L–LL and H–LL conditions was examined. This difference was significant only for the two-character skip data ($t(48) = .62$, $p > .05$; $t(48) = 3.69$, $p < .001$; $t(48) = 0.47$, $p > .05$ for one-, two- and three-character skips).

To test for word complexity effects, the difference between L–LL and L–HH was examined. This difference was significant at all word skip lengths ($t(48) = 3.02$, $p < .01$; $t(48) = 4.11$, $p < .001$; $t(48) = 2.84$, $p < .05$).

TABLE 10.5
Probabilities of Skipping at Least One, Two, and Three Characters for Different Types of Target Words Contained in All Sentences

Word type	H–LL	L–LL	L–LH	L–HL	L–HH
One character	91.2	88.8	87.7	85.9	82.1
Two characters	58.5	49.0	47.1	45.7	41.9
Three characters	16.0	14.6	13.5	12.6	10.5

Thus, the data support the hypothesis that, in reading Chinese, higher frequency words and words of lower complexity are most likely to result in characters being skipped. In addition, the pattern of data in Table 10.5 suggests that, if only one character in a two-character word is low complexity, then skipping may occur more frequently when that character is the first rather than the second character.

Frequency of Refixating Words. In the reading of English, it has been shown that higher frequency words are less likely than lower frequency words to receive an immediate second fixation, or refixation, following the initial fixation on the word (McConkie, Kerr, Reddix, Zola & Jacobs, 1989). Thus, we expect to observe this same pattern in the reading of Chinese text. In addition, words of lower complexity would be expected to have fewer cases of refixations, since visually simpler words should be more perceptible from a wider range of fixation locations in the word.

In the current data set, 21.0% of all target words received an immediate refixation, as did 19.8% of all two-character words in the passage. Table 10.6 shows the frequency of refixating target words of the five different types included in the study. An analysis of variance found a statistically significant difference in refixation rate among these different word types $F(4, 48) = 36.96$, $p < .0001$. In addition, the refixation frequency was lower for high- (H–LL) than for low- (L–LL) frequency words ($t(48) = 5.79$, $p < .0001$), and was lower for low- (L–LL) than for high- (L–HH) complexity words ($t(48) = 8.27$, $p < .0001$). Thus, both word frequency and word complexity have large effects on the likelihood of refixating words in reading Chinese.

Gaze Duration. The amount of time spent fixating a word during its first encounter (i.e., sum of fixation durations for the first consecutive sequence of fixations on the word) is referred to as the *gaze duration* on that word. This measure is a commonly used dependent variable in studying processing during reading (Just & Carpenter, 1987; Rayner & Pollatsek, 1989). Research on the reading of English found that gaze durations are greater on longer and on less frequent words. Thus, we would expect these same phenomena to be present in the reading of Chinese. In addition, assuming that words consisting of more complex characters are more difficult to process, gaze duration should also vary with word complexity.

TABLE 10.6
Probability of Refixating Different Types of Target Words

Word type	H–LL	L–LL	L–LH	L–HL	L–HH
Prob. of refix. (%)	7.5	15.4	21.1	24.8	31.2

TABLE 10.7
Mean Gaze Durations (ms) and Standard Deviations for All One-, Two-, Three-, and Four-Character Words in All Sentences

Word length	1-char.	2-char.	3-char.	4-char.
Mean gaze duration	306	362	472	698
Standard deviation	39	59	87	192

TABLE 10.8
Mean Gaze Durations (ms) and Standard Deviations for the Five Types of Target Words Differing in Usage Frequency and Structural Complexity

Word Type	H–LL	L–LL	L–LH	L–HL	L–HH
Mean gaze duration	321	355	359	384	401
Standard deviation	44	56	57	67	73

Table 10.7 presents means and standard deviations for gaze durations on all one-, two-, three-, and four-character words in all sentences in the study. As the table shows, mean gaze duration increases for longer words, as does the standard deviation. Since most Chinese words are one or two characters in length, the average reading time per word appears to be 300–400 ms.

Table 10.8 presents means and standard deviations for the gaze durations on the target words, for words of the five different types. An analysis of variance indicates that these means are significantly different $F(4,48) = 32.24$, $p < .0001$. Post hoc analyses show a significantly lower gaze duration for high- (H–LL) than for low- (L–LL) frequency words ($t(48) = 3.78$, $p < .002$), and a significantly higher gaze duration for high- (L–HH) than for low- (L–LL) complexity words ($t(48) = 5.12$, $p < .0001$). Thus, both word frequency and word complexity affect the amount of time that the eyes stay on words while reading Chinese.

DISCUSSION

The goal of this study was to examine some of the basic characteristics of eye behavior during the reading of Chinese text. The students who participated in this study showed an average fixation duration of 247 ms, with progressive saccades that were two characters in length. This is quite similar to the data reported by Sun, Morita, and Stark (1985), whose subjects also read text in a horizontal framework. They reported a mean fixation duration of 260 ms and progressive saccades averaging 2.6 characters in length. The individual differences observed among our subjects indicate that the differences between the results of these two studies is well within the range one could expect by taking different samples from a population of Chinese readers.

The eye movements of these readers show several characteristics previously observed in readers of alphabetic languages. Their regressive saccades are shorter than their progressive saccades, on average. Some words receive refixations following the initial fixation, with low-frequency words being refixated more often than high-frequency words. Some words are skipped during reading, and this occurs with high-frequency words more often than with low-frequency words. Low-frequency words and longer words receive longer gaze durations than do high-frequency and shorter words. The frequency of making regressive saccades, 18% in the current study, is only slightly greater than the 10% to 15% frequency reported for English by Rayner and Pollatsek (1989).

Chinese has one additional variable that has a substantial effect on the reader's eye behavior: character and word complexity. Even though the current study indexed this variable in only a crude way by counting the number of strokes in the character, its influence was still highly evident: It affected the likelihood of skipping characters, the likelihood of refixating words, and the gaze duration on a fixated word. It is clear that eye-movement studies investigating various aspects of processing during reading must take care in controlling for word and character complexity. Our experience in creating the materials for this study also suggests that there is a substantial relationship between character complexity and frequency, somewhat analogous to Zipf's Law in alphabetic languages, counting strokes instead of letters. As already described, additional complexity can arise in a number of different ways: by having nondecomposable characters with more strokes, or by having characters with additional components arranged in different ways. The effects of these differences on the reader's eye behavior, perhaps reflecting processing differences beyond the effects of simple complexity differences, have not yet been investigated.

We assume that visual complexity has two effects on eye movements. First, it was previously shown that more complex words take longer to identify (Yeh & Liu, 1972; but see Huang, Wang & Liou, 1991). This can account for some of the increase in gaze duration for more complex words. In some cases, this may be due to extra time required for decomposing a character into its components, or for the separate processing of the different components in arriving at a character identification. Second, more complex words can be identified less far into the periphery (Yang, 1994). This probably accounts for the effect of complexity on the frequency of skipping words and on the frequency of refixating. Related to this, we note that there is considerable variance in the average saccade lengths of the subjects who participated in this study. We wonder if some of this variance is related to differences in their ability to resolve complex patterns in their visual periphery.

The most striking difference between English and Chinese eye-movement data is the lack of a preferred viewing position phenomenon in Chinese.

This phenomenon is accepted as one of the primary indicators that eye-movement control in reading is word-based. Apparently, the fact that English readers are able to isolate words in the visual periphery based on the clearly marked blank space delimiters, thus enabling them to select a word to serve as the target for the following saccade, is critical for the existence of this phenomenon. The fact that this phenomenon is not present in Chinese readers' data suggests that words, or word-like units, are not functioning as saccade targets in Chinese as they do in English. Thus, there is no evidence that Chinese readers are able to parse the character string into words in their peripheral vision.

That observation raises two questions about reading in Chinese. First, when are characters combined into a word during reading? Are characters being considered sequentially, with word boundaries being located in a manner similar to that usually assumed in identifying constituent boundaries in the reading of English (e.g., Frazier & Fodor, 1978)? Or are characters being considered in parallel by a system having the characteristics of a neural network, which identifies character groups that constitute contextually appropriate lexical units, similar to letters in an English word? The issue of how the character string is parsed into word-like units, and when this takes place, is a fascinating and important theoretical question in understanding Chinese reading, particularly considering the potential parsing ambiguity that exists within the language.

The second question concerns the basis on which saccade length is determined. There are at least three ways that this might occur. First, the brain could set up the equivalent of a regression equation, enter the various factors that affect the solution, and derive a saccade length that is appropriate for the immediate situation. This is a *continuous control model*, with saccades varying continuously in length. Second, the brain could select some object in the visual array and send the eyes to that target. This is a *discrete control model*, with saccade length being determined by the distance of the particular target object selected. The preferred viewing position phenomenon has been used to argue for the discrete model of saccade control in opposition to the continuous model. Third, the eyes could be drawn to the "center of gravity" of the light distribution in some attended area, which would also be considered. The degree to which this type of model is discrete versus continuous depends on its details, particularly on the basis by which the attended region is determined.

If the saccade length calculation is not based on the locations of words in reading Chinese, this raises the question of whether some other unit, such as the character, is being selected as a saccade target, or whether the system here employs a more continuous control model. Our initial analyses failed to provide evidence for character-based control; the eyes were no more likely to go to a character than to go to the space between characters, and there was no indication of a multimodal pattern in the saccade length distribution.

It is possible that saccade guidance is character-based, but that we failed to observe it in the present analyses because the methods used were not sufficiently refined. McConkie et al. (1988) found evidence for two sources of error in saccadic landing position distributions that tend to spread out these distributions. Much narrower distributions on words are observed if data are plotted separately for cases where the eyes come from different distances and land on a word. Thus, the failure to observe character-related differences in fixation location frequencies may be due to the fact that Chinese characters are narrower than the lengths of all but the shortest English words, together with sources of oculomotor error that spread the fixation locations out. These factors may combine to hide an underlying character-unit basis in the control of saccades. Further analyses are required to explore this possibility.

The third hypothesis suggests that the eyes are sent to the center of gravity of the luminance pattern in some attended area that includes multiple characters (Vitu, 1991). Previous research indicates that the control of saccades is sensitive to the intensity, density, contrast, and size of stimuli (Deubel & Hauske, 1988; Findlay, 1992; Menz & Groner, 1987). Given the great variation in character complexity/density in Chinese, such a mechanism could show some of the characteristics previously described, such as greater likelihood of skipping less complex characters in favor of the eyes being drawn to more complex characters, and greater likelihood of refixating more rather than less complex characters. Testing such a hypothesis requires making claims about the control of attention, and, hence, about the attended area that influences saccade programming during fixations in reading.

In summary, this study failed to find the preferred viewing effect for words in Chinese reading, and thus suggested that saccades are not being sent to words as in the reading of English. At the same time, it clearly showed the effect of perceptual variables on saccade control, including word complexity and word frequency. Thus, it is clear that the eyes are not being moved simply on the basis of some oculomotor pattern, ignoring local text characteristics (Bouma & de Voogd, 1974; O'Regan, 1989, 1990; Shebilske, 1975; Tinker, 1958). However, it is not clear whether saccades are being sent to selected text units in a manner similar to the word-based control in English (McConkie et al., 1988; Morris, Rayner & Pollatsek, 1990; Morrison, 1984; Rayner & Pollatsek, 1996), or whether that control is based on center of gravity calculation in combination with attentional factors (Vitu, 1991).

Of course, these comments completely ignore the basis on which any selection may be taking place in determining where the eyes go, either with respect to selecting a saccade target or selecting a text region to which to attend. We assume that, at least to some degree, higher cognitive activities involved in the language processing are involved in these decisions (see Rayner and Pollatsek, 1989, for a review of studies showing influences of this type in English). The effects of word frequency observed in the present study indicate that language factors are having an influence, but just when and how are issues requiring additional studies.

ACKNOWLEDGMENTS

The data used in the analyses reported in this paper were collected at Beckman Institute, University of Illinois at Urbana-Champaign, as part of the first author's doctoral dissertation. The software for collecting and reducing data were written by Gary Wolverton. We acknowledge the assistance of Dr. C. C. Cheng in developing algorithms to access the Big Five character codes used for the display.

REFERENCES

Bouma, H., & de Voogd A. H. (1974). On the control of saccades in reading. *Vision Research, 14*, 273–284.

Cheng, C. M., & Wu, S. C. (1993, September). *Semantic satiation in Chinese*. Paper presented at the Sixth International Symposium on Cognitive Aspects of the Chinese Language, Taipei.

Coeffe, C., & O'Regan, J. K. (1987). Reducing the influence of non-target stimuli on saccade accuracy: Predictability and latency effects. *Vision Research, 27*, 227–240.

Coren, S., & Hoenig, P. (1972). Effect of non-target stimuli on the length of voluntary saccades. *Perceptual and Motor Skill, 34*, 499–508.

Deubel, H., & Hauske, G. (1988). The programming of visually guided saccades. In H. Marko, G. Hauske, & A. Struppler (Eds.), *Processing structures for perception and action* (pp. 67–74). Weinheim, Germany: Verlag Chemie.

Deubel, H., Wolf, W., & Hauske, G. (1984). The evaluation of the oculomotor error signal. In A. G. Gale & F. W. Johnson (Eds.), *Theoretical and applied aspects of oculomotor research*. Amsterdam: North-Holland

Findlay, J. M. (1981). Local and global influences on saccadic eye movements. In D. F. Fisher, R. A. Monty, & J. W. Senders (Eds.), *Eye movements, cognition and visual perception* (pp. 171–179). Hillsdale, NJ: Lawrence Erlbaum Associates.

Findlay, J. M. (1982). Global processing for saccadic eye movements. *Vision Research, 22*, 1033–1045.

Findlay, J. M. (1992). Programming of stimulus-elicited saccadic eye movements. In K. Rayner (Ed.), *Eye movements and visual cognition* (pp. 8–30). New York: Springer-Verlag.

Frazier, L., & Fodor, J. D (1978). The sausage machine: a new two-stage parsing model. *Cognition, 6*, 291–326.

Hoosain, R. (1991). *Psycholinguistic implications for linguistic relativity: A case study of Chinese*. Hillsdale, NJ: Lawrence Erlbaum Associates.

Huang, J.-T., & Wang, M. Y. (1992). From unit to Gestalt: Perceptual dynamics in recognizing Chinese characters. In H.-C. Chen & O. J. L. Tzeng (Eds.) *Language processing in Chinese* (pp. 3–35). North-Holland: Elsevier.

Huang, J.-T., Wang, M. Y., & Liou, S. N. (1991). *Component diagnosticity and pattern complexity in object recognition*. Unpublished manuscript.

Inhoff, A. W., & Rayner, K. (1986). Parafoveal word processing during eye fixations in reading: Effects of word frequency. *Perception & Psychophysics, 40*, 431–439.

Just, M. A., & Carpenter, P. A. (1987). *The psychology of reading and language comprehension.* Newton, MA: Allyn & Bacon.

McConkie, G. W. (1983). Eye movements and perception during reading. In K. Rayner (Ed.), *Eye movements in reading: Perceptual and linguistic aspects* (pp. 65–96). New York: Academic Press.

McConkie, G. W., Kerr, P. W., Reddix, M. D., & Zola, D. (1988). Eye movement control during reading: I. The locations of initial eye fixations in words. *Vision Research, 28*(10), 1107–1118.

McConkie, G. W., Kerr, P. W., Reddix, M. D., Zola, D., & Jacobs, A. M. (1989). Eye movement control during reading: II. Frequency of refixating a word. *Perception & Psychophysics, 46*(3), 245–253.

Menz, C., & Groner, R. (1987). Saccadic programming with multiple targets under different task conditions. In K. O'Regan & A. Lévy-Schoen (Eds.), *Eye movements: From physiology to cognition* (pp. 95–103). Amsterdam: North-Holland.

Morris, R. K., Rayner, K., & Pollatsek, A. (1990). Eye guidance in reading: The role of parafoveal letter and space information. *Journal of Experimental Psychology: Human Perception and Performance, 16*, 268–281.

Morrison, R. E. (1984). Manipulation of stimulus onset delay in reading: Evidence for parallel programming of saccades. *Journal of Experimental Psychology: Human Perception and Performance, 10*, 667–682.

O'Regan, J. K. (1990). Eye movement and reading. In E. Kowler (Ed.), *Review of oculomotor research: Vol. 4. Eye movements and their role in visual and cognitive processes* (pp. 395–453). Amsterdam: Elsevier.

O'Regan, J. K., & Jacobs, M. A. (1992). Optimal viewing position effect in word recognition: A challenge to current theory. *Journal of Experimental Psychology: Human Perception and Performance, 18*, 185–197.

O'Regan, J. K., & Lévy-Schoen, A. (1987). Eye movement strategy and tactics in word recognition and reading. In M. Coltheart (Ed.), *Attention and performance XII: The psychology of reading* (pp. 363–383). Hillsdale, NJ: Lawrence Erlbaum Associates.

Peng, D. L., Orchard, L. N., & Stern, J. A. (1983). Evaluation of eye movement variables of Chinese and American readers. *Pavlovian Journal of Biological Science, 18*, 94–102.

Rayner, K., & Pollatsek, A. (1989). *The psychology of reading.* Hillsdale, NJ: Lawrence Erlbaum Associates.

Rayner, K., & Pollatsek, A. (1996). Reading unspaced text is not easy: Comments on the implications of Epelboim et al (1994) study for models of eye movement control in reading. *Vision Research, 36*, 461–470.

Shebilske, W. (1975). Reading eye movements from an information-processing point of view. In E. W. Massaro (Ed.), *Understanding language* (pp. 291–311). New York: Academic Press.

Sun, F., Morita, M., & Stark, L. W. (1985). Comparative patterns of reading eye movement in Chinese and English. *Perception & Psychophysics, 37*, 502–506.

Tinker, M. A. (1958). Recent studies of eye movements in reading. *Psychological Bulletin, 55*, 215–231.

Vitu, F. (1991). The existence of a center of gravity effect during reading. *Vision Research, 31*, 1289–1313.

Yang, H. M. (1994). *Word perception and eye movements in Chinese reading.* Doctoral dissertation, The University of Illinois at Urbana-Champaign.

Yeh, J. S., & Liu, I. M. (1972). Factors affecting recognition thresholds of Chinese characters. *Acta Psychologica Taiwanica, 14*, 113–117.

Yu, B. L., Hu, L. H., & Pan, Y. J. (1993, September). *Context and space effects on target characters' cognition.* Paper presented at the Sixth International Symposium on Cognitive Aspects of the Chinese Language, Taipei.

Zhang, W., & Feng, L. (1992). A study on the unit of processing in recognition of Chinese characters. *Acta Psychologica Sinica, 24*, 379–385.

CHAPTER

11

Use of Prelexical and Lexical Information During Chinese Sentence Reading: Evidence From Eye-Movement Studies

Albrecht Werner Inhoff
Weimin Liu
Department of Psychology
State University of New York at Binghamton

ZhiHua Tang
Department of Psychology
Hangzhou University, China

Irrespective of whether Chinese or English text is read, reading requires that a spatially ordered sequence of visual symbols is recognized. Furthermore, irrespective of text type, reading is a dynamic task in which the perception and recognition of a sequence of spatially ordered linguistic symbols is accomplished by executing a sequence of spatially ordered eye movements, also referred to as *saccades*. Execution of saccades is necessary because high acuity vision is limited to a small area of text projected onto the fovea and the immediately adjacent parafovea. Readers thus need to look at—or *fixate*—characters and words before they can be recognized and discriminated from visually similar alternatives. In this chapter, we first review work that measured eye movements during the reading of alphabetic text. General issues related to oculomotor control and word recognition are of particular interest. After this, we report the results of several recent eye-movement studies with Chinese text, which pursued similar issues. Alphabetic (English) script and Chinese script differ in a number of key characteristics, making it likely that the control of eye movements during reading and the nature of linguistic analyses applied to visible text are codetermined by script type.

Alphabetic English text is composed of a relatively small number of visuospatially simple letters, each of which denotes a phonemic unit. There is a relatively strong relationship between the graphemic structure of a sequence of letters and its pronunciation, as graphemically similar letter sequences have generally similar pronunciations. With few exceptions, letters cannot stand alone as words; instead, they are concatenated until a word boundary is reached. The boundaries of successive words in running text are well defined, consisting of visuospatially distinct blank spaces.

In contrast to this, Chinese script is composed of a relatively large number of spatially distinct characters, with modern (simplified) Chinese script containing approximately 7,000 such characters (Yin & Rohsenow, 1994). Each character corresponds to a syllable and, in the vast majority of instances, to a (root) morpheme. The relationship between the graphemic structure of a character and its pronunciation is poor, as graphemically similar characters can have radically different pronunciations. Most characters can stand alone as one-character words, though most are constituents of common two-character words and of less common three- and four-character words in running text. In contrast to characters, words are not visuospatially distinct, and successive characters are concatenated until a punctuation mark is reached. Consequently, the lexical status of a character, that is, whether it constitutes a word or part of a word, often remains ambiguous.

EYE MOVEMENTS AND WORD RECOGNITION DURING THE READING OF ALPHABETIC TEXT

Fixation durations and saccade sizes during the reading of alphabetic (English) text are fairly normally distributed, with modal values of approximately 220 ms and 9 character spaces, respectively (e.g., Rayner & Pollatsek, 1989, p. 117; Vitu, O'Regan, Inhoff, & Topolski, 1995). There is considerable variability in both measures, with fixation durations ranging from less than 100 ms to more than 500 ms. Similarly, saccades extend across less than 1 character space to more than 20 character spaces. Readers of alphabetic text also execute a considerable number of negative saccades, that is, regressions to previously read text, with typical regression rates amounting to 15% to 20% (Rayner & Pollatsek, 1989).

Individual fixation durations and saccades during the reading of alphabetic text were shown to be determined by on-line processing demands. The duration of the first fixation on a selected (target) word and the cumulated target viewing duration (also referred to as *gaze duration*) are, for instance, determined by the target's temporal visibility (Morrison, 1984; Rayner & Pollatsek, 1981; Sereno & Rayner, 1992). In these studies, eye-movement-contingent display changes were used to present a foil stimulus at the onset of a fixation and to replace it with the to-be-identified target after a prespecified interval. For instance, in Sereno and Rayner's (1992) study, a contextually inappropriate word foil was shown immediately after the fixation of the target area. After a delay of 30 ms, 45 ms, or 60 ms, it was replaced with the target (the target was shown at fixation onset in a control condition). The results showed increased first fixation durations and gaze durations on the target area when the foil was presented, with approximate increases in the two viewing duration measures (first fixation durations and gaze durations) of 50/60 ms, 74/80 ms, and 80/100 ms for the 30 ms, 45 ms, and 60 ms delay conditions, respectively. Hence, any delay in the visibility of a meaningful (alphabetic) word resulted in an immediate increase in viewing durations.

First fixation durations and gaze durations on target words are also determined by the ease of linguistic analyses, showing that readers of alphabetic text complete the lexical analysis of a word–including the assignment of word meaning–while it is fixated. For instance, readers spent less time viewing common high-frequency words than length-matched low-frequency words (Henderson & Ferreira, 1990; Inhoff & Rayner, 1986; Kennison & Clifton, 1995; Rayner & Duffey, 1986), and less time viewing a meaningful, context-consistent word than a graphemically and phonologically matched context-inconsistent word or pseudoword (Inhoff & Topolski, 1994). Effect sizes were quite substantial in these studies, amounting to approximately 30 ms to 50 ms when high-frequency words were compared to low-frequency words and to more than 100 ms when contextually appropriate legal words were compared to context-inappropriate words and pseudowords.

EYE MOVEMENTS AND LEXICAL PROCESSES DURING THE READING OF CHINESE SENTENCES

Do readers of Chinese text use similar eye-movement patterns as readers of alphabetic text, and do they recognize characters and words while they are fixated? Figures 11.1 and 11.2 depict frequency distributions for fixation durations and the size of saccades during the reading of Chinese sentences. The data were obtained from 11 skilled readers who read sentences taken from a popular Chinese magazine. All text was written in simplified script, as used in mainland China, and all readers were native speakers of Chinese who studied at the State University of New York at Binghamton. A commonly used word-processing program, Chinese Star 2.0, was used to generate and display characters, each of which occupied a 24 × 24 pixel grid, with strokes occupying a 24 (vertical) by 23 (horizontal) pixel area. A blank 1-pixel vertical column thus separated successive characters. In this and all other reported

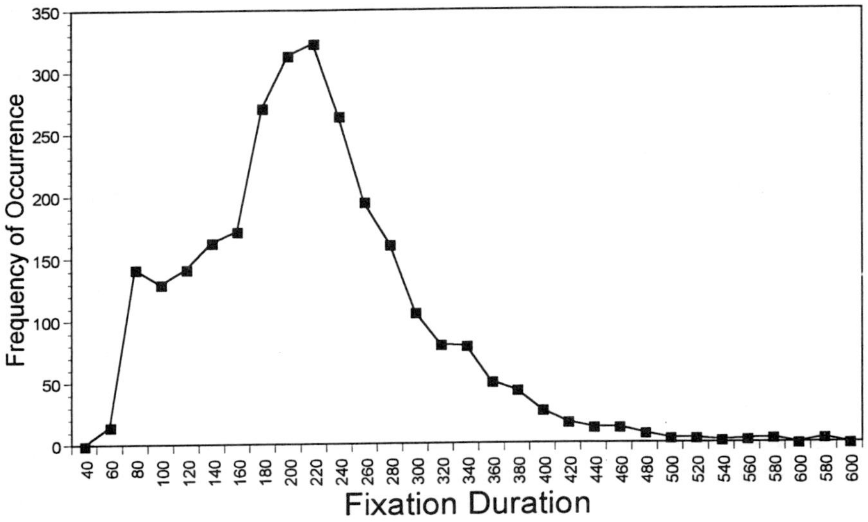

FIG. 11.1. The frequency distribution of fixation durations during the reading of sequences of unrelated sentences. Eleven readers contributed to the distribution, which is based on approximately 2,700 observations.

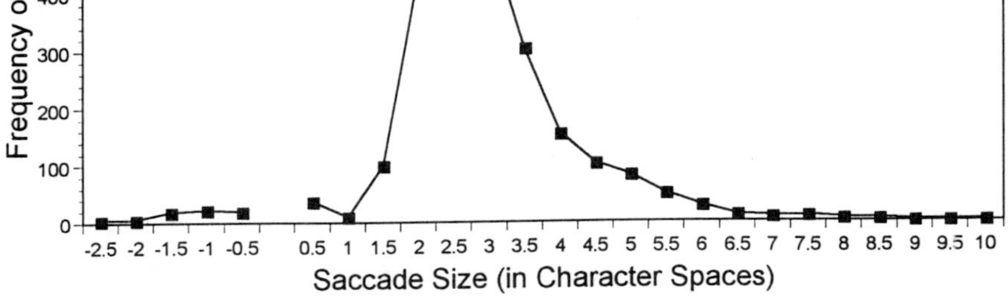

FIG. 11.2. The frequency distribution of saccade during the reading of unrelated sentences. Eleven readers contributed to the distribution, which is based on approximately 2,700 observations.

experiments, eye movements were recorded with a fifth-generation dual-Purkinje SRI eye-tracking system (Crane & Steele, 1985). The system is extremely accurate with an absolute visual resolution of <10 minutes of arc. Fixation locations were determined at 2-ms intervals and were used to determine whether the eyes were fixated or in motion (saccading). To distinguish saccades from fixations, a saccade was assumed to have occurred when two successive recordings of fixation locations revealed that the eyes had moved across one third or more of a character space. Following the convention adopted for the measurement of saccades during the reading of alphabetic text, saccade size was measured in (Chinese) character spaces. Additional technical details can be obtained from Inhoff and Liu (1998).

Similar to the corresponding frequency distributions for alphabetic text, the modal fixation duration amounted to 220 ms during the reading of Chinese text. However, brief fixation durations of 160 ms or less were considerably more common. Hence, mean fixation durations are shorter when Chinese text is read (approximately 230 ms) than when alphabetic (English) text is read (approximately 250 ms). Examination of saccade sizes also revealed a distinct paucity of regressive saccades (indicated by the small number of negative values in Fig. 11.2). As noted before, 15% to 20% of all saccades were directed to previously read text when alphabetic text was read, but only 4% of saccades were directed to the left when similar Chinese text was read.

Global differences in script type and differences in the pattern of eye movements during the reading of alphabetic and Chinese text suggest that oculomotor control and word recognition processes may also be a function of script type. Fixation duration could, for instance, be less closely linked to the onset and completion of lexical analyses when Chinese text is read than when alphabetic text is read due to the ambiguity in the lexical status of individual characters.

To examine this possibility, we (Inhoff & Liu, 1998) delayed the onset of a Chinese two-character word by either 50 ms or 150 ms after the onset of the first fixation on the target area and measured effects of target delay on the ensuing first fixation duration and the cumulated viewing duration (gaze duration). The target word was visible upon its fixation in a control condition. The results yielded large and robust effects of target delay. Specifically, presentation of two foil characters at fixation onset for 50 ms increased first fixation durations and gaze durations by approximately 100 ms and 200 ms, respectively. Increasing the delay by another 100 ms to 150 ms added an additional 30 ms and 65 ms to first fixation durations and gaze durations, respectively. Delay of the target thus led to an immediate increase in first fixation durations and gaze durations during the reading of Chinese text, as it did for alphabetic English text. Irrespective of script type, eye movements thus respond immediately to the availability of meaningful text.

Furthermore, readers of Chinese text determine a character's or word's meaning while it is fixated. Similar to earlier studies with alphabetic text, in which the word frequency of target words was manipulated, we (Inhoff & Liu, 1998) embedded two types of target words, consisting of high- and low-frequency two-character words, in simple declarative sentences and measured target viewing duration as a function of word frequency.

The mean word frequency of high- and low-frequency two-character words amounted to 477 and 2 per million, respectively (Beijing Institute of Language Teaching and Research, 1986). High- frequency target words were also more likely to contain familiar character constituents than low-frequency targets, with constituent character frequency counts of 379 and 353 per million for the first and second constituent characters of high-frequency targets, respectively, and of 64 and 250 per million for low-frequency targets. Our results showed robust effects of word frequency. Specifically, first fixation and gaze durations amounted to 195 ms and 212 ms, respectively, when a high-frequency word was fixated, and to 223 ms and 249 ms, when a low-frequency word was fixated. Lexical analyses are thus executed on-line when a Chinese word or character is fixated.

Direct evidence for the extraction of character meaning during, and perhaps even prior to, a character's fixation also accrued in Liu and Inhoff's (1998) study, where readers were required to determine whether a sentence was meaningful after it had been read. Four sentence conditions were created (see Fig. 11.3). Sentences in each of the four conditions were identical except for a single target character: In the *base condition*, the target was consistent with preceding and following sentence context. The target did not fit prior and subsequent sentence context in three other conditions that consisted of a *homophone condition*, in which the target was homophonically related to the base; a *graphemic condition*, in which the target was graphemically similar to the base; and an *unrelated* (or *dissimilar*) *condition*, in which the target was neither phonologically nor graphemically related to the base.

Examination of sentence classification accuracies indicated that virtually all base sentences were classified as meaningful and that approximately 80% of the sentence in the homophone, graphemic, and dissimilar conditions were classified as nonmeaningful. Hence, successful assignment of meaning was made in these conditions after sentence reading had been completed. More important, inspection of first fixation durations and gaze durations on the target character of correctly classified sentences revealed profound effects of target type. Specifically, first fixation durations and gaze durations on a target character were shorter in the base condition (240 ms and 276 ms, respectively) than in either the homophone condition (330 ms and 342 ms), the graphemic condition (269 ms and 325 ms), or the dissimilar condi-

对该事　　件悼捶丅约°__它Ä的讨论仍然可以继续。

*

(Thought and discussion to this event can still keep on).

	Character	Pronunciation	Translation
base Word	对	"dui"	to
homophonic	队	"dui"	row
graphemic	付	"fu"	pay
unrelated	立	"li"	stand

Note. * indicates the sentence location which was occupied by one of the four types of target character

FIG. 11.3. The four target characters that were embedded in sentence context. Inclusion of the base character resulted in a meaningful expression. Inclusion of the homophonic (phonologically similar), graphemically similar, or unrelated (dissimilar) character resulted in a nonmeaningful expression.

tion (340 ms and 349 ms). The results from recent experiments thus converge, showing that eye movements are controlled on-line by the ease of linguistic processes and that linguistic processes during the reading of Chinese text include word recognition and the assignment of character meaning.

FLUENT READING OF ALPHABETIC TEXT ENTAILS THE USE OF THE PARAFOVEALLY VISIBLE WORD

Although full recognition of an alphabetic word generally requires that it be fixated, its linguistic analysis appears to begin before the eyes are moved toward it. There is converging evidence that the range of effective vision during the reading of alphabetic text extends up to 15 character spaces to the right of fixation (see Rayner & Pollatsek, 1989, for a review) and that it generally includes the fixated word and the next (*parafoveal*) word in the text (Inhoff, 1990). When parafoveally available word information is denied, so that skilled readers of alphabetic text obtain useful information only from the fixated word, reading becomes dysfluent and reading rate declines by approximately one third (Blanchard, Pollatsek, & Rayner, 1989; Rayner, Well, Pollatsek, & Bertera, 1982).

Recent findings suggest that parafoveal preview of alphabetic text provides two functionally distinct types of information: visuospatial cues, in the form of parafoveally visible word length, and linguistic specifications. Parafoveal word length is sought relatively early in a fixation and appears to be used for the planning of saccades. This view is supported by results showing that the masking of parafoveal word spaces disrupts eye movements primarily when the masking is applied during the initial 50 ms of a fixation (Morris, Rayner, & Pollatsek, 1989; Pollatsek & Rayner, 1982), and that the size of interword saccades increased with the length of to-be-fixated words (Inhoff, 1989; see also Rayner & Pollatsek, 1981, for a review). Furthermore, saccades generally position the eyes near the visuospatial center of the parafoveally visible word (McConkie, Kerr, Reddix, Zola, & Jacobs, 1989; Radach & Kempe, 1993), presumably because this location constitutes an optimal location for subsequent word identification (O'Regan, 1992; Vitu, O'Regan, & Mittau, 1990).

Parafoveal word preview also provides linguistic information, and some researchers argue that it is used for the planning of saccades. In support of this view, we (Inhoff, Briihl, & Schwartz, 1996; see also Hyona, Niemi, & Underwood, 1989) noted that readers executed larger saccades to compound words than to matched monomorphemic controls, and that readers executed somewhat larger saccades to parafoveally visible words than to length-matched sequences of random characters (Inhoff, 1989). Other studies showed that readers tend to execute larger saccades to words with an informative letter sequence than to words with an uninformative letter sequence (Beauvillain, Dore, & Baudouin, 1996; Underwood, Bloomfield, & Clews, 1988; Underwood, Clews, & Everatt, 1990). However, so far these claims remain controversial. Effects of the informative value of beginning letter sequences were completely absent in Rayner and Morris's (1992) experiment; other studies that manipulated linguistic properties of parafoveally visible text failed to discern effects on saccade size (e.g., Inhoff & Rayner, 1986).

There is, however, considerable agreement that parafoveally visible linguistic information is used for word recognition. In some instances, the parafoveally visible word may be fully recognized; more often parafoveally obtained linguistic information is used to facilitate word recognition when the previewed word is subsequently fixated (e.g., Balota, Pollatsek, & Rayner, 1985; Inhoff & Rayner, 1986; Kennison & Clifton, 1995; Pollatsek, Lesch, Morris, & Rayner, 1992; Rayner, 1975). Most studies examining the nature of the linguistic information that can be obtained from the parafovea used an eye-movement-contingent display change method in which the parafoveal visibility of a target word was manipulated. In Rayner's (1975) original study, sentence frames containing a target word were created and eye movements were recorded during sentence reading. Prior to the fixation of the target, the reader was shown one of several target foils in the right parafovea, including the target word itself (base preview), a length-matched graphemically similar sequence of characters (similar foil), or a graphemically dissimilar sequence of characters (dissimilar foil). During the saccade to the parafoveal stimulus, the previously visible foil was replaced with the contextually appropriate base word, and base word viewing durations were measured as a function of foil type. Rayner's (1975) original finding—and the results of several subsequent studies—indicated that readers often obtain useful graphemic information, expressed in shorter target viewing durations, when parafoveal previews were graphemically similar to the target than when they were dissimilar.

Subsequent studies showed that readers also obtain useful phonological and lexical information from the parafovea. Pollatsek et al. (1992) obtained shorter target viewing durations when a parafoveally previewed foil was homophonically related to the subsequently fixated target than when it was graphemically similar, indicating that readers of alphabetic text obtained phonological information in addition to graphemic information from the parafoveal word. Two other studies (Inhoff & Rayner, 1986; Kennison & Clifton, 1995) showed that parafoveal preview was more useful when it revealed a high-frequency word than when it revealed a low-frequency word, presumably because parafoveal preview was used to activate a set of lexical candidates, with more activation accruing in the representations of high-frequency candidates than of low-frequency candidates.

Finally, there is now some evidence indicating that readers of alphabetic text obtain some useful linguistic information *after* a word was fixated (Inhoff, Radach, & Starr, 1998; see also Binder, Pollatsek, & Rayner, submitted). In our recently completed experiment, sentences were used that contained a short target word. Eye-movement-contingent display changes were used to change the target after the eyes moved to the right of it (e.g., *hat* was then changed to *mat*). After sentence reading was completed, readers were asked to identify the word shown in the sentence by choosing among three visually similar candidates (e.g., *hat, mat, cat*). In this task, the target was selected on approximately 60% of the trials, and the word that had never been shown was selected on approximately 10% of the trials. The foil was selected on approximately 30% of the trials, indicating that readers obtained some useful information from the word to the left of fixation.

Together, these results indicate that recognition of an alphabetic word is spatially distributed, being initiated before the word is fixated, that is, when it is visible in the right parafovea, being continued when it is fixated, and being completed—at last on occasion—after its fixation.

LINGUISTIC ANALYSES ARE ALSO SPATIALLY DISTRIBUTED DURING THE READING OF CHINESE TEXT

Fluent reading of Chinese text, just like the fluent reading of alphabetic text, requires the acquisition of useful information from parafoveally visible text. This conclusion emerged from several recent experiments (Inhoff & Liu, 1997, 1998). In two of the experiments, eye-movement-contingent display changes were used to create 11 viewing conditions. Two baseline conditions were created in which the reader either saw the full line of text during each fixation, as occurs during normal sentence reading, or only the fixated character, that is, all text to the right and left of the character was masked by phonologically and graphemically dissimilar characters. The visibility of text to the right and left of the fixated character was manipulated in the remaining conditions. In seven right-directed preview conditions, one to seven characters of text were shown to the right of the fixated character, and in two left-directed preview conditions, an additional one or two characters were shown to the left of fixation (the window of legible text moved with the eyes in all conditions). As can be seen in Fig. 11.4, the perceptual span, determined by selecting the smallest viewing area with normal gaze durations, encompassed the fixated character, one to two characters to its left, and up to three characters to its right. Since the majority of words contained two character constituents, readers of Chinese text thus obtained useful information from the fixated word and from the next word(s) in the text. Furthermore, since reading was hampered when characters to the left of fixation were masked, readers of Chinese text appear to obtain useful linguistic information to the left of a fixated linguistic unit, just as do readers of alphabetic text.

In English texts, words vary greatly in length, with word lengths ranging from 1 character space to more than 20 character spaces, and, as noted before, all words are preceded and followed by blank spaces, which makes them visuospatially distinct and well suited for the planning of saccades. In Chinese text, by contrast, successive characters of text are separated by small blank spaces and there is no spatial marking of word units. Consequently, word length appears ill suited for saccade planning.

This view is consistent with results of our analysis in which we (Inhoff & Liu, 1998) examined the size of a right-directed saccade off a fixated character as a function of the length of the next word in the text. Three groups of parafoveal word lengths were discriminated, those consisting of one character, of two characters, and those containing more characters. The results showed that saccades to the right extended across approximately 2.6 character spaces (when the full line of text was visible during each fixation), irrespective of the length of the next word in the text.

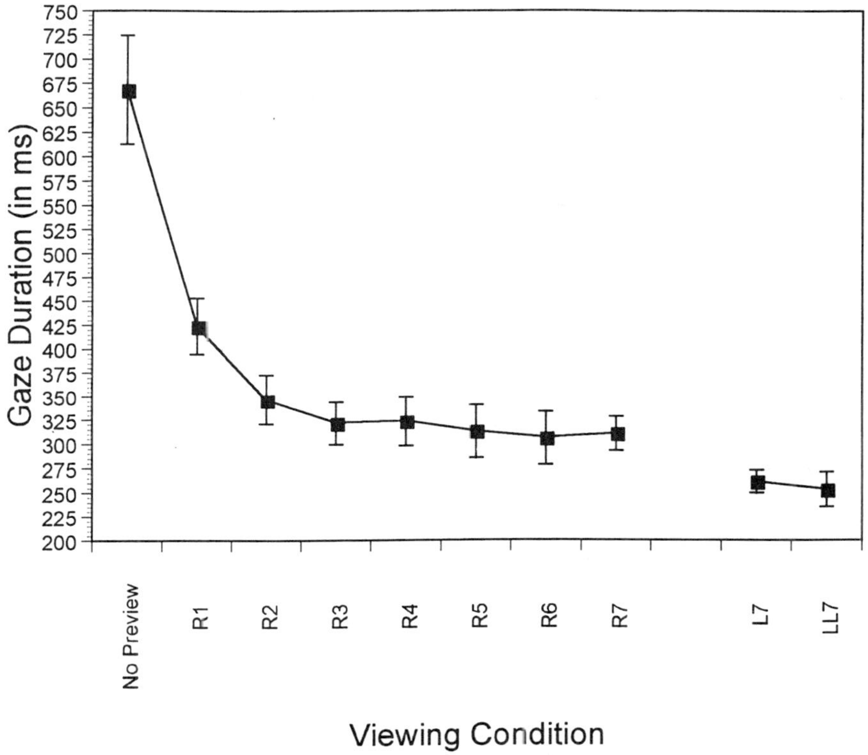

FIG. 11.4. Cumulated word viewing durations (gaze durations) as a function of the visibility of text during each fixation. The no-preview condition refers to a viewing condition in which only the fixated character was visible. R1 to R7 refer to viewing conditions in which one to seven characters of text, respectively, were visible to the right of fixation. L1R7 and L2R7 refer to two viewing conditions in which the fixated character, seven characters to its right, and either one or two characters to its left were visible during each fixation.

Linguistic properties of to-be-fixated text did, however, determine saccade size. In Liu and Inhoff's (1998) study, readers made slightly larger saccades to a parafoveally visible context-consistent target (2.4 character spaces) than to an inconsistent character (2.0 character spaces). Similarly, in Inhoff and Liu's (1998) study, saccades to two-character words were larger when the target was visible in the parafovea (2.6 character spaces) than when it was masked by a visually similar (but context-inconsistent) two-character foil prior to its fixation (2.2 character spaces). Though some of these findings are still preliminary, they converge showing that character and word meaning, rather than the visuospatial layout of parafoveally visible words, are used for the planning of saccades when Chinese text is read.

Saccade planning may occur relatively late during the reading of Chinese text, after linguistic information has been extracted from parafoveally visible characters.

What type of linguistic information do readers of Chinese text obtain from parafoveally visible characters? To answer this question, we (Liu & Inhoff, submitted) manipulated the linguistic properties of a parafoveally visible target character by creating four target preview conditions: a *control condition*, in which the contextually consistent target (base) character was shown in the parafovea prior to its fixation; a *homophone condition*, in which the target was replaced with a homophonic foil prior to its fixation; a *graphemically similar* condition, in which the target was replaced with a graphemically similar foil; and a *dissimilar condition*, in which the target was replaced with a visually and phonologically dissimilar foil. In all conditions, the target was visible upon the fixation of its location. In the main analyses, first fixation durations and gaze durations on the target character were analyzed as a function of its parafoveal preview.

The results showed a robust effect of parafoveal preview. First fixation durations and gaze durations on a target character were shorter when the base or a graphemically similar character had been visible in the parafovea (249 ms/253 ms and 257 ms/272 ms, respectively) than when a homophonic or dissimilar foil had been visible (300 ms/313 ms and 300 ms and 315 ms, respectively). These results imply that parafoveally visible graphemic information was used during the following identification of the target character and the assignment of meaning. Parafoveally visible phonological information, by contrast, is of little or no use for character recognition, perhaps because the preponderance of homophony would generally result in the accessing of a large number of inappropriate character meanings.

Parafoveal preview of a character thus appears to activate a set of graphemically similar candidates. Activation levels are codetermined by linguistic properties, being higher for high-frequency candidates than for low-frequency candidates. This conception emerged from the comparison of parafoveal target previews that were high- and low-frequency words (Inhoff & Liu, 1998). In a standard reading condition, these words were visible in the parafovea prior to their fixation, as occurs in normal reading. In a parafoveal masking condition, the two types of target words were replaced with graphemically similar foils until they were directly fixated. The results revealed considerably shorter first fixation and gaze durations on high-frequency targets when parafoveal previews had been provided, with first fixation durations and gaze durations amounting to 174 ms and 226 ms, respectively, for high-frequency words and to 182 ms and 272 ms for low-frequency words. Effects of word frequency were reduced when the targets had been masked prior to their fixation, with first fixation durations and gaze durations amounting to 211 ms and 357 ms for high-frequency words and to 191 ms and 371 ms for low-frequency words. Hence, the frequency of occurrence determined the usefulness of its parafoveal preview. Furthermore, posttarget fixations were 20 ms longer when a low-frequency two-character word had been masked prior to its fixation, but no corresponding posttarget spillover of word frequency effects occurred when the target had been visible in the parafovea.

Together the results from these experiments indicate that readers of Chinese text obtain useful graphemic, phonological, and lexical information from a parafoveally visible character or word prior to its subsequent fixation. Use of this information leads to a dramatic reduction in the subsequent character and word viewing duration, presumably because parafoveal previews can be used to activate lexical candidates prior to a word's (character's) fixation. Furthermore, the results revealed considerably larger saccades to parafoveally visible words (2.6 character spaces) than parafoveally visible graphemically similar foils (2.2 character spaces).

TO WHAT EXTENT ARE ON-LINE LINGUISTIC ANALYSES A FUNCTION OF TEXT TYPE?

Readers of Chinese text, just like readers of alphabetic text, need to obtain useful information from an effective area of vision (the perceptual span) that can be used for linguistic computations and the planning of saccades. Our results show that irrespective of text type, the perceptual span is relatively large and encompasses the fixated linguistic unit (the word in alphabetic text and the character or word in Chinese text) and adjacent unit(s) to the right and left of fixation. Irrespective of text type, each individual fixation duration is determined by the temporal visibility of a target and by the ease of its linguistic analysis. Furthermore, the linguistic analysis of a target extends across several fixations. Analyses are initiated before the target is fixated, they are executed while it is fixated, and they may be continued (at least on occasion) after it has been fixated.

Yet, reading of alphabetic and Chinese text also appears to entail the development of script-specific reading skills. Examination of eye movement patterns indicate that short fixation durations of 160 ms or less are considerably more common when Chinese text is read than when alphabetic text is read. Furthermore, regressive saccades to previously read text are less common when Chinese text is read. The time course of information usage also appears to be a function of script type. This applies, in particular, to the use of phonological information, which appears to be used for the accessing of meaning when alphabetic text is read but not when Chinese text is read.

Finally, script type has a profound effect on the specification of saccade size. The length of a parafoveally visible word determines saccade size when alphabetic text is read, but it has virtually no effect when Chinese text is read, presumably because of its lack of visuospatial distinctiveness. Conversely, linguistic properties of a parafoveally visible word (character) determine saccade size when Chinese text is read, but it remains unclear whether it is used when alphabetic text is read.

ACKNOWLEDGMENTS

The research was supported by NIMH grant 503870 and by the Center for Cognitive and Psycholinguistic Studies at SUNY Binghamton.

REFERENCES

Balota, D. A., Pollatsek, A., & Rayner, K. (1985). The interaction of contextual constraints and parafoveal visual information in reading. *Cognitive Psychology, 17*, 364–390.

Beauvillain, C. D., Dore, K., & Baudouin, V. (1996). The "center of gravity" of words: Evidence for an effect of word initial letters. *Vision Research, 36*, 589–603.

Beijing Institute of Language Teaching and Research. (1986). *A frequency dictionary of modern Chinese*. Beijing: Beijing Language Institute Press.

Binder, K. F., Pollatsek, A., & Rayner, K. (1998). *Extraction of information to the left of the fixated word in reading*. Manuscript submitted for publication.

Blanchard, H. E., Pollatsek, A., & Rayner, K. (1989). The acquisition of parafoveal word information in reading. *Perception & Psychophysics, 46*, 85–94.

Crane, H., & Steele, C. M. (1985). Generation V dual Purkinje-image eyetracker. *Applied Optics, 24*, 527–537.

Henderson, J. M., & Ferreira, F. (1990). Effects of foveal processing difficulty on the perceptual span in reading: Implications for attention and eye movement control. *Journal of Experimental Psychology: Learning, Memory, and Cognition, 17*, 417–429.

Hyona, J., Niemi, P., & Underwood, G. (1989). Reading long words embedded in sentences: Informativeness of word halves affects eye movements. *Journal of Experimental Psychology: Human Perception and Performance, 15*, 142–152.

Inhoff, A. W. (1989). Parafoveal processing of words and saccade computation during eye fixations in reading. *Journal of Experimental Psychology: Human Perception and Performance, 15*, 544–555.

Inhoff, A. W. (1990). Integrating information across eye fixations in reading: The role of letter and word units. *Acta Psychologica, 73*, 281–297.

Inhoff, A. W., Briihl, D., & Schwartz, D. (1996). Compound word effects differ in reading, on-line naming, and delayed naming. *Memory & Cognition, 24*, 466–476.

Inhoff, A. W., & Liu, W. (1997). The perceptual span during the reading of Chinese text. In H.-C. Wang (Ed.), *The cognitive processing of Chinese and related Asian languages*. Hong Kong: The Chinese University Press.

Inhoff, A. W., & Liu, W. (1998). The perceptual span and oculomotor activity during the reading of Chinese sentences. *Journal of Experimental Psychology: Human Perception and Performance, 24*, 20–34.

Inhoff, A. W., & Liu, W. (1998). *Eye movements and the use of lexical information during the reading of Chinese high- and low-frequency words.* Manuscript submitted for publication.

Inhoff, A. W., Radach, R., & Starr, M. (1998). *Readers obtain useful linguistic information to the left of fixation.* Manuscript in preparation.

Inhoff, A. W., & Rayner, K. (1986). Parafoveal word processing during eye fixations in reading: Effects of word frequency. *Perception & Psychophysics, 40*, 431–439.

Inhoff, A. W., & Topolski, R. (1994). Use of phonologial codes during eye fixations in reading and in on-line and delayed naming tasks. *Journal of Memory and Language, 33*, 689–713.

Kennison, S. M., & Clifton, C. (1995). Determinants of parafoveal preview benefit in high and low working memory capacity readers: Implications for eye movement control. *Journal of Experimental Psychology: Learning, Memory, and Cognition, 21*, 68–81.

Liu, W., & Inhoff, A. W. (1998). *Eye movements and the use of graphemic and phonological character information during the reading of Chinese sentences.* Manuscript submitted for publication.

McConkie, G. W., Kerr, P. W., Redix, M. D., Zola, D., & Jacobs, A. M. (1989). Eye movement control during reading: II. Frequency of refixating a word. *Perception & Psychophysics, 46*, 245–253.

Morris, R. K., Rayner, K., & Pollatsek, A. (1989). Eye movement guidance in reading: The role of parafoveal letter and space information. *Journal of Experimental Psychology: Human Perception and Performance, 16*, 268–281.

Morrison, R. E. (1984). Manipulation of stimulus onset delay in reading: Evidence for parallel programming of saccades. *Journal of Experimental Psychology: Human Perception and Performance, 10*, 667–682.

O'Regan, J. K. (1992). Optimal viewing position in words and the strategy-tactics theory of eye movements in reading. In K. Rayner (Ed.), *Eye movements and visual cognition: Scene perception and reading.* New York: Springer-Verlag.

Pollatsek, A., Lesch, M., Morris, R., & Rayner, K. (1992). Phonological codes are used in integrating information across saccades in word identification and reading. *Journal of Experimental Psychology: Human Perception and Performance, 18*, 148–162.

Pollatsek, A., & Rayner, K. (1982). Eye movement control in reading: The role of word boundaries. *Journal of Experimental Psychology: Human Perception and Performance, 8*, 817–833.

Radach, R., & Kempe, B. (1993). An individual analysis of initial fixation positions in reading. In G. d'Ydewalle & J. van Rensbergen (Eds.), *Perception and cognition: Advances in eye movement research* (pp. 213–225). Amsterdam: North-Holland.

Rayner, K. (1975). The perceptual span and peripheral cues in reading. *Cognitive Psychology, 7*, 65–81.

Rayner, K., & Duffey, S. D. (1986). Lexical complexity and fixation times in reading: Effects of word frequency, verb complexity, and lexical ambiguity. *Memory & Cognition, 14*, 191–201.

Rayner, K., & Morris, R. (1992). Eye movement control in reading: Evidence against semantic preprocessing. *Journal of Experimental Psychology: Human Perception and Performance, 18*, 163–172.

Rayner, K., & Pollatsek, A. (1981). Eye movement control during reading: Evidence for direct control. *The Quartery Journal of Experimental Psychology, 33a*, 351–373.

Rayner, K., & Pollatsek, A. (1989). *The psychology of reading.* Englewood Cliffs, NJ: Prentice-Hall.

Rayner, K., Well, A. D , Pollatsek, A., & Bertera, J. B. (1982). The availability of useful information to the right of fixation in reading. *Perception & Psychophysics, 31*, 537–550.

Sereno, S. C., & Rayner, K. (1992). Fast priming during eye fixations in reading. *Journal of Experimental Psychology: Human Perception and Performance, 18*, 163–172.

Underwood, G., Bloomfield, R., & Clews, S. (1988). Information influences the pattern of eye fixations during sentence comprehension. *Perception, 17*, 267–278.

Underwood, G., Clews, S., & Everatt, J. (1990). How do readers know where to look next? Local information distribution influences eye fixations. *Quarterly Journal of Experimental Psychology, 42a*, 39–65.

Vitu, F., O'Regan, J. K., Inhoff, A. W., & Topolski, R. (1995). Mindless reading: Eye movement characteristics are similar in scanning strings and reading text. *Perception & Psychophysics, 57*, 352–365.

Vitu, F., O'Regan, J. K., & Mittau, M. (1990). Optimal landing position in reading isolated and continuous text. *Perception & Psychophysics, 47*, 583–600.

Yin, B., & Rohsenow, J. S. (1994). *Modern Chinese characters*. Beijing, China: Sinolingua.

CHAPTER

12

Word Recognition During the Reading of Chinese Sentences: Evidence From Studying the Word Superiority Effect

Jenn-Yeu Chen
National Chung-Cheng University, Taiwan

ABSTRACT

Word segmentation and word recognition were investigated by an examination of the word superiority effect in the context of a Chinese sentence. The same basic task was employed in four experiments, in which, on each trial, subjects searched for a predesignated character in a sentence. The target character was embedded in a two-character word, a two-character nonword, a string of asterisks, or a string of scrambled characters. Where there was a word or a nonword, it was manipulated to fit or not fit the meaning of the sentence. The results show that detection of a character was fastest when it was embedded in a string of asterisks, was faster when it was part of a word, whether or not the word fit the meaning of the sentence, and was slowest when it was part of a nonword or was among a string of scrambled characters. The word superiority effect was evident in the comparison of the word conditions with the scrambled-character condition. The presence of a word superiority effect, whether or not the word fit the meaning of the sentence, indicated that the sentence was not normally read; word segmentation and recognition nevertheless took place during character search. Further, such processes must have occurred fairly automatically, given that they were never the primary goal of the task.

One pressing question for students of Chinese reading is how words are recognized during the reading of a Chinese sentence. For our purpose, the question does not relate so much to the matter of lexical access, as is typically the case in English reading, as to the matter of word segmentation that logically should precede lexical access. The reason that word segmentation is an issue in the reading of Chinese sentences has to do with the way a Chinese sentence is printed. A Chinese sentence is printed by characters, not by words. That is, there are spaces between characters, but no extra spaces between words. As a result, a Chinese reader must perform word segmentation before she or he can proceed with lexical access and other higher level processes in sentence reading.

In this chapter, we address the question by studying the word superiority effect in a sentence, as was first demonstrated by H.-C. Chen (1987). Based on a series of experiments that examine the nature of this effect, we conclude that word segmentation takes place automatically during character reading.

The *word superiority effect* and its sister effect, the *word inferiority effect*, are two seemingly contradictory phenomena that coexist in the reading literature. In one, letter search is found more accurate or faster when a target letter appears in a word than when it appears in a nonword. This is the word superiority effect (Kreuger, 1992; Reicher, 1969; Wheeler, 1970). In the other, the opposite result is observed, that is, letter search tends to be less accurate or slower when the target letter is part of a word than when it is part of a nonword. This is the word inferiority effect (Healy, 1976, 1980; Healy & Drewnowski, 1983).

Similar effects were demonstrated in Chinese with the same procedures. For example, employing the tachistoscopic presentation, Cheng (1981) showed that a character target could be detected more accurately when it was part of a word than when it was part of a nonword (word superiority). In addition, a radical was better detected when it was embedded in a legal character than when it was embedded in a pseudocharacter or an illegal character (character superiority). By contrast, H.-C. Chen (1984) observed a character inferiority effect when a radical was to be detected in a text. That is, more errors were found when the radical formed part of a legal character than when it formed part of an illegal character.

An interesting finding by H.-C. Chen (1987) is a word superiority effect, not an inferiority effect, when he asked his subjects to search for a target character in a sentence. Because a sentence also involves multiple words much like a text, a word inferiority effect should be expected. Therefore, H.-C. Chen's finding constitutes a surprise, and it serves as a point of departure in our investigation.

A major theory of the word superiority effect and the word inferiority effect is the unitization hypothesis formulated by Healy and colleagues (Drewnowski & Healy, 1977; Healy, 1994; Healy & Drewnowski, 1983). According to this hypothesis, during reading of a text, several units are being processed simultaneously at different levels. Partial information from the subordinate units can add up to activate a unit so that the unit may be processed to its completion before its subordinate units. Once a unit at a given level is completely processed (i.e., identified), the reader moves on to the next unit at that level, leaving incomplete the processing of the units at its subordinate levels. Because the comprehension processes in text reading typically "pull" the reader onward, searching a letter in a text suffers from a word inferiority effect. On the other hand, if only a word is to be read and searched, processing of the letters can be completed and it can benefit from the processing of the word, resulting in a word superiority effect.

The unitization hypothesis, in its present form, does not appear capable of explaining the word superiority effect in the context of a Chinese sentence. The reason is that processing of a sentence should presumably invoke comprehension that will pull the reader's attention onward once a word has been identified. Processing of the characters that make up the word will, then, be terminated prematurely and, according to the unitization hypothesis, a word inferiority effect should result.

There are a number of procedural differences that the unitization hypothesis may resort to in order to account for H.-C. Chen's (1987) finding. For example, it may argue that a word superiority effect is typically observed with a tachistoscopic presentation method, and since this method was used in H.-C. Chen's study to show the sentences, the same kind of effect was observed. However, this argument cannot be correct. Employing the tachistoscopic presentation method, Healy, Oliver, and McNamara (1987) presented a text, either one or four words at a time, while asking their subjects to detect a predesignated letter. They observed a word superiority effect with one-word presentation, but a word inferiority effect with four-word presentation. Because a sentence display is also a multiple-word display, the prior argument fails to explain why a word superiority effect was found in H.-C. Chen's study, whereas a word inferiority effect was observed in Healy et al.'s study. Thus, the tachistoscopic presentation per se cannot be the sole cause of the word superiority effect found in H.-C. Chen's study.

The unitization hypothesis may also argue that because the word inferiority effect is typically observed with a text, whereas the word superiority effect is typically found with a word, perhaps it is the size of the display that matters. However, in Healy, Oliver, and McNamara's (1987) study, both the paper condition (where the entire text was in view) and the four-word condition gave rise to a word inferiority effect. Furthermore, Kreuger (1970) observed a word superiority effect even though he used long word lists, each containing 25 words arranged in a 5×5 matrix.

In an attempt to reconcile Kreuger's finding, Healy (1994, Note 6) recently proposed that the critical factor was the emphasis of the task. In her task, reading was emphasized, whereas in Kreuger's (1970) task, searching was stressed. However, since the goal of both tasks was to detect a predesignated letter, therefore both requiring searching of the stimulus display, it is not clear how the two tasks could differ in emphasis and how the alleged differential emphases could be independently assessed and verified. The trouble is heightened when H.-C. Chen's (1987) finding is placed under similar consideration. Unlike the word lists used by Kreuger, sentences were the stimuli in H.-C. Chen's task. It is conceivable that reading could have taken place when H.-C. Chen's subjects were trying to detect a target letter in the sentence. Perhaps the most troublesome for this explanation, though, is that Healy (1976) herself observed a word inferiority effect when a scrambled-words text was used, where normal reading could not have occurred.

Instead of appealing to the differential emphases in tasks, the unitization hypothesis may be modified to adopt the following explanation to account for H.-C. Chen's (1987) finding. In H.-C. Chen's study, a different target character was designated before each sentence was shown. This procedure is the same as that typically used to demonstrate a word superiority effect, but it is different from the procedure typically used to demonstrate a word inferiority effect. In the latter case, a single predesignated target is searched in a text. It is possible that when there is only a single target to be searched in an entire text, the momentum of reading forward can easily be induced, thus creating a situation ideal for demonstrating a word inferiority effect. On the other hand, when a different target is to be searched for each sentence, showing the targets can easily disrupt the "pull" of the sentences, and the momentum of reading forward cannot be maintained. Under this circumstance, the subjects in H.-C. Chen's study may have adopted the same search strategy as they would do should they search a word only. Remember that in a typical experiment that demonstrates the word superiority effect, a different target is also designated each time a word or a nonword is to be searched, exactly the same procedure as used in H.-C. Chen's study. This explanation implies that H.-C. Chen's subjects did not really read a sentence. Rather, they only scanned it for the purpose of finding the designated target character, just as they would have done should a word or a nonword alone have been shown and have to be searched. In other words, the word superiority effect in a sentence context might have been driven by the same mechanism as the word superiority effect in a word-alone context.

Although this explanation seems plausible, one puzzling point remains. If the subjects did not read a sentence, but only scanned it character by character, how could there be a word superiority effect? Bear in mind that words are not visually marked in a Chinese sentence, so that the word giving rise to the superiority effect was "hidden" and must be recognized first. The fact that there was the word superiority effect indicates that the word must have been recognized. If we are to accept the prior argument, we must be prepared to entertain the possibility that words in a sentence can be recognized even though the sentence is never really read. Later in this chapter, we argue, based on our experimental investigation, that this may well be the case. But, before presenting our experimental findings, we should like to evaluate H.-C. Chen's (1987) own explanation of his finding.

H.-C. Chen (1987) explained his result by pointing to the characteristic difference between an English letter and a Chinese character. An English letter is not a meaningful unit, whereas a Chinese character is. Because reading a sentence invokes semantic processes of all the meaningful units in the sentence, processing of a word is carried out at the same level (semantic) as processing of the characters in Chinese, and therefore, the latter can benefit the former. In English, on the other hand, processing of a word is carried out at a different level than processing of the letters; therefore, letter processing is not aided by word processing. To the contrary, it is hurt by the dominating comprehension processes in sentence reading.

H.-C. Chen's (1987) account, which may be called "the shared processing hypothesis," can explain why, under similar conditions (i.e., a text and a sentence), a word inferiority effect was observed in an English letter detection task, whereas an opposite effect was found in the Chinese character detection. The hypothesis, however, may be challenged on several grounds. First, as H.-C. Chen himself demonstrated in another study (1984), there is a character inferiority effect in Chinese when a radical is the target of detection in a running text. The radical used in that study ("日") can be a stand-alone character and, like many other radicals in Chinese, carries meaning (*sun* or *day*). Thus, the explanation used to account for the finding in character detection contradicts the finding in radical detection.

It can be argued that, even though the radical "日" has meaning, that meaning is not shared by its superordinate unit, the character "是," which is equivalent to the verb *be* in English. Accordingly, the processing of "日" cannot benefit from the processing of "是." However, the shared processing hypothesis is not salvaged with this modification. This is because a similar relationship also exists between the words and their composing characters, which were used to demonstrate the word superiority effect in H.-C. Chen's (1987) study. For example, "牡丹" (a type of flower) bears no obvious semantic relationship with "牡" or with "丹." Similarly, the meaning of "消息" is opaque with respect to the meanings of its constituents "消" and "息." Yet, detection of "丹" was faster and more accurate when it appeared in "牡丹" than when it appeared in "空丹," a nonword in Chinese.

The shared processing hypothesis can also be challenged on the ground that it makes the unverified assumption that sentences are automatically processed and processed to the semantic level, even though the task is to search for a character. Because the sentences in H.-C. Chen's (1987) study were presented in the same way as the single words in a typical experiment that demonstrates a word superiority effect, it is possible, as was alluded to earlier, that the same strategy was adopted by H.-C. Chen's subjects to simply search the sentence character by character without attempting to comprehend it.

Last, the shared processing hypothesis may be challenged on two methodological grounds. First, the nonwords used in H.-C. Chen's (1987) experiment inevitably disrupted the well-formedness of the sentences so that the effect of word was confounded with the effect of sentence. In other words, the observed word superiority effect may, in part or in whole, be a sentence superiority effect. Second, the word superiority effect was calculated against a nonword condition. There was no neutral condition in the study. This makes the word–nonword effect dubious. It is not clear whether the effect is due to the lengthening of processing time in the nonword condition, or due to the shortening of processing time in the word condition. In other words, the effect could be a word superiority effect, as H.-C. Chen interpreted it, or it could be a nonword inferiority effect. These two methodological problems put in doubt the validity of the so-called word superiority effect and directly affect the integrity of the shared processing hypothesis.

To sum up, we hypothesized that when subjects attempted to detect a character in a sentence, they did not read the sentence, but searched it character by character. This occurred not so much because the task emphasis was different as because the task included multiple trials, each of which involved a different character and a different sentence. Further, although the primary processing was at the character level, word segmentation and recognition nevertheless occurred, producing the word superiority effect. The major evidence we sought in the experiments was the presence of word superiority effect even when the word did not fit the sentence.

We started by dealing with the methodological problems in H.-C. Chen's (1987) study. Experiment 1 was a replication of that study to make sure that the effect could be repeated. Experiment 2 included a neutral condition (a character embedded in a string of asterisks) to see whether it was a word superiority effect or a nonword inferiority effect. Experiment 3 modified the word condition of H.-C. Chen's experiment in such a way that the word was semantically incongruent with the sentence in which it was embedded. This word/nonsentence condition should eliminate the word superiority effect if the sentence confounding had indeed existed in H.-C. Chen's study. This condition would also provide the evidence for testing our hypothesis. Experiment 4 included a different neutral condition (a target character embedded in a string of scrambled characters) to further clarify the nature of the word superiority effect.

EXPERIMENT 1

Because the word superiority effect that H.-C. Chen (1987) observed is an unexpected finding, we need to make sure that it is repeatable. Thus, Experiment 1 was an exact replication of H.-C. Chen's experiment. However, only the whole-sentence presentation condition from his experiment was included. The method is described again here because the original experiment was described in Chinese.

Method

Subjects. Twenty-eight students from the National Chung-Cheng University participated in this experiment. They all had normal or corrected-to-normal vision.

Materials. The materials consisted of 48 Chinese two-character words, half of which were high-frequency words (at least 125 per million occurrences, according to Liu, Chuang, & Wang, 1975; the average frequency was 238), and the other half low-frequency words (1 per million). One sentence was constructed from one of these words. Each sentence from the high-frequency list contained 9 to 12 characters, with an average of 10.6 characters per sentence. The high-frequency words were located, on the average, at the 3.7th character position of the sentences. The sentences from the low-frequency list were matched with the same characteristics as the high-frequency sentences. The two sets of sentences were also equated in readability. This was accomplished in H.-C. Chen's (1987) study through the judgment of six subjects.

For each word, a nonword was also constructed by replacing the first or the second character of a word with the corresponding character in another word. These nonwords were inserted into the same set of sentence frames used by the words. Thus, there were a total of 48 sentences with words and 48 sentences with nonwords. Half of these sentences were designated as positive sentences, that is, sentences containing a target character, and the other half as negative sentences. These materials can be found in Appendixes A and B (from H.-C. Chen, 1987).

Equipment. The experiment was conducted on a 80486 IBM-compatible personal computer. Each character was made of 16×16 white pixels, each measured 0.4×0.9 cm in area, and 0.35 cm separated two adjacent characters. During the experiment, the subjects sat 50 cm in front of the computer monitor. The room light was turned off, with some illumination reflected from the hallway.

Procedure. The experiment employed a completely within-subjects design. The two variables were: (a) wordness (word or nonword), and (b) word frequency (high or low). The 96 sentences were randomly ordered and presented one at a time. Each trial started with an asterisk shown for 500 ms. A target character followed and stayed for 1500 ms. Then, a string of asterisks appeared for 500 ms, followed by a sentence. The subjects' task was, on seeing the sentence, to decide if the target character appeared in the sentence and to press a key for yes or another key for no. The response keys were counterbalanced across subjects. Each subject received 10 practice trials before proceeding to the experimental trials.

Results and Discussion

Percentage of accuracy was first computed for each subject in each of the four conditions, for the positive trials only. These percentages were then subjected to a repeated-measures analysis of variance. None of the main effects or the interaction effect were significant: $F(1, 27) = 1.03$, $MSe = .00626$, $p = .3201$ for the wordness main effect; $F(1, 27) = .03$, $MSe = .00671$, $p = .8625$ for the frequency main effect; $F(1, 27) = 3.47$, $MSe = .00591$, $p = .0734$ for the interaction effect. The means and standard deviations of these percentages are presented in Table 12.1.

Next, a median response time (RT) was computed for each subject in each condition, for the positive and correct trials. Medians, and not means, were used to avoid the influence of extreme values that were sometimes present in reaction times. These median RTs were also subjected to a repeated-measures analysis of variance. The wordness main effect was significant, showing a word superiority effect: $F(1, 27) = 7.56$, $MSe = 4301.52$, $p = .0105$. The frequency main effect was also significant; that is, character detection was faster in a high-frequency word than in a low-frequency word: $F(1, 27) = 5.08$, $MSe = 5104.54$, $p = .0326$. The Wordness × Frequency interaction

TABLE 12.1
Means and Standard Deviations (in parentheses) of the Proportion of Accuracy (1st row) and the Median Character Search Times (in ms; 2nd row) for the Wordness × Frequency Conditions of Experiment 1

	High Frequency		*Low Frequency*		*Marginal*	
	Mean	*(STD)*	*Mean*	*(STD)*	*Mean*	*(STD)*
Word/Sentence	0.89	(0.11)	0.92	(0.08)	0.90	(0.09)
	573.71	(142.23)	584.20	(125.68)	578.96	(133.09)
Nonword/	0.93	(0.07)	0.90	(0.09)	0.92	(0.08)
Pseudosentence	588.27	(131.89)	638.21	(178.47)	613.24	(157.52)
Marginal	0.91	(0.09)	0.91	(0.09)		
	580.99	(136.10)	611.21	(155.35)		

effect was not significant: $F(1, 27) = 2.42$, $MSe = 4419.44$, $p = .1318$. The means and standard deviations of these median RTs are also displayed in Table 12.1.

The RT results from Experiment 1 replicated H.-C. Chen's (1987) findings, indicating that the word superiority effect with a sentence context is a robust effect. The next few experiments were, then, designed to clarify the nature of this effect. These experiments all employed the same equipment and very similar design and procedures as Experiment 1. They all included two within-subjects variables: wordness and word frequency. The same materials were used, but different wordness conditions were created for different experiments. Moreover, instead of a mixed presentation as in Experiment 1, the different wordness conditions were presented in blocks. The order of presentation was always counterbalanced across subjects.

EXPERIMENT 2

Experiment 2 included a neutral condition for the wordness variable in order to find out whether the observed effect was a word superiority effect or a nonword inferiority effect. H.-C. Chen (1987) calculated the word superiority effect against a nonword condition. It is not clear whether the effect was due to the lengthening of processing time in the nonword condition, or due to the shortening of processing time in the word condition. A neutral condition that serves as a control is needed for a proper interpretation of the effect. In the Reicher (1969) and the Wheeler (1970) studies, a letter-only condition served as the control. In the present experiment, instead of a character only, we embedded the character in a string of asterisks so that search might be equally required in the neutral condition as in the word and the nonword conditions.

Method

Subjects were 24 college students from the National Chung-Cheng University. They all had normal or corrected-to-normal vision. The neutral condition was created by replacing all the nontarget characters in a sentence with asterisks. The three conditions of the wordness variable were presented in separate blocks. Each sentence in Experiment 1 was used twice here for each condition, once for the positive and once for the negative trials. Thus, for the word condition there were 24 positive high-frequency sentences, 24 positive low-frequency sentences, 24 negative high-frequency sentences, and 24 negative low-frequency sentences. The 96 sentences were randomly ordered and presented one at a time to the subjects. The sentences in the nonword condition and the asterisk strings in the neutral condition were arranged similarly.

Results and Discussion

Analyses focused on the RT data, although the means and standard deviations of both accuracy and RT are presented (see Table 12.2). The RT data were analyzed in the same way as those in Experiment 1. There was a significant wordness main effect: $F(2, 46) = 35.85$, $Mse = 8497.75$, $p < .0001$. The frequency main effect was not significant: $F(1, 23) = 1.90$, $MSe = 1607.73$, $p = .1812$; nor was the two-way interaction effect: $F(2, 46) = .79$, $MSe = 2855.69$, $p = .4587$. Post hoc comparisons showed that the mean RT for the word condition was shorter than that for the nonword condition: $F(1, 23) = 4.18$, $p = .0526$, and that the mean RT for the neutral condition was shorter than that for the word condition: $F(1, 23) = 39.34$, $p < .0001$.

The word superiority effect in a sentence context was replicated once more in this experiment. However, this effect was computed against the nonword condition. When the neutral condition served as the control, there was no longer evidence of a word superiority effect. Whether this word–nonword difference reflected a facilitatory effect or an inhibitory effect or a combination of both could not be determined yet in this experiment. This is because the response times for the word and the nonword conditions were both longer than those for the neutral condition. The longer response times in the nonword condition relative to the neutral may be interpreted as an inhibitory effect. However, the longer response times in the word condition relative to the neutral may not be interpreted the same way. It is possible that the longer response times in the word condition reflected both an inhibitory effect and a facilitatory effect. The inhibitory effect may have resulted from the delay in searching through a multiple-word string as opposed to a string of asterisks. The former provided a heterogeneous context, the latter a homogeneous context. The facilitatory effect, on the other hand,

TABLE 12.2
Means and Standard Deviations (in parentheses) of the Proportion of Accuracy (1st row) and the Median Character Search Times (in ms; 2nd row) for the Wordness × Frequency Conditions of Experiment 2

	High Frequency		*Low Frequency*		*Marginal*	
	Mean	*(STD)*	*Mean*	*(STD)*	*Mean*	*(STD)*
Word/	0.96	(0.05)	0.96	(0.04)	0.96	(0.05)
Sentence	666.63	(121.44)	665.85	(133.67)	666.24	(126.34)
Nonword/	0.93	(0.06)	0.96	(0.04)	0.95	(0.05)
Pseudosentence	725.15	(92.85)	700.08	(111.86)	712.61	(102.48)
Neutral/Asterisks	0.96	(0.03)	0.96	(0.03)	0.96	(0.03)
	558.31	(85.88)	556.50	(75.93)	557.41	(80.20)
Marginal	0.95	(0.05)	0.96	(0.04)		
	650.03	(121.67)	640.81	(124.55)		

may truly be an effect of the word in which a target character was embedded. A neutral condition using a heterogeneous context that matches the word condition may be a more appropriate control than the one used in the present experiment. This issue is pursued again in Experiment 4.

Aside from the significant wordness main effect, the present experiment observed no significant frequency main effect or significant wordness by frequency interaction. A discussion of these null results is deferred to the general discussion section, where the results related to these two effects from the other experiments can be considered at the same time.

In Experiment 3 we dealt with the sentence confound that might have contaminated the observed word superiority effect in a sentence context.

EXPERIMENT 3

We noted earlier that the nonwords in H.-C. Chen's (1987) study inevitably disrupted the well-formedness of the sentences so that the word–nonword effect was confounded with the sentence–nonsentence effect. In the present experiment, we aimed to find out whether the word–nonword effect could still be observed when the sentence confound had been removed. Since a nonword is necessarily associated with a nonsentence, we modified the word condition so that a word was also embedded in a nonsentence. This was accomplished by simply replacing the target word in one sentence with the target word in another sentence. In addition, because the sentences originally designed for the nonword condition are disrupted by only one character, we replaced the target nonword in one sentence with the target nonword in another sentence to ensure equivalent "ill-formedness" of the sentences in the word and the nonword conditions.

Method

Subjects were 24 college students from the National Chung-Cheng University. They all had normal or corrected-to-normal vision. Except for the changes made to the word and the nonword conditions, the rest of Experiment 3 was identical to Experiment 2. The three wordness conditions were: word/nonsentence, nonword/nonsentence, and neutral.

Results and Discussion

Analyses focused on the RT data, although the means and standard deviations of both accuracy and RT are presented (see Table 12.3). The RT data were analyzed in the same way as in previous experiments. Three subjects' data were removed from analyses because of extreme values (more than three standard deviations away from the mean) in the word and the nonword conditions. The analyses revealed a significant wordness main effect: F(2,

TABLE 12.3
Means and Standard Deviations (in parentheses) of the Proportion of Accuracy (1st row) and the Median Character Search Times (in ms; 2nd row) for the Wordness × Frequency Conditions of Experiment 3

	High Frequency		*Low Frequency*		*Marginal*	
	Mean	*(STD)*	*Mean*	*(STD)*	*Mean*	*(STD)*
Word/	0.97	(0.03)	0.98	(0.03)	0.97	(0.03)
Nonsentence	689.98	(127.36)	710.69	(143.95)	700.33	(134.65)
Nonword/	0.95	(0.06)	0.94	(0.04)	0.95	(0.05)
Nonsentence	769.26	(147.05)	709.81	(145.59)	739.54	(147.62)
Neutral/Asterisks	0.98	(0.03)	0.96	(0.04)	0.97	(0.04)
	586.10	(95.17)	586.95	(105.13)	586.52	(99.04)
Marginal	0.97	(0.04)	0.96	(0.04)		
	681.78	(144.38)	669.15	(143.25)		

40) = 34.46, MSe = 7698.63, $p < .0001$. The frequency main effect was not significant: $F(1, 20) = 1.37$, $MSe = 3666.52$, $p = .2556$. The Wordness × Frequency interaction effect was significant: $F(2, 40) = 5.65$, $MSe = 3241.46$, $p = .0069$. Post hoc comparisons showed that the mean RT for the word condition was shorter than that for the nonword condition: $F(1, 20) = 3.54$, $p = .0744$, and that the mean RT for the neutral condition was shorter than that for the word condition: $F(1, 20) = 42.97$, $p < .0001$. Analyses of the simple interaction effects indicated an equivalent word-neutral difference in the high- and the low-frequency conditions: $F(1, 20) = 1.06$, $MSe = 7842.73$, $p = .3164$. The word–nonword differences, however, were not equivalent in the high- and the low-frequency conditions: $F(1, 20) = 7.84$, $MSe = 17204.58$, $p = .0110$. An examination of the simple effect indicated that the word–nonword difference was significant in the high-frequency condition, but not in the low-frequency condition.

The results from this experiment show that the word–nonword difference persisted even when the sentence confound had been removed. Thus, the word superiority effect in a sentence context is truly an effect of the word, and not an effect of the sentence. Further evidence supporting this conclusion was obtained when the results of Experiment 2 and Experiment 3 were compared. These comparisons revealed no significant differences between the two experiments in any of the effects. Thus, although character search was carried out in a sentence, the nature of the sentence had little influence on the search; at the same time, it was found that the immediate context surrounding the character (i.e., the word) had a significant effect on the search. This suggests that the subjects did not attempt to comprehend the sentence; nonetheless, because a Chinese sentence is printed by characters, not by words, given that words had an effect on characters, word segmentation and word recognition must have occurred automatically in the course of character search.

A remaining puzzle to solve is why character search in the neutral condition was faster than that in the word condition. This was dealt with in Experiment 4.

EXPERIMENT 4

The puzzle may be solved by invoking two equally plausible explanations. One is that an asterisk was easier to process than a character. Since most of the string in the neutral condition was comprised of asterisks, overall search time would be less in this condition than in the word condition where the entire string to be searched contained characters. The other explanation is that, whereas the entire string of characters needed to be searched in the word condition, only two needed to be searched in the neutral condition. This is because the asterisks in the neutral condition formed a homogeneous context that made it possible for the subjects to quickly home in on the characters in the string. In both explanations, differential perceptual complexity was assumed to be the cause of the different search times in the neutral and the word conditions.

The perceptual complexity hypothesis predicts that character search would be slowed down in the neutral condition if asterisks were replaced by scrambled characters. More specifically, search time in this modified neutral condition would be slower than that in the word condition, but would be as slow as that in the nonword condition. The latter prediction was based on the findings of Experiments 2 and 3, which indicated that detection of the target character was affected by the local context (i.e., the word) rather than by the global context (i.e., the entire string). Since the modified neutral condition and the nonword condition were identical with respect to the local context that surrounded the target, the time for detecting the target should also be identical in the two conditions.

In the present experiment, the word/sentence condition from Experiment 2 and the word/nonsentence condition and the nonword/nonsentence condition from Experiment 3 were included. In addition, a major change was made to the neutral condition. Instead of asterisks, scrambled characters were used. This guaranteed equivalent perceptual complexity in the neutral and the other conditions. Accordingly, the results of this experiment should be consistent with the prediction of the perceptual complexity hypothesis explicated in the last paragraph. In addition, based on the findings of Experiments 2 and 3, no difference was expected between the two word conditions.

Method

Subjects were 32 college students from the National Chung-Cheng University and the National Cheng-Kung University. They all had normal or corrected-to-normal vision. Four conditions were included in the wordness variable:

word/sentence, word/nonsentence, nonword/nonsentence, and neutral. The neutral condition was created by taking the sentences in the word/sentence condition and randomly rearranging the characters within a sentence. The position of the target character was kept intact. The rest of the experiment was identical to Experiments 2 and 3.

Results and Discussion

Analyses focused on the RT data, although the means and standard deviations of both accuracy and RT are presented (see Table 12.4). These data were analyzed the same way as in previous experiments. Two subjects' data had to be discarded because of a corruption in their data files. An additional subject's data were removed from further analyses due to extreme values (more than three standard deviations away from the mean). The analyses revealed a significant wordness main effect: $F(3, 84) = 4.64$, $MSe = 13467.32$, $p = .0047$. The frequency main effect was not significant: $F(1, 28) = .06$, $MSe = 1963.19$, $p = .8160$. Neither was the Wordness × Frequency interaction effect: $F(3, 84) = 1.78$, $MSe = 2490.42$, $p = .1569$. Post hoc comparisons showed that the mean RT for the two word conditions was shorter than that for the nonword and the neutral condition: $F(1, 28) = 16.86$, $MSe = 20972.14$, $p = .0003$, whereas the mean RTs between the two word conditions or the mean RTs between the nonword and the neutral condition were not significantly different from each other: Fs < 1.

The results of this experiment are consistent with the perceptual complexity hypothesis. That is, searching for a target character in a string of asterisks was easier than in a string of characters because the former was perceptually simpler than the latter. When perceptual complexity was

TABLE 12.4
Means and Standard Deviations (in parentheses) of the Proportion of Accuracy (1st row) and the Median Character Search Times (in ms; 2nd row) for the Wordness × Frequency Conditions of Experiment 4

	High Frequency		*Low Frequency*		*Marginal*	
	Mean	*(STD)*	*Mean*	*(STD)*	*Mean*	*(STD)*
Word/	0.95	(0.05)	0.96	(0.04)	0.95	(0.05)
Sentence	674.00	(156.75)	709.60	(228.68)	691.80	(195.20)
Word/	0.96	(0.05)	0.97	(0.04)	0.96	(0.05)
Nonsentence	683.05	(185.98)	680.27	(163.55)	681.66	(173.64)
Nonword/	0.90	(0.16)	0.93	(0.17)	0.92	(0.17)
Nonsentence	714.92	(163.77)	722.28	(166.44)	718.60	(163.75)
Neutral/Scrambled	0.94	(0.07)	0.95	(0.06)	0.95	(0.06)
characters	749.60	(151.05)	727.17	(159.25)	738.38	(154.30)
Marginal	0.94	(0.10)	0.95	(0.10)		
	705.39	(165.54)	709.83	(180.36)		

equated by replacing the asterisks in the neutral condition with scrambled characters, character detection became equally fast in the neutral and the nonword conditions, and both were slower than character detection in the two word conditions. Thus, it may be concluded that the word superiority effect observed in H.-C. Chen's (1987) study, as well as in the present study, reflected a facilitatory effect, not an inhibitory effect.

The results also replicated the findings of Experiments 2 and 3. Together, they suggest that while searching for a character in a sentence, subjects did not read the sentence in a normal way, but processed it character by character. Word segmentation nevertheless occurred automatically as a result of processing the characters. Furthermore, both the words and the characters are processed to the point of identification. It appears, then, that the word superiority effect in the sentence context is governed by the same mechanism as the word superiority effect in the word alone context. The only difference between the two situations is the amount of perceptual processing required of each. With the sentence context, more characters are to be searched and the perceptual demand is high. With the word-alone context such as used in Cheng's (1981) experiment, only two characters need to be searched, and the perceptual demand is minimal.

GENERAL DISCUSSION

In this chapter we investigated word segmentation and word recognition in a Chinese sentence by examining the word superiority effect in the context of a sentence, as originally documented by H.-C. Chen (1987). With proper controls, the results of four experiments confirmed that the effect was facilitatory rather than inhibitory and that it was an effect of the word, not the sentence. These results, in turn, revealed that processing of the characters in a sentence could lead to an automatic word segmentation and word recognition.

Whether a higher level of processing (i.e., syntactic and semantic processing at the sentence level) might also be possible during character search is a question that is worth further investigation. If it is possible, it will present a serious problem for the unitization model because the model assumes that the word inferiority effect is the consequence of normal reading, whereas the word superiority effect is the consequence of letter or character search in the absence of normal reading.

This conclusion and discussion are based on the findings of the wordness main effect. Our experiments also manipulated the frequency variable. But, except for Experiment 1, no reliable frequency effect was found; in addition, aside from Experiment 3, no reliable wordness by frequency effect was observed. The literature shows clearly that word frequency has an effect on letter or character detection, producing a high-frequency advantage comparable to the word superiority effect and a high-frequency disadvantage comparable to the word inferiority effect. The frequency-related effects are also an important prediction of Healy's unitization theory (Drewnowski & Healy, 1977; Healy, 1994; Healy & Drewnowski, 1983) and McClelland and Rumelhart's (1981) interactive activation model. So, why did we fail to find the effects?

One reason might be the procedural difference between Experiment 1 and the other experiments. Trials from the different wordness conditions were intermixed in Experiment 1, but were separated in blocks in the others. Another reason might have to do with the fact that word frequency and character frequency were confounded in the experiments. That is, high-frequency words were made up of high-frequency characters, and low-frequency words were made up of low-frequency characters. This confound was noted in H.-C. Chen (1987). Unfortunately, the exact mechanisms that produced the inconsistent frequency-related effects across experiments remain unclear at this point and must await future investigation.

APPENDIX A

A Sample of Stimulus Sentences Used in the Experiments

High Frequency	Word/ Sentence	1st Position	Positive	地	牛頓發現地球有特殊引力	Exp. 1, 2, 4
			Negative	運	他是衛生組織的負責人	
		2nd Position	Positive	準	模特兒都有標準的身段	
			Negative	師	運動員有體育精神是應該的	
	Nonword/ Pseudo-sentence	1st Position	Positive	歷	這古物很有歷女價值	Exp. 1, 2
			Negative	山	他是一個理婦主義者	
		2nd Position	Positive	界	和平是全稅界的人都渴望的	
			Negative	動	他是本牡司的總經理	
	Word/ Non-sentence	1st Position	Positive	精	在香港賣精神是違法的	Exp. 3, 4
			Negative	標	他是一個山莊主義者	
		2nd Position	Positive	莊	那位山莊快要結婚了	
			Negative	然	我的精神是一名留學生	
	Nonword/ Non-sentence	1st Position	Positive	老	玩砌老檬是很有趣的遊戲	Exp. 3, 4
			Negative	同	用手很難捉公鮮因爲它很滑	
		2nd Position	Positive	司	適量的牡司可增強心臟功能	
			Negative	式	他把幾個稅界放在冰箱中	
	Neutral/ Asterisks	1st Position	Positive	世	＊＊＊＊世＊＊＊＊＊＊＊	Exp. 2, 3
			Negative	老	＊＊方＊＊＊＊＊＊＊＊＊	
		2nd Position	Positive	德	＊＊＊＊＊德＊＊＊＊＊	
			Negative	然	＊＊＊＊然＊＊＊＊＊＊＊	
	Neutral/ Scrambled characters	1st Position	Positive	同	居又是學既同鄰是和他我	Exp. 4
			Negative	事	球現殊有地牛發頓力特引	
		2nd Position	Positive	師	的是一師學生名我老留	
			Negative	準	美個元一戶他儲口有蓄	

Note. From Chen (1987). Reprinted with permission.

APPENDIX B

A Sample of Stimulus Sentences Used in the Experiments

Low Frequency	Word/ Sentence	1st Position	Positive	口	女士們把口紅塗在嘴唇上	Exp. 1, 2, 4
			Negative	奇	捉拿盜匪是捕快的職責	
		2nd Position	Positive	數	他把所有的奇數塡在空格上	
			Negative	木	捉拿盜匪是捕快的職責	
	Nonword/ Pseudo-sentence	1st Position	Positive	奇	他把所有的奇料塡在空格上	Exp. 1, 2
			Negative	吧	那位伴準快要結婚了	
		2nd Position	Positive	數	他把所有的太數塡在空格上	
			Negative	稿	在香港賣精竹是違法的	
	Word/ Non-sentence	1st Position	Positive	捕	這幾天的捕快有很大差距	Exp. 3, 4
			Negative	積	人在伴郎之前一律平等	
		2nd Position	Positive	流	他認爲暖流是要自己創造	
			Negative	枝	運動員有體育吧女是應該的	
	Nonword/ Non-sentence	1st Position	Positive	奇	在古時候奇料是一種藥材	Exp. 3, 4
			Negative	洋	這幾天的檀驗有很大差距	
		2nd Position	Positive	班	這些美麗的故班都很名貴	
			Negative	女	他愛在空丹界中尋找樂趣	
	Neutral/ Asterisks	1st Position	Positive	荔	＊＊＊＊＊荔＊＊＊＊＊＊	Exp. 2, 3
			Negative	暖	＊＊爆＊＊＊＊＊＊＊	
		2nd Position	Positive	盤	＊＊＊＊＊盤＊＊＊＊＊	
			Negative	快	＊＊＊＊＊＊流＊＊＊	
	Neutral	1st Position	Positive	郎	那要婚郎結伴快位了	Exp. 4
			Negative	奇	涯已厭倦生女她的吧	
		2nd Position	Positive	鮮	頭邊石海鮮在遍佈苔上	
			Negative	爆	空數有上的塡他在格把所奇	

Note. From Chen (1987). Reprinted with permission.

REFERENCES

Chen, H.-C. (1984). Detecting radical component of Chinese characters in visual reading. *Chinese Journal of Psychology*, *26*, 29–34.

Chen, H.-C. (1987). Character detection in reading Chinese: Effects of context and display format. *Chinese Journal of Psychology*, *29*, 45–50.

Cheng, C.-M. (1981). Perception of Chinese characters. *Chinese Journal of Psychology*, *23*, 137–153.

Drewnowski, A., & Healy, A. F. (1977). Detection errors on *the* and *and*: Evidence for reading units larger than the word. *Memory & Cognition*, *5*, 636–647.

Healy, A. F. (1976). Detection errors on the word *the*: Evidence for reading units larger than letters. *Journal of Experimental Psychology: Human Perception and Performance*, *2*, 235–242.

Healy, A. F. (1980). Proofreading errors on the word *the*: New evidence on reading units. *Journal of Experimental Psychology: Human Perception and Performance*, *6*, 45–57.

Healy, A. F. (1994). Letter detection: A window to unitization and other cognitive processes in reading text. *Psychonomic Bulletin & Review*, *1*, 333–344.

Healy, A. F., & Drewnowski, A. (1983). Investigating the boundaries of reading units: Letter detection in misspelled words. *Journal of Experimental Psychology: Human Perception and Performance*, *9*, 413–426.

Healy, A. F., Oliver, W. L., & McNamara, T. P. (1987). Detecting letters in continuous text: Effects of display size. *Journal of Experimental Psychology: Human Perception and Performance*, *13*, 279–290.

Kreuger, L. (1970). Search time in a redundant visual display. *Journal of Experimental Psychology*, *83*, 391–399.

Kreuger, L. (1992). The word-superiority effect and phonological recoding. *Memory & Cognition*, *20*, 685–694.

Liu, I.-M., Chuang, C.-J., & Wang, S.-C. (1975). *Frequency count of 40,000 Chinese words*. Taipei, Taiwan: Lucky Books.

McClelland, J. L., & Rumelhart, D. E. (1981). An interactive activation model of context effects in letter perception: Part I. An account of basic findings. *Psychological Review*, *88*, 375–407.

Reicher, G. M. (1969). Perceptual recognition as a function of meaningfulness of stimulus material. *Journal of Experimental Psychology*, *81*, 275–280.

Wheeler, D. D. (1970). Processes in word recognition. *Cognitive Psychology*, *1*, 59–85.

CHAPTER

13

How Do Readers of Chinese Process Words During Reading for Comprehension?

Hsuan-Chih Chen
The Chinese University of Hong Kong

If we give a sentence, either in written or spoken form, to a group of English speakers (or speakers of any other European language) and ask them to count the number of words in the sentence, what would happen? In most cases, the task would be considered very easy, and there would be no difference in the results of counting, even among different people. This result would not surprise anyone, because, after all, words are units that are widely used in daily life. For instance, individual words are one of the fundamental blocks on which reading acquisition is built and are listed as basic entries in dictionaries, let alone the fact that word boundaries are clearly marked by spaces in writing. Hence, it is quite obvious that words are transparent language units for speakers of European languages. It is, therefore, no wonder that the question of how words are processed has been a major focus of psychological research ever since Cattell (1886).

It is also interesting to note that contemporary theories of human language processing typically consider words as the basis of complex comprehension processes. In fact, it was suggested, not only by reading researchers (e.g., Rayner & Pollatsek, 1989), but also by those who study speech (e.g., Marslen-Wilson & Tyler, 1980), that the process of normal language comprehension is essentially word-by-word. However, the arguments and conclusions related to word processing and the tendency of treating this issue as central to the understanding of the processes behind language comprehension are based almost exclusively on psycholinguistic studies with English and/or with a small group of closely related European languages such as

Dutch, French, and German. It is by no means self-evident whether (or the extent to which) the conclusions that have been reached can be generalized to other languages, such as Chinese, in which words are not transparent units, especially in written form.

The purpose of this chapter is to consider the way in which words are processed during the reading of Chinese text. There are five main parts to the chapter. The first one reviews two general hypotheses proposed to explain how and when each successive word in a text is comprehended as a person reads through the text. The second part presents a short review of a number of peculiar properties Chinese characters possess (especially those related to the script). The third summarizes the notable characteristics of words in Chinese. The fourth part describes some recent studies on how words are processed during reading for comprehension in Chinese. The final section discusses universal and language-specific mechanisms associated with word processing during the reading of different writing systems.

IMMEDIATE VERSUS DELAYED PROCESSING IN READING

Two general hypotheses are proposed to account for the way in which words are processed during reading for comprehension. The first is the delayed comprehension hypothesis. According to this hypothesis, in order to maximize the likelihood of correctly interpreting a given word, readers postpone related comprehension analyses of the word and wait for the context (e.g., Marcus, 1980). A particularly strong form of delayed comprehension assumes that readers do not start to interpret each successive word until reaching the end of phrase or sentence boundaries (Bouma & de Voogd, 1974).

An alternative view is that all levels of comprehension (including, for example, encoding, interpretation, and integration) occur immediately in reading. According to this proposition, rather than waiting for the context to make the interpretation, readers try to process each piece of the text as soon as it is in view. An extreme form of this hypothesis, as Just and Carpenter (1980) proposed, postulates that readers do not move on to new information until all processing of the currently fixated word is complete.

There is evidence in the literature to reject the two most extreme positions just described (see, e.g., Ehrlich & Rayner, 1983; Just & Carpenter, 1987). It appears that intermediate views, especially those in line with the immediate processing position, are more generally accepted. However, we will not discuss in detail the evidence for and against different versions of these hypotheses because that is not the focus of the present chapter. Rather, because these different hypotheses are all developed from research on reading in English or other European languages, our main concern here is to explore the extent to which these positions can help us to understand reading comprehension in a distinctively different writing system, namely, Chinese. Before moving on to consider the processing role of words in Chinese, let us first look at major properties of the Chinese script, especially those related to characters and words.

MAIN PROPERTIES OF THE CHINESE SCRIPT

The Chinese script has a number of peculiar properties that have interesting and important psychological implications (for reviews, see, e.g., Chen, 1992, 1996). To begin, let us look at its physical layout. Chinese text is constructed by arrays of little square symbols of equal size called characters. Each character is made up of single or multiple strokes, radicals (i.e., certain combinations of strokes), simple characters, or a certain combination of these components in a constant box-form region. Because different characters can vary in number of strokes (e.g., 籲, meaning *to call*, has 32 strokes, whereas 一, *one*, has only one), in number of radicals (e.g., 木, *wood*, and 森, *forest*), or in manner of construction (e.g., 杏, *almond*, is formed by two simple characters 木 and 口, while 呆, *stupid*, is made up by the same components but in reverse order), the complexity of the characters can be very high. In contrast to the symmetrical layout of written Chinese, text written in an alphabetic script is typically arranged by strings of words of various lengths and shapes. Thus, as compared to the amount of visual spatial cues that are available in the alphabetic systems, the physical layout of Chinese script provides very limited spatial cues to its readers (for recent research on visual spatial analysis in Chinese reading, see, e.g., Chen & Tang, in press; Inhoff & Liu, 1997; Tang, Au Yeung, & Chen, 1997).

Note that characters are not only distinct orthographic units of the language well known to and commonly used by Chinese people, they are also the ones that are marked by spaces in text. Characters, therefore, serve as basic perceptual units in the reading of Chinese and actually function as encoding units similar to the role of words in alphabetic reading (Chen, 1986, 1987). However, this does not mean that characters are indeed words. To clarify this, let us look at the linguistic properties of the characters.

The Chinese script is constructed mainly on the logographic (i.e., morphographic) principle. That is, characters basically represent morphemes (basic units of meaning). This contrasts drastically with the alphabetic principle that symbols (i.e., letters) are made mainly to correspond to basic units in speech, that is, phonemes. Adopting a logographic script, as opposed to a sound-based system, is indeed a fairly ideal choice in the case of Chinese. This is because the Chinese are a large population spreading over a wide geographical area and speaking quite a number of distinctively different dialects. Since Chinese characters are logographic in nature, they have been continuously used not only by the Chinese people for several thousand years, but also by the Japanese and the Korean for about 2,000 years. In fact, there are about 1,800 to 2,000 Chinese characters currently used in Korea and Japan (Taylor, 1997). The continuing and common use of a sizable number of logographic characters has contributed greatly to the exchange of culture and knowledge in East Asia.

Furthermore, although the Chinese script is not made to represent spoken Chinese, the characters do correspond to speech at the syllabic level (i.e., the characters are typically monosyllabic). Since there are much fewer possible syllables (about 400 in Mandarin, excluding tones, or about 1,300 with tones) than there are characters that are regularly used (about 4,600), the degree of homophony in the language is fairly high. In other words, while each character usually just corresponds to one single syllable (e.g., 易 corresponds to *yi*, with tone 4 in Mandarin), each syllable often corresponds to a number of different characters (e.g., *shi* in Mandarin, with tone 4, corresponds to more than 30 different characters). Given that Chinese has a high degree of homophony and a highly opaque script–speech relationship, it is not clear whether the recognition processes that have often been found in identifying alphabetic words will also appear in the recognition of Chinese characters (for recent research and discussion on this issue, see, e.g., Chen, Flores d'Arcais, & Cheung, 1995; Chen & Shu, 1998; Leung, Tang, & Chen, 1995; Liu, 1995; Perfetti & Tan, 1995).

We now turn to look at the semantic properties of Chinese characters and their relationship with Chinese words. Generally speaking, the meaning of a Chinese character is not transparent by itself and can be highly context-dependent. This is because the majority of characters can join others to form multiple-character words with distinctively different meanings. Indeed, we need only 2,460 different characters to form 15,708 different words (Cheng, 1982), and this number of words can account for about 99% of the material in a Chinese corpus of about one million words (Liu, Chuang, & Wang, 1975). Although there are many characters that can be used alone as single-character words in text, these characters can also be used with others to form very different words (and this is quite often the case). For instance, 生, in the following sentence 他生在北京但在香港長大, meaning *he (was born) in Beijing but grew up in Hong Kong*, is a single-character word. However, the same character can also be used in many other words such as 生意, *business*, 生氣, *anger*, 學生, *student*, 花生, *peanut*, and so on.

It was shown that the properties of the character, as reviewed above, have interesting and important implications for the memory and processing of the characters (e.g., Chen et al., 1995; Chen & Juola, 1982; Chen & Tsoi, 1990; Peng, Guo, & Zhang, 1985). For instance, Chen and Juola (1982) demonstrated that graphemic information and processing plays the most prominent role among various dimensions of lexical coding in Chinese, whereas visual representation plays a less important role for English lexical items than phonological and semantic representations. These results are striking, but are in line with the properties of Chinese characters previously noted, indicating that the graphic aspects of each character are much more distinctive and informative, in comparison with phonological and semantic aspects.

THE CHARACTERISTICS OF WORDS IN CHINESE

Thus far we have examined a number of salient attributes of Chinese characters. It is important to note that characters are typically morphemes. Their major function is to form words. However, it is the word, rather than the morpheme, that is commonly considered to be the basic unit of language processing and is used as the unit of analysis to study such questions as how sentences are structured and comprehended (e.g., Marslen-Wilson & Tyler, 1980; Rayner & Pollatsek, 1989). We now turn to discuss issues more specifically related to the processing of words in Chinese.

As noted earlier, unlike speakers of various European languages, the notion of the word is not very clear to most, if not to all, users of Chinese (Chao, 1968). Let us consider a few reasons why this is the case. Foremost, morphological complexity in Chinese is relatively low compared to that in many European languages (e.g., Li & Thompson, 1981). Chinese words generally have no inherently marked lexical categories. They do not usually possess explicit markers that specify their grammatical categories, and they can often be used in multiple lexical categories because the syntactic role of a word is typically determined by sentential context. In addition, they have no inflectional markings to indicate various kinds of grammatical attributes (e.g., the number, gender, and case for nouns, or the tense and aspect for verbs), let alone intrasentence concordance rules such as subject–verb agreement. As a matter of fact, among various ways to form polysyllabic or multimorphemic Chinese words (e.g., reduplication, affixation, etc.), compounding is by far the most important and productive method. (This is very different from how words are commonly formed in English and other European languages.) Consequently, a great majority of Chinese words can be analyzed into separate morphemes, just as sentences can be broken down to words, and paragraphs to sentences. To put it another way, the word in Chinese does not seem to have salient and distinctive characteristics compared to other structural units of the language (e.g., the orthographically distinctive unit, the character).

Furthermore, the word is not simply a frame unit in modern Chinese. This is because the size of a word is not constant and can be formed by one or more characters (or one or more syllables in the case of spoken Chinese). Also, word boundaries are not explicitly marked by extra spacing in written Chinese (or typically specified by a pause in the spoken form). For instance, in the sentence 他喜歡看連環圖, meaning *he likes to read comics*, it is impossible to figure out, on the basis of spacing in the sentence, that there are actually two single-character words (他 & 看), one two-character word (喜歡), and one three-character word (連環圖). In other words, identifying individual words in Chinese reading is not as simple and straightforward as is reading text written in English or any other alphabetic system.

The obvious effect of all these properties of Chinese words is that Chinese readers have only vague ideas about what a word is. This is supported by the finding that there is considerable disagreement concerning what are word boundaries in simple sentences, even among skilled Chinese readers (Hoosain, 1992).

Given these characteristics of Chinese words, there are a number of interesting and important questions one can ask concerning the processing of words in Chinese. For instance, how are multimorphemic words represented and processed in the mind of Chinese readers? Are they stored as a whole word, or are they represented at the morpheme level in terms of their constituent morphemes? Also, can idiomaticity affect the mental storage and processing of the word? However, these questions are typically studied with isolated lexical items as stimuli and are topics of other chapters in this volume and thus are not discussed here. Rather, we focus in this chapter on the processing of Chinese words in sentence or text.

PROCESSING THE WORD IN CHINESE READING

We reviewed striking properties of Chinese characters and words. The question to consider now is: How do Chinese readers process words during reading for comprehension? Two sets of experiments are reviewed in the following sections.

The Basic Experimental Paradigm

The basic manipulation in the experiments to be described is to change a critical word in a short passage so that certain combinations of lexical, syntactic, and semantic information are violated. Different types of violations can be formed by replacing a genuine word with a nonword (i.e., a lexical violation; note that in this condition the syntactic and semantic information is also violated; nevertheless the label is used for simplicity), making changes in semantic category (i.e., a semantic violation), or substituting a semantically anomalous and syntactically inappropriate word for the original word (i.e., a syntactic+semantic violation). By observing and comparing the patterns of disruptions caused by different types of violations on character/word viewing times, various questions with respect to how words are processed in reading can then be assessed.

The violation paradigm was originally developed by Danks and his colleagues and proved to be a valid research tool to study comprehension processes in oral reading (Danks, Bohn, & Fears, 1983). This paradigm was also found to be equally effective in studying comprehension in a silent reading condition where character or word viewing times, rather than oral reading protocols, were measured as on-line indicators of comprehension processes. For instance, I adopted a specially designed, self-paced display technique to collect the viewing time of individual words in English passages and successfully replicated two major findings of Danks et al. (Chen, 1992). These results include immediate disruption effects (i.e., all different types of violations produced longer word viewing times than those in the intact, controlled condition at the critical word position where those violations were introduced) and a lexical over syntactic/semantic advantage effect (i.e., larger disruption for lexical violation than for syntactic and/or semantic violations).

The focus of the following experiments is on three interrelated aspects of the question how do readers of Chinese process words during reading for comprehension? First, are readers of Chinese sensitive to various types of linguistic information carried or implied by different types of words? Second, can Chinese readers immediately activate and use the linguistic information of individual words, or do they have to wait until reaching the end of a phrase or sentence boundary to start relevant semantic and syntactic analyses? Third, how do different types of words contribute to Chinese comprehension processing?

Processing of Nouns and Verbs in Chinese Reading

Chen and Au Yeung (1995) adopted the aforementioned violation paradigm to investigate the processing of content words in the reading of Chinese. In this study, nouns and verbs were chosen as the target stimuli because they represent two major categories of content words in Chinese and serve quite different functions in sentence processing. The target words were modified to form three types of violations (i.e., lexical, semantic, and syntactic+semantic). By contrasting patterns of violations for nouns and verbs, we attempt to investigate the way in which the readers of Chinese gain access to various properties of words in reading for comprehension.

A stimulus passage used in the experiment (and the corresponding manipulations) is presented in Fig. 13.1. To prevent possible confoundings due to word length and other morphological factors, all critical stimuli (i.e.,

上星期警方據報在河底發現了一具屍體。調查結果顯示，這
個人在事前綁了幾塊鉛在自己身上。這樣他死後身體就不會浮
起來。雖然警方仍未能確定死者的身份，但相信他是自殺的。

	Control	Lexical	Semantic	Syntactic+Semantic
Noun	人 (person)	[illegible]	力 (force)	上 (up)
Verb	綁 (bind)	[illegible]	喝 (drink)	蛾 (moth)

FIG. 13.1. Example of a stimulus passage used in the experiment. The two potential targets (the first one is a noun and the second a verb) are underlined. The lexical, semantic, and syntactic+semantic violations are illustrated in the lower half of the figure (their corresponding meanings are indicated in parentheses).

target words and their replacements) were embedded in an unmarked subject-verb-object (SVO) sentence and contained either a single-character word or a noncharacter. (Note that similar experiments have also been carried out using multiple-character words as critical stimuli and the results are basically the same; see, e.g., Chen, 1992.) Also, all the target words and their replacements were matched as closely as possible in terms of complexity (i.e., number of strokes) and frequency.

Twenty-four skilled readers of Chinese participated, with each subject receiving 38 short stories (including two practice ones) for comprehension. All stimulus passages were displayed on a computer screen using a self-paced, moving-window technique, in which a mechanical mouse was linked to the computer for the control of stimulus display (Chen, 1992). The position of the mouse was represented on the screen by a pointer that would move underneath horizontally arranged line segments and punctuation marks. A character corresponding to a certain position was displayed right above the line segment when the pointer came to that position. Consequently, by moving the mouse from left to right (within a line) and right to left (at the end of a line to the beginning of the next), a subject viewed successive characters of a text displayed in spatially appropriate locations on the screen. The character-by-character exposure times were taken and recorded to calculate the dependent variable, namely, the average character viewing time per word.

To see disruption profiles for the stimuli with nouns as critical words, we can plot the differences in the average viewing times between the three experimental conditions and their controls across a fixed range of word positions (i.e., from one word before the critical one to three words after it

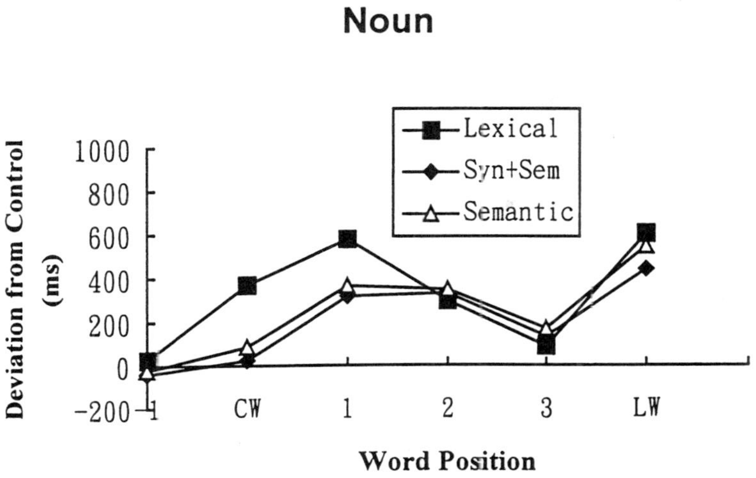

FIG. 13.2. Differences in mean viewing times between the lexical, syntactic+semantic (Syn+Sem), and semantic violations and their controls for those passages with nouns as critical words (CW). LW refers to the last word in the sentence.

plus one additional word located at the end of sentence). Figure 13.2 shows these difference scores. Lexical violation reveals immediate disruption at the critical word position, whereas both semantic and syntactic+semantic violations show patterns of delayed disruption at the word+1 position. Furthermore, all three types of violations produce reliable interference at the end of the sentence, indicating that the subjects attempted to use contextual information to resolve the violations introduced in the sentence.

Figure 13.3 shows the disruption profiles for the verb stimuli. Again, lexical disruption starts without delay at the critical word position, whereas syntactic+semantic disruption does not appear immediately, but at the word+1 position. Furthermore, reliable disruption effects are shown at the last word position for all types of violations. These results are highly compatible with those for the noun stimuli.

However, a very different pattern is shown for the semantic violation: No semantic disruption can be seen until three words after the critical one, indicating that a considerable amount of contextual information is needed until a purely semantic violation for verbs can be detected. We suspect that this extreme delay in semantic violation for verbs is due to the fact that the target word was embedded in a SVO sentence in which a target verb is typically followed by a noun phrase that contains three words (i.e., a numeral or demonstrative or quantifier, a classifier, and the object). Since the main function of a verb is to relate the nouns in a sentence, in order for a reader to comprehend that the verb provided in the semantic violation

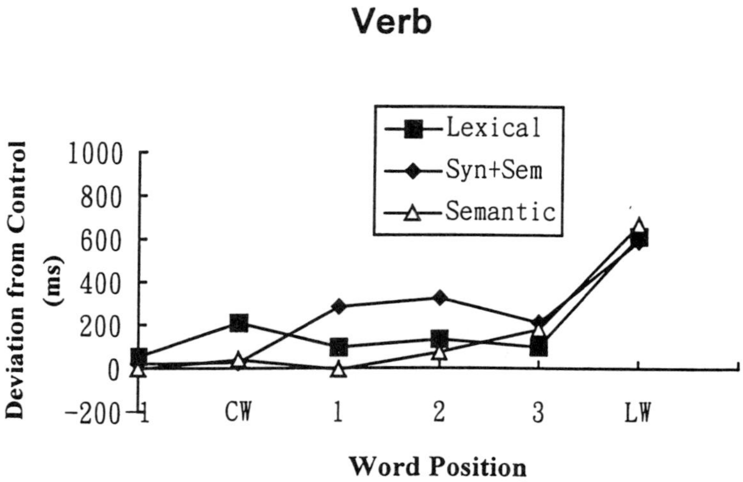

FIG. 13.3. Differences in mean viewing times between the lexical, syntactic+semantic (Syn+Sem), and semantic violations and their controls for those passages with verbs as critical words (CW). LW refers to the last word in the sentence.

condition is anomalous, a reader has to know what the nouns of the sentence are.

To verify that hypothesis with respect to the specific pattern of semantic disruption for verbs and to investigate possible effects of word order on the processing of verbs, we collected additional data using two modified sets of stimuli. The modified stimuli were constructed by changing the word order of the sentence with the critical word embedded from the original SVO structure to subject-object-verb (SOV) and object-subject-verb (OSV) structures. Consequently, in either a SOV or OSV sentence, the reader can see the main participants of the sentence before reaching the verb. Since the disruption profiles for the SOV and OSV stimuli are very similar, the results are therefore collapsed across the two types of stimuli and are shown in Fig. 13.4.

The most striking finding that can be seen in Fig. 13.4 is that, in contrast to the much delayed disruption effects for semantic violation in the SVO sentences, the semantic violation in the SOV/OSV stimuli can be detected quickly at the word+1 position. It is also interesting to note that unlike the strong boundary effects found in the SVO sentence, the boundary effect in the SOV/OSV sentences is either relatively weak, as in the syntactic+semantic condition, or totally absent. It appears that more integration is done at the end of the sentence in SVO sentences, but less in SOV/OSV sentences. This could be because the latter sentences are marked by either the object marker, *ba*, or the passive marker, *bei*, whereas the former are unmarked.

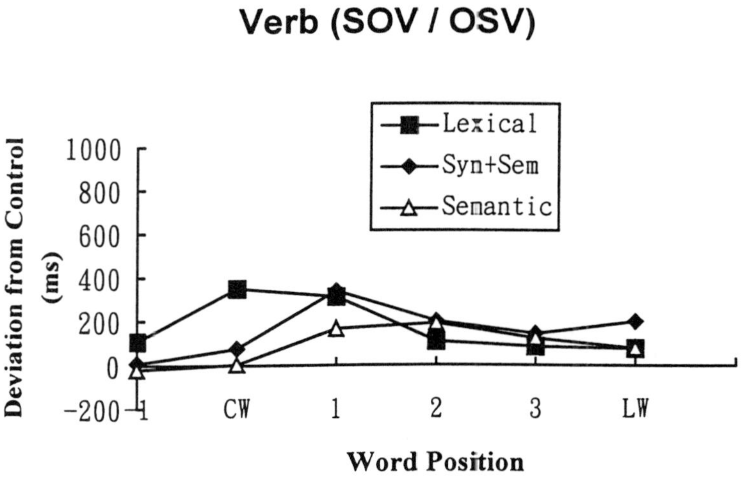

FIG. 13.4. Differences in mean viewing times between the lexical, syntactic+semantic (Syn+Sem), and semantic violations and their controls for those passages with verbs as critical words (CW). LW refers to the last word in the sentence.

In other words, it seems plausible that because additional information is provided in the marked SOV/OSV sentences to specify the semantic roles played by various participants in the sentence, comprehension analyses can be carried out relatively on-line before reaching the end of the sentence. Other results for SOV/OSV sentences (i.e., the patterns of lexical and syntactic+semantic violations) are generally compatible with those for SVO sentences.

Taken together, the semantic violation results for verbs in the three types of sentences suggest that (a) the semantic processing of a verb relies, to a considerable extent, on information such as the relative positioning of main participants in a sentence, and (b) the way in which a verb is processed in a Chinese sentence can clearly be affected by its word order.

Furthermore, we also manipulated word order to assess its effects on the processing of nouns. The results are shown in Fig. 13.5. However, except for the fact that the boundary effect was not found in SOV/OSV sentences, word order did not produce any other notable effect on the disruption profiles for the noun stimuli, indicating that (a) the effect of word order is probably greater on certain types of comprehension analysis such as integration, and (b) verbs play a more important role than nouns on sentence integration for Chinese.

Clearly, there are similarities and differences between the results for nouns and verbs. Figure 13.6, for instance, shows the overall pattern of the disruption effects for nouns and verbs. However, before we start to discuss

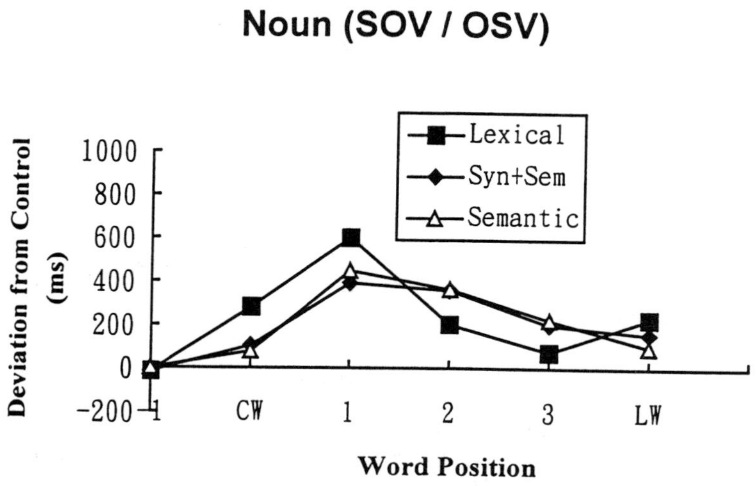

FIG. 13.5. Differences in mean viewing times between the lexical, syntactic+semantic (Syn+Sem), and semantic violations and their controls for those passages with nouns as critical words (CW). LW refers to the last word in the sentence.

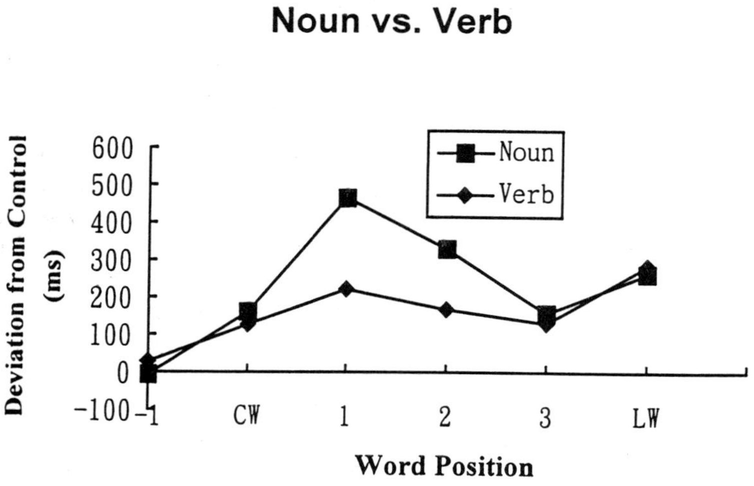

FIG. 13.6. Mean disruption as a function of type of critical word (CW). LW refers to the last word in the sentence.

these findings and their implications for the understanding of Chinese reading, let us look at another study that adopted the same violation paradigm to explore and contrast the processing of content words and function words in Chinese reading.

Processing of Classifiers and Prepositions in Chinese Reading

This research (Chen, 1994) focuses on classifiers and two special types of prepositions (i.e., a pretransitive one, *ba*, and an agentive one, *bei*). Classifiers are words that occur almost obligatorily with numerals (and/or demonstratives or certain quantifiers) before nouns. Since there are quite a few classifiers (more than 300 according to Chao, 1968), the choice of which is determined very much by the noun it modifies, the use of classifiers is considered to be one of the most peculiar characteristics of Chinese.

Being an isolating language, Chinese is well known for its general lack of inflectional morphology. Indeed, unlike European languages, Chinese does not rely much on the use of affixes in the expression of grammatical relationships. Rather, other devices such as word order play a more important role in the expression of grammatical relationships among the elements of a sentence in Chinese (Norman, 1988). One way to change the word order in a sentence is to use such prepositions as *ba* and *bei*. For instance, *ba* can be used to construct the following word order: subject-*ba*-direct object-verb. Thus, *ba* is also called the object marker. *Bei*, on the other hand, can be adopted as the passive marker to form another word order: object-*bei*-subject-verb.

Twenty-four Chinese subjects participated in this study. The design and materials were the same as those used in the study mentioned before, except that classifiers and the prepositions *ba* and *bei* were the critical words in the present study. Examples of the critical words and related manipulations are shown in Fig. 13.7.

Figure 13.8 shows the disruption profiles for the classifier stimuli. Similar to the pattern for the noun stimuli observed in the aforementioned study, lexical violation appears earlier and stronger than either the syntactic+semantic violation or the semantic one: Lexical violation shows up at the critical word position, whereas the others occur at the word+1 position. However, unlike the results with SVO sentences, but similar to those with SOV/OSV sentences, no sentence boundary effect is found for the classifier stimuli.

The disruption profiles for *ba* and *bei* are very similar. Thus, the data are collapsed across the two types of stimuli and are shown in Fig. 13.9. Again, lexical disruption occurs immediately at the critical word position. The syntactic+semantic disruption effect shows up at the word+1 position and

Subject-ba-Object-Verb sequence:

調查結果顯示，這個人在事前把幾塊鉛綁在自己身上。

Object-bei-Subject-Verb sequence:

調查結果顯示，幾塊鉛在事前被這個人綁在自己身上。

	Control	Lexical	Semantic	Syntactic+Semantic
Classifier	塊 (piece)	荅扌	群 (group)	煩 (bother)
Ba	把	斤土	叫 passive marker	吹 (blow)
Bei	被	瓜石	將 object marker	能 (able)

FIG. 13.7. Examples of the three types of critical words and their corresponding sentences used in the experiment (the critical words are underlined). The related manipulations are illustrated in the lower half of the figure (the corresponding meanings are indicated in parentheses).

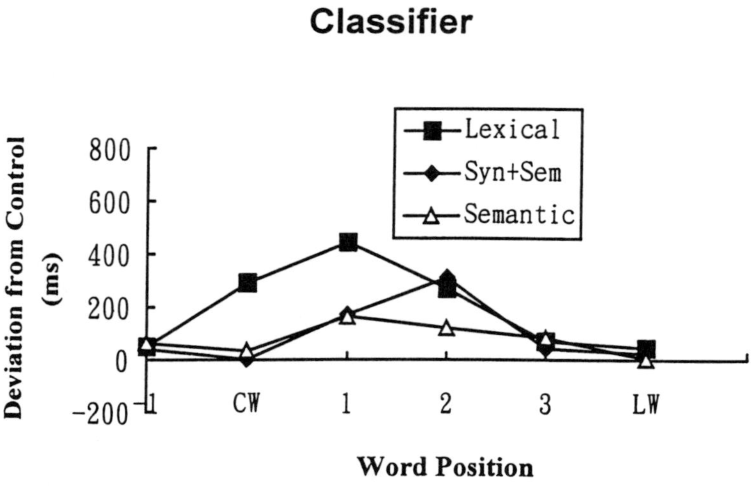

FIG. 13.8. Differences in mean viewing times between the lexical, syntactic+semantic (Syn+Sem), and semantic violations and their controls for those passages with classifiers as critical words (CW). LW refers to the last word in the sentence.

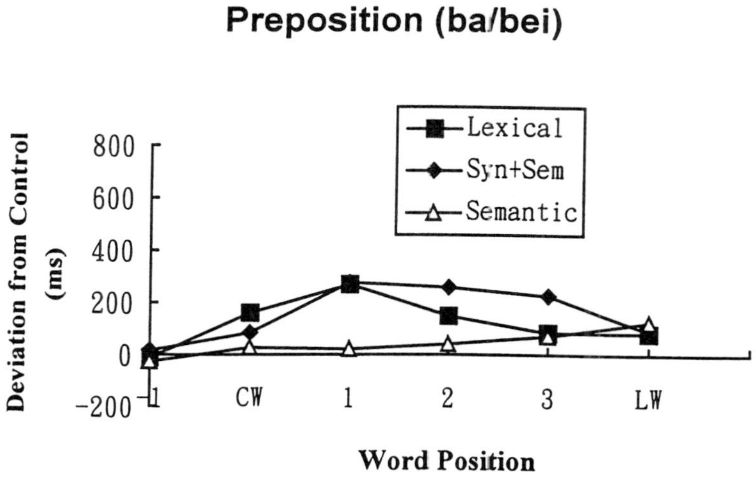

FIG. 13.9. Differences in mean viewing times between the lexical, syntactic+semantic (Syn+Sem), and semantic violations and their controls for the passages with prepositions (i.e., *ba* or *bei*) as critical words (CW). LW refers to the last word in the sentence.

continues through to the word+3 position. In contrast, the semantic disruption is generally missing. But, at the end of sentence, a small but nevertheless reliable effect of semantic disruption was found, indicating that either the subjects needed to have the maximal amount of sentential context to detect a deviant preposition or the effect could only be detected at the sentence wrap-up stage.

Finally, Fig. 13.10 shows the mean disruption for classifiers and prepositions. Note that classifiers produced clear and strong disruption effects around the critical word area (i.e., starting at the critical word position and continuing through the word+2 position), whereas prepositions resulted in weaker but more long-lasting effects across a wider range of the sentence. This is again in line with the idea that the role of classifiers in comprehension is to facilitate the processing of the noun (and thus have more local effects), whereas *ba* and *bei* contribute more generally to the word sequence of a sentence (and hence have more global effects).

Implications for the Nature of Word Processing in Chinese Reading

What can we learn from the results we've described with respect to the question of how readers of Chinese process words during reading for comprehension? Let us first look at the issue of whether Chinese readers are sensitive to various types of linguistic information from different types of words.

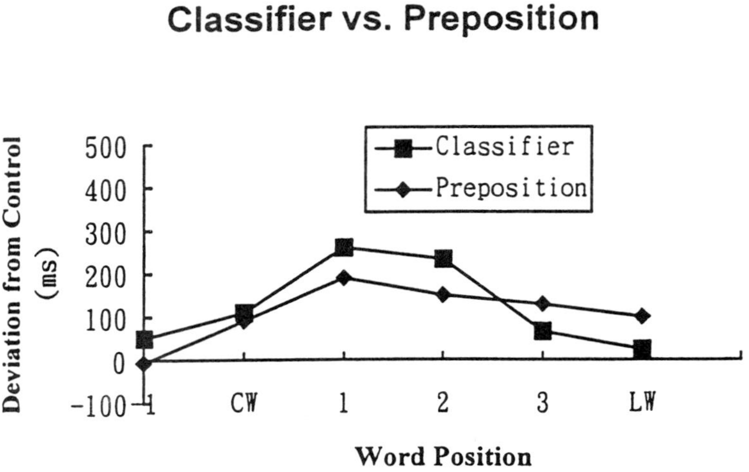

FIG. 13.10. Mean disruption as a function of type of critical word (CW). LW refers to the last word in the sentence.

Traditional linguists divide Chinese words into two major categories: full and empty words. *Full words* are those that have semantic content and can further be separated into body words (e.g., nouns, pronouns, classifiers, etc.) and predicates (i.e., verbs and adjectives). *Empty words* are those that usually have minimal semantic substance and are typically used to indicate grammatical relationships, including, for instance, prepositions, adverbs, conjunctions, and so on (e.g., Norman, 1988; Zhu, 1982). But do these linguistic categorizations of Chinese words have any psychological implications? This question can be tested by comparing patterns of disruption results for different kinds of words. Overall, the results showed that various types of words produced considerably different patterns of results. For instance, when nouns or classifiers were used as the critical words, both the semantic and syntactic+semantic violations yielded compatible patterns of strong disruption, suggesting that these two types of content words contribute equally to both syntactic and semantic operations in the process of reading Chinese. When the critical words were verbs, however, the syntactic+semantic violations resulted in clear and stable disruptions, whereas the disruptive effects created by the semantic violations depended very much on the relative position of the verb in the sentence. The results for verbs are in line with the notion that verbs serve to relate various participants of a sentence to describe an action or state (and are thus sensitive to the word sequence of sentence). Furthermore, when using the prepositions *ba* and *bei* as critical words, the syntactic+semantic violations resulted in reliable and long-lasting disruptive effects, whereas semantic violations did not produce a significant effect until the end of the sentence was reached. This finding confirms the notion that as function words, *ba* and *bei* play a more important role in syntactic analyses than in semantic analyses. Together, these results indicate that the linguistic properties of various types of words have important psychological implications. Indeed, readers of Chinese are clearly sensitive to the linguistic information carried by different types of words.

The next issue of concern is immediate versus delayed processing in Chinese reading. The results noted before indicate that various types of violations result in distinctively different profiles of disruption in Chinese reading. With respect to the relative positioning resulting from each type of violation, although the lexical disruption typically appeared at exactly where the violation was introduced, the disruption effects created by the syntactic+semantic violation usually occurred slightly delayed at the word+1 position. In contrast, the position of semantic disruption varied considerably from the word+1 position to the last word in the sentence. Obviously, these results are not in line with the delayed processing view of Chinese reading (cf. Aaronson & Ferres, 1986). However, the findings are also not consistent with the most radical view of immediate processing (Just & Carpenter, 1980), according to which various types of violation effects should be observed right at the critical word position, regardless of the type of critical word used. Although there is clear evidence indicating that Chinese readers start to process lexical, syntactic, and semantic information for different types of words as soon as they can, these different sorts of linguistic information for individual words obviously do not become available simultaneously, possibly because, in Chinese, different sorts of linguistic information are not directly marked by the word itself, but often need to be inferred from attributes of sentential context such as word order. Indeed, the disruptive effects produced by various sorts of manipulation often last for more than one word, indicating clearly that the Chinese readers rely heavily on a more diffused, context-dependent strategy during reading for comprehension as opposed to a more focused, word-dependent strategy often adopted by readers of alphabetic systems (Chen, 1992).

What would the time course of lexical, syntactic, and semantic processing in Chinese reading be, then? The result that lexical violations unexceptionally produce reliable and immediate disruption provides clear and strong support for the notion that lexical access precedes any other form of comprehension analysis. What is less clear, however, is the time course of syntactic and semantic processing during Chinese reading. This complex but important issue cannot be answered unambiguously by the violation results described earlier and needs to be explored in future research.

Another issue that emerged from the violation research described here concerns the extent to which different types of words contribute to the process of Chinese comprehension. It appears that the linguistic distinction between body words (e.g., nouns and classifiers) and predicates does have notable implications in the process of Chinese reading. Whereas both nouns and classifiers produce highly similar patterns of disruption profiles, there are interesting differences between the results for nouns and those for verbs. For instance, as mentioned before, the profiles for verbs were sensitive to the manipulation of word order in a sentence in two different ways: the relative position of the semantic violation and the sentence boundary effect. However, for nouns, word order affected only the appearance of the sentence boundary effect. These results are consistent with the notion that nouns mainly help in the analysis of the constituents of a clause/sentence into different semantic roles (e.g., actor, object, etc.) or the building of basic structures, whereas verbs play a more important role in integrating various parts of a sentence to describe an action. Because the integration part of comprehension probably relies heavily on the verb, the sequence of words in a sentence may therefore have a clearer impact on the processing of verbs than of nouns.

Moreover, some researchers who study language comprehension in English propose that the verb plays a critical role in analyzing the configuration of roles to which the various participants of a sentence are assigned (e.g., Fillmore, 1971; Schank, 1975; for arguments against this view, see Just & Carpenter, 1987). This, however, does not seem to apply to the verb in Chinese, because there are clearly other devices that indicate semantic roles of various parts of a sentence, such as word order and grammatical markers (e.g., *ba* and *bei*). Furthermore, there are at least two lines of evidence to indicate that if we compare the relative importance of the noun and the verb in Chinese comprehension, it is the noun rather than the verb that seems to be more relied on in comprehension analyses. First, the violation studies described earlier demonstrated a clear noun-over-verb advantage: The magnitude of disruption was significantly larger when various types of violations were introduced to a noun compared to when these violations were introduced to a verb. The second line of evidence is related to the use of classifiers in Chinese. The grammar of Chinese is well known for relying more on semantic and pragmatic factors than on explicit rules and/or formal structure, such as inflections, agreement, and so on (e.g., Li, 1992; Lu, 1986). Given the structural simplicity of the language, it is interesting to note that there are so many classifiers and that the classifiers are so heavily used in Chinese. Is this a simple accident? Perhaps not! Rather, it seems much more reasonable to think that the use of classifiers actually reflects the need of the processing system. In other words, the language has probably evolved to suit a processing mechanism that relies on the noun more heavily for the purpose of comprehension. Thus, the use of classifiers is to facilitate and provide support for the processing of nouns. This interesting possibility warrants systematic investigation in future research.

Finally, let us consider the relative importance of such prepositions as *ba* and *bei* in comprehension. Both word order and prepositions have often been assumed to play an important role in determining the grammatical relationship among various constituents of a sentence (e.g., Norman, 1988). We saw earlier how word order affects sentence processing through the appearance of sentence boundary effects and the semantic processing of verbs. But what about prepositions? The results obtained from the violation experiments reviewed in this chapter indicate that although reliable disruptions could be found through introducing different types of violations to *ba* and *bei*, the magnitude of the disruptions were relatively small, especially when an inappropriate marker was used to replace the original one (i.e., the semantic violation condition). These results suggest that prepositions such as *ba* and *bei* do contribute to the comprehension analysis involved in Chinese reading. However, the grammatical markers themselves do not seem to play as critical a role in comprehension as do other linguistic devices such as word order. These observations with respect to the preposition are consistent with the idea that Chinese empty words or markers are not very critical for comprehension, probably because they are not always present in sentences, nor do they always provide constant and reliable syntactic information (for further discussion, see Chao, 1968; Li & Thompson, 1981).

CONCLUSIONS

There are many varieties of written, as well as spoken, languages currently used by the people around the world. Since different writing and speech systems are constructed on distinctive principles and present linguistic information in diverse forms, it is not self-evident that the same processes are utilized to comprehend information expressed in different systems. It is therefore important and useful to conduct cognitive research across languages to explore both universal and language-specific processes associated with the processing of different systems. This in turn can help us to build a genuinely comprehensive theory of human language processing.

The research reviewed in this chapter suggests that despite the fact that Chinese words are not transparent units and lack explicit marking and inflections, they are used as important processing units for various sorts of comprehension analyses, just as words in alphabetic systems are used. The results of the violation studies further indicate that syntactic and semantic operations can be conducted on-line while reading Chinese text without much delay. It appears that readers of Chinese are sensitive to the linguistic properties of different types of words, and can use them in a timely, if not immediate, fashion. Therefore, it seems reasonable to conclude that the word as a prominent processing unit, and the tendency for immediacy processing, are good candidates for universal principles in reading and comprehension.

However, we have also seen evidence to suggest that Chinese readers, unlike readers of alphabetic writing systems, rely more heavily on context and less on individual words. This reminds us not to underestimate the difference between various writing systems (and spoken languages as well) while looking for universal principles underlying various kinds of language processing and comprehension activities. There is indeed good evidence to indicate that readers are sensitive to the differences in the cognitive demands imposed by different orthographic and linguistic structures in reading comprehension and that they adapt their processing strategies to meet the cognitive demands of the orthography and the language.

ACKNOWLEDGMENTS

This work was supported by Earmarked Grants CUHK2/92H, 153/94H, and 144/96H from the Hong Kong Research Grants Council.

REFERENCES

Aaronson, D., & Ferres, S. (1986). Sentence processing in Chinese-American bilinguals. *Journal of Memory and Language, 25*, 136–162.

Bouma, H., & de Voogd, A. H. (1974). On the control of eye saccades in reading. *Vision Research, 14*, 273–284.

Cattell, J. M. (1886). The time it takes to see and name objects. *Mind, 11*, 63–65.

Chao, Y. R. (1968). *A grammar of spoken Chinese*. Berkeley: University of California Press.

Chen, H.-C. (1986). Component detection in reading Chinese characters. In H. S. R. Kao & R. Hoosain (Eds.), *Linguistics, psychology, and the Chinese language* (pp. 1–10). Hong Kong: University of Hong Kong.

Chen, H.-C. (1987). Character detection in reading Chinese: Effects of context and display format. *Chinese Journal of Psychology, 29*, 45–50.

Chen, H.-C. (1992). Reading comprehension in Chinese: Some implications from character reading times. In H.-C. Chen & O. J. L. Tzeng (Eds.), *Language processing in Chinese* (pp. 175–205). Amsterdam: North-Holland (Elsevier).

Chen, H.-C. (1994, November). *Comprehension processes in reading Chinese*. Paper presented at the 35th Annual Meeting of the Psychonomic Society, St. Louis, MO.

Chen, H.-C. (1996). Chinese reading and comprehension: A cognitive psychology perspective. In M. H. Bond (Ed.) *Handbook of Chinese psychology* (pp. 43–62). Hong Kong: Oxford University Press.

Chen, H.-C., & Au Yeung, L. H. (1995, November). *The time course of lexical, syntactic, and semantic processing in Chinese reading*. Paper presented at the 36th Annual Meeting of the Psychonomic Society, Los Angeles, CA.

Chen, H.-C., Flores d'Arcais, G. B., & Chueng, S.-L. (1995). Orthographic and phonological activation in recognizing Chinese characters. *Psychological Research/Psychologische Forschung, 58*, 144–153.

Chen, H.-C., & Juola, J. F. (1982). Dimensions of lexical coding in Chinese and English. *Memory & Cognition, 10*, 216–224.

Chen, H.-C., & Shu, H. (1998). *Lexical activation during the recognition of Chinese characters*. Manuscript submitted for publication.

Chen, H.-C., & Tang, C. K. (in press). The effective visual field in reading Chinese. *Reading and Writing*.

Chen, H.-C., & Tsoi, K.-C. (1990). Symbol-word interference in Chinese and English. *Acta Psychologica, 75*, 123–138.

Cheng, C.-M. (1982). Analysis of present day Mandarin. *Journal of Chinese Linguistics, 10*, 282–358.

Danks, J. H., Bohn, L., & Fears, R. (1983). Comprehension processes in oral reading. In G. B. Flores d'Arcais & R. J. Jarvella (Eds.), *The process of language understanding* (pp. 193–223). Chichester, Sussex: Wiley.

Ehrlich, K., & Rayner, K. (1983). Pronoun assignment and semantic integration during reading: Eye movements and immediacy of processing. *Journal of Verbal Learning and Verbal Behavior, 22*, 75–87.

Fillmore, C. J. (1971). Verbs of judging: An exercise in semantic description. In C. J. Fillmore & D. T. Langendoen (Eds.), *Studies in linguistic semantics*. New York: Holt, Rinehart & Winston.

Hoosain, R. (1992). Psychological reality of the word in Chinese. In H.-C. Chen & O. J. L. Tzeng (Eds.), *Language processing in Chinese* (pp. 111–130). Amsterdam: Elsevier.

Inhoff, A. W., & Liu, W. (1997). The range of effective vision during the reading of Chinese sentences. In H.-C. Chen (Ed.), *Cognitive processing of Chinese and related Asian languages* (pp. 243–265). Hong Kong: Chinese University Press.

Just, M. A., & Carpenter, P. A. (1980). A theory of reading: From eye fixation to comprehension. *Psychological Review, 87*, 329–354.

Just, M. A., & Carpenter, P. A. (1987). *The psychology of reading and language comprehension*. Boston: Allyn & Bacon.

Leung, C. H., Tang, S. L., & Chen, H.-C. (1995, November). *Phonological processing in the recognition of Chinese characters: Effects of task*. Paper presented at the 36th Annual Meeting of the Psychonomic Society, Los Angeles, CA.

Li, C., & Thompson, S. (1981). *Mandarin Chinese: A functional reference grammar*. Berkeley: University of California Press.

Li, J. X. (1992). *Xin Zhu Guo Yu Wen Fa* [The new version of Chinese grammar]. Beijing: Commercial Press.

Liu, I.-M. (1995, December). *The issue of prelexical phonology in reading Chinese characters*. Paper presented at the 7th International Conference on the Cognitive Processing of Chinese and other Asian Languages, Hong Kong.

Liu, I.-M., Chuang, C. J., & Wang, S. C. (1975). *Chang youg zhong wen ci de chu xian ci shu* [Frequency count of 40,000 Chinese words]. Taipei, Taiwan: Liu Guo Chu Ban She.

Lu, S. X. (1986). Flexibility in Chinese syntax. *Zhongguo Yuwen, 190*, 1–9

Marcus, M. P. (1980). *A theory of syntactic recognition for natural language*. Cambridge, MA: MIT Press.

Marslen-Wilson, W. D., & Tyler, L. K. (1980). The temporal structure of spoken language understanding. *Cognition, 8*, 1–71.

Norman, J. (1988). *Chinese*. Cambridge, England: Cambridge University Press.

Peng, D. L., Guo, D. J., & Zhang, S. L. (1985). The retrieval of information of Chinese characters in making similarity judgment under recognition condition. *Acta Psychologica Sinica, 3*, 227–234.

Perfetti, C. A., & Tan, L. H. (1995, December). *The time course of written Chinese word identification*. Paper presented at the Seventh International Conference on the Cognitive Processing of Chinese and Other Asian Languages, Hong Kong.

Rayner, K., & Pollatsek, A. (1989). *The psychology of reading*. Englewood Cliffs, NJ: Prentice-Hall.

Schank, R. C. (1975). *Conceptual information processing*. Amsterdam: North-Holland.

Tang, C. K., Au Yeung, L. H., & Chen, H.-C. (1997). The effective visual field in Chinese reading. In H.-C. Chen (Ed.), *Cognitive processing of Chinese and related Asian languages* (pp. 267–286). Hong Kong: Chinese University Press.

Taylor, I. (1997). Psycholinguistic reasons for keeping Chinese characters in Korean and Japanese. In H.-C. Chen (Ed.), *Cognitive processing of Chinese and related Asian languages* (pp. 299–319). Hong Kong: Chinese University Press.

Zhu, D. X. (1982). *Yu Fa Jiang Yi* [Lecture notes on grammar]. Beijing: Commercial Press.

CHAPTER

14

Sentence Understanding in Chinese

Miao Xiaochun
East China Normal University

Sentence understanding is an important part of language comprehension. In some sense, it plays a central role in language comprehension in that sentences constitute the elementary form of information transmission in language communication. However, in the study of Chinese language comprehension, sentence understanding received much less attention than character and word recognition. As far as I know, studies of sentence understanding, only a few in number, are chiefly focused on the roles of factors affecting comprehension. In this chapter, therefore, I mainly introduce and review the studies in this respect.

THE ROLES OF SYNTACTIC AND SEMANTIC FACTORS IN UNDERSTANDING SIMPLE CHINESE SENTENCES

Generally speaking, syntactic, semantic, pragmatic, phonological, and even suprasegmental factors all contribute to sentence understanding. However, research has shown that while understanding sentences, people speaking different languages rely heavily on different information and use different strategies because of the difference of languages. For example, several studies (Bates et al., 1984; Bates, McNew, MacWhinney, Devescovi, & Smith, 1982; MacWhinney, Bates, & Kliegl, 1984) found that English speakers determined the subject of a sentence mainly according to word order and native Italians on the basis of lexical meaning and morphological markers. Chinese is a language distinct from Indo-European languages. An evident peculiarity of Chinese is the lack of inflection morphology. There are no case markers, no tense suffixes, and no agreement markers in Chinese. The grammatical relations and function of words in Chinese sentences are, to a large degree, determined by word order and function words. Word order seems to be an important, even the essential, syntactic device in Chinese. The basic word order of the Chinese sentence is SVO (subject-verb-object). Therefore, problems such as how Chinese speakers understand a sentence, what the contribution of syntactic and semantic factors to Chinese sentence understanding is, and whether the word order is the essential cue used by Chinese speakers in comprehension aroused the interest of researchers in this field.

Two earlier studies (Miao, 1981; Miao, Chen, & Ying, 1984) examined the relative roles of word order and word meaning in understanding simple Chinese sentences. The researchers presented adult Chinese subjects with a series of simple sentences composed of three words: two nouns and one verb (no articles, prepositions, or other words were introduced). The nouns used in the sentences comprised animate names and inanimate names; all verbs were transitive. The sentences were divided into three types, each type reflecting a different word order: noun-verb-noun (NVN), verb-noun-noun (VNN) and noun-noun-verb (NNV).

As two kinds of nouns were introduced in the sentences, the test materials were composed of nine types of sentences altogether: AVA, AVI, IVA, VAA, VAI, VIA, AAV, AIV, and IAV (A stands for animate noun, I stands for inanimate noun, V stands for verb). Examples of the nine types of sentences are listed in Table 14.1.

The subjects were asked to act out the sentence meaning, by using toys or other objects, after listening to each sentence. Table 14.2 and Table 14.3

TABLE 14.1
Nine Types of Sentences With Examples

Type	*Example*		
AVA	骆驼推乌龟	(Luotuo tui wugui)	(Camel push turtle)
AVI	骆驼拍皮球	(Luotuo pai piqiu)	(Camel pat ball)
IVA	盒子推乌龟	(Hezi tui wugui)	(Box push turtle)
VAA	推骆驼乌龟	(tui luotuo wugui)	(Push camel turtle)
VAI	摸骆驼调羹	(Mo luotuo tiaogeng)	(Touch camel spoon)
VIA	拍铅笔乌龟	(pai qianbi wugui)	(Pat pencil turtle)
AAV	骆驼乌龟推	(Luotuo wugui tui)	(Camel turtle push)
AIV	骆驼铅笔拍	(Luotuo qianbi pai)	(Camel pencil pat)
IAV	铅笔乌龟摸	(Qianbi wugui mo)	(Pencil turtle touch)

Note. In order to retain the characteristics of Chinese, the examples of Chinese sentences are literally translated into English in this chapter.

TABLE 14.2
Percentages of First Noun Choice in Miao's (1981) Study

Word Meaning	Word Order			Average
	NVN	VNN	NNV	
AA	77.1	41.7	68.8	62.5
AI	97.9	75	85.4	86.1
IA	29.2	10.4	14.6	18.1
Average	68.1	42.4	56.3	

TABLE 14.3
Percentages of First Noun Choice in Miao et al.'s (1984) Study

Word Meaning	Word Order			Average
	NVN	VNN	NNV	
AA	96.7	37.5	57.5	63.9
AI	100	81.7	88.3	90
IA	35.8	0.8	8.3	15
Average	77.5	40	51.4	

illustrate the percentages of responses that chose the first noun as the subject of the sentences in Miao's (1981) study and Miao et al.'s (1984) study, respectively.

The general trends of subjects' responses in the two studies coincided with each other. The subjects frequently assigned the agent[1] to the animate noun when one noun was animate and the other was inanimate despite their position in the sentence, because only the animate could initiate an action, whereas the inanimate could not. In the first study, the main effect of word order and the two-way factor interaction did not prove to be significant. The main effect of animacy was of significance. Although significant in the second study, the main effect of word order was weaker than that of animacy, and was displayed mainly in the responses to the sentences composed of two animate nouns, especially to AVA and VAA. These results indicated that Chinese speakers relied more heavily on the semantic factor (word meaning) than word order in sentence understanding. They used semantic strategy primarily. More surprisingly, only 29.2% and 35.8% respectively of the first nouns were selected as the agent in IVA, which, as the standard word order in Chinese, corresponded to the agent–action–patient sequence. In

[1]Agent and subject are used in different concepts. However, in the test sentences of the studies introduced here, the agent is usually the subject of a sentence so I use them interchangeably in this chapter.

most cases, the animate noun following the verb was chosen as the agent, rather than the inaminate one preceding the verb. This means that when the information of word order conflicts with the semantic information, word meaning, or lexical constraints in the present studies, usually win in Chinese sentence understanding. Native Chinese rely on both word order and word meaning, but the role of word meaning, namely, semantic plausibility, is predominant.

These two studies were the first attempt to reveal the roles of syntactic and semantic factors in comprehending Chinese sentences. They were rudimentary because the subject sample was small and only the choice behavior in acting out was used as the indicator of understanding. Nevertheless, the results were, on the whole, supported by some subsequent studies.

In a study of sentence comprehension in Chinese aphasia patients (Chen, Tzeng, & Bates, 1990), using the enactment task similar to Miao's, Chen et al. found the patients relied both on word order and animacy as cues in understanding NVN, NNV, and VNN sentences. However, the effect of animacy was much stronger than that of word order.

Later, Li et al. studied the problem more intensively and systematically (Li, Bates, Liu, & MacWhinney, 1992; Li, Bates, & MacWhinney, 1993; Li, 1994). In addition to the choice response, they analyzed the reaction time data and used the gating technique to examine Chinese sentence comprehension, including the time course of the comprehension. The task of their subjects was to identify which of the two objects in the picture presented was the agent of the action. In understanding sentences composed of two names and one verb, similar to those used by Miao, both the choice response and reaction time data showed that the effects of word order, noun animacy, and their interaction were significant. However, in comparison with noun animacy, the effect of word order was relatively weak. It manifested itself mainly in the phenomenon that when there was no animacy contrast, that is, on the AA items, the subjects chose the first noun as the agent in the standard NVN order most of the time and took a second-noun strategy in the nonstandard NNV and VNN orders. In addition, the reaction time data indicated that when word order and noun animacy were consistent, namely both pointing to the same noun as the agent, the response behavior was facilitated. In contrast, if the two cues conflicted and competed with each other, the sentence processing was inhibited and the response speed was slowed down. In brief, their results were, on the whole, compatible with those obtained by Miao (1981) and Miao et al. (1984) though they found that the effect of word order was even more evident than that in Miao et al.'s findings.

In sum, several studies demonstrated unanimously the predominance of word meaning in Chinese sentence comprehension. This finding is contrary to people's expectation because it is generally believed that when understanding a sentence, Chinese speakers have to rely almost exclusively on word order to determine which word is the subject or agent in the sentence and which is the object or patient because there is no inflection in Chinese. Nevertheless, if we investigate the Chinese language carefully, we find that besides the standard word order SVO, other types of word order are available, especially in spoken Chinese. For example, SOV (*ba* construction), OSV (object topicalization), and VOS (subject as an afterthought supplement) are frequently encountered in oral communication. The word order of Chinese sentences is far more variable and much less powerful than was assumed. According to some grammarians, this is also associated with the basic peculiarity that there is no inflection in the Chinese language. On one hand, no inflection leads to a relatively fixed word order; on the other hand, it makes the word order relatively flexible, because Chinese permits the change of word order of a sentence without any marker and the meaning of the sentence remains the same. For example, a sentence in SVO order like Sentence 1 can be changed into three other forms (Sentences 2, 3, and 4) of different orders, which basically express the same meanings.

1. 我写完了两封信 (Wo xie wan le liang feng xin) (I complete two letter SVO)
2. 两封信我写完了 (Liang feng xin wo xie wan le) (two letter I complete OSV)
3. 我把两封信写完了 (Wo *ba* liang feng xin xie wan le) (I *ba* two letter complete SOV)
4. 两封信写完了我 (Liang feng xin xie wan le wo) (Two letter complete I OVS)

Thus, the event probability, instead of word order, must be the essential cue for sentence comprehension. Also, because the topic is a very important notion in Chinese, some grammarians argue that Chinese is a topic-prominent language. The first position in a Chinese sentence may be strongly associated with the topic, which may not necessarily be the subject or the agent. Moreover, even a subject may be absent from a sentence, as the subject is allowed to be omitted. Chang (1992) argued that the fact that Chinese adults do not rely heavily on word order to assign the subject relation has more to do with the topic–comment structure of Chinese sentences in which the subject is determined by word meaning rather than by word order. Consequently, when Chinese speakers are required to identify the subject of a sentence, they do not merely rely on word order, but more frequently on lexical meaning.

In their studies, Li et al. (1992) investigated the roles of the object marker *ba* and passive marker *bei* in sentence comprehension. The sentence with *ba* is a sentence type peculiar to Chinese. *Ba* is an object marker that should precede the object of a sentence and, in turn, the object is followed by a verb. Therefore, the typical word order of the sentence with *ba* is SOV. Li et al. inserted *ba* in front of the second noun of the simple sentence, for example:

5. 小鸭把大树扔掉 (Xiaoya *ba* dashu rengdiao) (Duckling *ba* tree throw away A *ba* IV)
6. 猴子吃掉把香蕉 (Houzi chi diao *ba* xiangjiao) (Monkey eat-up *ba* banana AV *ba* I)
7. 放走风筝把绵羊 (Fang zou fengzheng *ba* mianyang) (Let go kite *ba* sheep VI *ba* A)

The result suggested the contribution of *ba* was mainly shown in the nonstandard orders, especially in AAV. The presence of *ba* in simple sentences raised the rate of the first noun choice from 34% in AAV to 70% in A *ba* AV, but it did not raise the first noun choice to an even higher level. According to traditional grammar, however, the *ba* construction is strongly associated with SOV order. More importantly, in *ba* sentences, the subjects' choice response pattern as seen in the simple sentences did not change fundamentally. Animacy was still a predominant cue in assigning the subject. The reaction time data coincided with the choice response data. Both demonstrated that the contribution of the marker *ba* is limited in Chinese sentence comprehension. Native Chinese do not rely heavily on *ba* in determining the subject of a sentence.

The passive marker *bei* was also inserted before the second noun to construct such sentences as the following:

8. 汽球被女孩扔掉 (Qiqiu *bei* nuhai ren-diao) (balloon *bei* girl throw-away I *bei* AV)
9. 小猪咬烂被大树 (Xiao zhu yao-lan *bei* dashu) (pig bite-mash *bei* tree AV *bei* I)
10. 打破小狗被苹果 (Da-po xiao gou *bei* pingguo) (hit-break dog *bei* apple VA *bei* I)

Bei is a passive marker, and the word order of the sentence with *bei* is S *bei* OV, in which S is the object or patient and O is the agent or doer of an action. The results showed that their subjects chose the second noun following *bei* as the agent most of the time, in spite of the word order and noun animacy. This means that Chinese speakers rely more heavily on the passive marker *bei* than on word order and animacy. *Bei* often wins over other cues when they come under competition. The researchers attributed the different strength of the object marker *ba* and the passive marker *bei* to the difference of their functionality, in that *bei* in Chinese sentences functions exclusively as a passive marker, whereas *ba* can be multifunctional.

From their results, Li et al. (1992) drew the conclusion that in general the hierarchy of cue strength is passive marker > animacy > word order > object marker. They applied their results in support of the competition model of sentence processing proposed by MacWhinney and Bates (Bates & MacWhinney, 1987; MacWhinney, 1987), which emphasizes that cues compete and converge to determine sentence meaning and that the weight of cues varies from language to language.

In brief, Chinese speakers make use of a variety of cues in sentence understanding. Each plays its role to some degree, and complex interactions are observed. The role of a cue is affected and restricted by other cues. Among them, some are generally more important. Though function words such as *bei*, called a semimorphological marker by Li et al. (1992), play a very important role in Chinese sentence comprehension, their presence is infrequent, particularly in informal situations. Therefore, semantic information or lexical meaning is probably the most important cue to Chinese speakers in simple sentence understanding. This contrasts with the finding that English speakers rely more on word order to determine the meaning of a sentence.

UNDERSTANDING SENTENCES WITH PRONOUNS

The pronoun is a frequently used linguistic means to keep the sentences coherent. When coming across a pronoun, one has to find its referent in the context; otherwise the sentence is not understood. Sometimes a pronoun is ambiguous; two or more words in the context may be considered the antecedent of the pronoun. In this case, the listener or reader must make the decision by means of a number of cues—syntactic, semantic, pragmatic, and nonlinguistic Consequently, there are a number of factors affecting pronoun assignment and comprehension of sentences with pronouns.

Miao examined the roles of several factors in pronoun assignment such as sentence semantics, verb meaning, the parallel function, and the distance between a pronoun and its antecedent. In a recent study, Miao and Song (1995) asked the subjects to complete sentence fragments ended with the pronoun *he*, which could refer to either the first or the second noun in the fragment, to observe subjects' pronoun assignment. Examples of the sentence fragments are as follows:

11. 王标批评张新，他 (Wang Biao piping Zhang Xin, *ta*) (Wang Biao criticize Zhang Xin, he)

12. 李明追赶陈龙，因为他 (Li Min zuigan Chen Long, yinwei *ta*) (Lin Min chase Chen Long, because he)
13. 徐良帮助吴奇，但是他 (Xu Liang bangzhu Wu Qi, danshi *ta*) (Xu Liang help Wu Qi, but he)
14. 汤珊被戴宗殴打，他 (Tang Shan *bei* Dai Zhong ouda, *ta*) (Tang Shan *bei* Dai Zhong hit, he)

They found that if the sentence fragment contained the conjunction *because*, the assignment of most pronouns was biased to either the first noun or the second noun, depending on the verb in the sentence. This suggested that verb meaning influenced pronoun assignment and sentence comprehension. For some actions, people are inclined to attribute them to the agents, whereas for other actions, people are inclined to ascribe them to the patients. If the conjunction in the sentence was *but*, the preference of the subjects' response changed in its direction or even disappeared. This result is consistent with the hypothesis proposed by Garvey, Caramazza, and Yates (1976) and Caramazza, Grober, and Garvey (1977) that some verbs possess implicit meaning of causality that determines pronoun processing and consequently sentences with pronouns. Chen and Hung's (1995) study obtained similar results. In completing the sentence fragments containing some verbs, most subjects took the first noun as the antecedent of the pronoun *he*; for other materials, most subjects took the second noun; and for still others, no obvious preference was found. It indicated that the subjects were very sensitive to semantic factors when they processed pronouns and sentences.

Sheldon (1974) proposed that parallel function is a factor affecting pronoun processing. He found the subjects were inclined to think that a pronoun referred to the noun that had the same grammatical function as the pronoun in the sentence. If a pronoun was the subject of a clause, the subject of the preceding clause was more often chosen as the antecedent of the pronoun. This hypothesis was supported by Grober, Beardsley, and Caramazza (1978), though they found some semantic factors influenced the use of the parallel function strategy. In Miao and Song's (1995) study, when the fragment did not contain any conjunction, it seemed that the subjects used the parallel function strategy in completing sentences most of the time. The subject of the preceding clause was frequently taken as the antecedent of the pronoun *he*, whatever verbs were used. This tendency was more obvious when the preceding clause was passive.

In Miao's other two studies (1996a, 1996b), he further examined the role of several factors in pronoun processing and sentence comprehension. The subjects were asked to determine which was the antecedent of the pronoun just after the whole sentence was presented on the computer screen. The test sentences were classified into three types in terms of their meaning, and the classification was made according to college students' rating. The first type was composed of sentences like Sentence 15, in which the antecedent of the pronoun *he* should be the first noun of the first clause, according to semantic probability. In the second type of sentence, like Sentence 16, the antecedent should be the second noun of the clause. Sentence 17 is an example of the third type of sentence, in which there is no semantic bias for pronoun assignment.

15. 小梁终于找到了迷路的小孩，他谢绝接受一笔报酬。 (Xiao Liang zhongyu zhao dao le milu de xiaohai, ta xiejue jieshou yibi baochou.) (Xiao Liang finally found lost child, he refused to accept the reward.)
16. 经理在办公室当众表扬了小钱，他挽回了一大笔损失。 (Jingli zai bangongshi dangzhong biaoyang le Xiao Qian, tawanhuile yi dabi sunshi.) (The manager in office in the presence of all praised Xiao Qian, he retrieved a great loss.)
17. 小梁终于找到了迷路的小孩，他当时累得站立不住。 (Xiao Liang zhongyu zhao dao le milu de xiaohai, ta dangshi lei de zhan li buzhu.) (Xiao Liang finally found lost child, he was tired and could not stand.)

In addition, the sentences varied in the distance between the pronoun and its antecedent, the agreement of grammatical function between them, the number of clauses preceding the pronoun, and the voice of the clause (whether it is active or passive). The results showed that semantic probability was the main factor affecting pronoun processing. Most of the time, the subjects determined the antecedent of a pronoun and comprehended the sentence based on the semantic information provided, regardless of the parallel grammatical function of and the distance between the pronoun and its antecendent (see Table 14.4). Even when there was no obvious semantic bias implied in the sentence, the subjects did not respond on the basis of

TABLE 14.4
Percentages of Choosing the First Noun (Subject) as the Antecedent of the Pronoun *he*

		Active Clause	*Passive Clause*
	subject-subject	91.67	84.17
strong semantic bias	object-object	13.33	29.17
	subject-object	87.5	81.67
	object-subject	19.17	16.67
no obvious semantic bias	pronoun is subject	54.17	75.83
	pronoun is object	43.33	65.87

Note. The first word is the function of the antecedent, the second word is the function of the pronoun *he*.

the parallel function principle. The parallel function found expression only in the following:

1. When there was no obvious semantic bias in the sentence, the first clause was a passive clause and the pronoun *he* was the subject of a clause, the parallel function strategy was used most of the time. 75.83% of the responses took the subject of the first clause as the antecedent of *he*.

2. The parallel function facilitated the pronoun assignment and sentence understanding in certain conditions. When a strong semantic bias existed and both the pronoun and its antecedent were the subjects of the clause, the determination of antecedent was speeded up, especially in sentences in which the first clause with the antecedent was active. Nevertheless, the average reaction time in the conditions of parallel and nonparallel function did not differ significantly (see Table 14.5). Moreover, when both the pronoun *he* and its antecedent were the object in function and the first clause was a passive clause, the subjects took the most reaction time. Thus it can be seen that the effect of parallel function is limited and is only partially supported by these studies.

In comparison with semantic information, the effect of distance between the pronoun and its antecedent is also less important. Both the subjects' choice and reaction time depended more on semantic probability than on distance. The results coincided with Hirst and Brill's (1980), but were different from those of Clark and Sengul (1971) or Ehrlich and Rayner (1983). When the test sentences consisted of two clauses and the antecedent was in the first clause, the position of an antecedent, that is, the distance between the pronoun and its antecedent, did not affect the subjects' choice and reaction time. Only when the sentences consisted of three clauses did the distance affect the processing rate in some degree. If the antecedent was in the clause immediately preceding the pronoun according to semantic

TABLE 14.5
Reaction Time (ms) of Pronoun Assignment

			Active Clause	*Passive Clause*
strong semantic bias	parallel function	subject-subject	2929.6	3280.4
		object-object	3300.3	3902.5
		average	3114.9	3591.4
	nonparallel function	subject-object	3230.9	3501.2
		object-subject	3245.3	3656.5
		average	3238.1	3578.8
no obvious semantic bias	pronoun is subject		3668.7	3616.1
	pronoun is object		3640.7	3896.3
	average		3654.7	3757.2

information, the reaction time was shorter than that of the antecedent in the first clause. When there was no semantic information suggesting which noun was the antecedent of the pronoun, the subjects preferred to choose the noun in the immediately preceding clause as its antecedent. This showed that the distance between the pronoun and its antecedent contributed a bit to pronoun assignment and sentence comprehension, and the distance should be regarded as the number of clauses rather than characters. It suggests that the clause is the unit of storage; two nouns within a clause are the elements of the identical mental representation, so they shared the same difficulty of access. In contrast, two nouns in different clauses belong to two different mental representations. When the whole sentence is presented, the representation of the first clause has become the background, whereas the representation of the second clause is still in the foreground. Consequently, the retrieval of an element from the latter is easier than from the former.

Chen and Hung (1995) investigated how the semantic and structural factors influenced pronoun assignment by means of the moving-window technique, in which each character was sequentially displayed by a mouse-controlled device and the reading time of each character was measured as an indicator of the comprehension process. They presented the subjects with sentences containing verbs that led to first noun choice or second noun choice as the antecedent of a pronoun. Each sentence was composed of four parts. The first part included two possible antecedents. The second part included a pronoun; according to the information provided in this part, the reader could not determine the referent of the pronoun. The third part was the clarifying area that gave the information for determining the antecedent. The information sometimes was consistent and sometimes inconsistent with the preference suggested by the verb used in the first part. The fourth part was irrelevant to pronoun assignment. Examples of the materials are as follows:

18. 甲杀乙的时候(1)/他真的害怕 (2)/
事后会被人发现 (3)/但这忧虑现已消除了 (4)
(Jia sha Yi de shihou (1)/ta zhen de haipa (2)/shihou hui bei ren faxian (3) /dan zhe youlu xian yi xiaochu le (4))
(When A killed B (1)/he was really afraid of (2)/to be exposed afterwards (3) /but the anxiety has been dispelled (4))

19. 甲杀乙的时候(1)/他真的害怕 (2)/
生命会就此完结 (3)/但这忧虑现已消除了 (4)
(Jia sha Yi de shihou (1)/ta zhen de haipa (2)/shengming hui jiuci wanjie (3) /dan zhe youlu xian yi xiaochu le (4))
(When A killed B (1)/he was really afraid of (2)/the end of his life (3) /but the anxiety has been dispelled (4)

20. 王太太邀请了林小姐担任晚会的表演嘉宾（1）/她很兴奋因为（2）/林小姐自认是城里最有名的歌星（3）/ 晚会嘉宾还有影星超人。(4)
(Wang tai yaoqing le lin xiaojie danren wanhui de biaoyan Jiabin(1)/ta hen xingfen yinwei (2)/lin xiaojie zi ren shi chengli zui you ming de gexing (3)/wanhui jiabin haiyou yingxing chaoren (4))
(Mrs. Wang invited Miss Lin to be the honored performing guest in the evening party (1)/she was very excited because (2)/Miss Lin considered herself the most famous singing star in the city (3)/in the party there was another honored guest, movie star Superman (4))
21. 王太太邀请了林小姐担任晚会的表演嘉宾（1）/她很兴奋因为（2）王太太给的报酬很高(3)/ 晚会嘉宾还有影星超人。(4)
(Wang tai yaoqing le Lin xiaojie danren wanhui de biaoyan Jiabin (1)/ta hen xingfen yinwei (2)/Wang tai gei de baochou hen gao (3)/wanhui jiabin haiyou yingxing chaoren (4))
(Mrs. Wang invited Miss Lin to be the honored performing guest in the evening party (1)/she was very excited because(2)/Mrs. Wang paid high remuneration (3)/in the party there was another honored guest, movie star Superman (4))

They found that when the information in the clarifying area indicated that the first noun in the preceding clause was the antecedent of the pronoun, like Sentences 18 and 20, the average reading time for every character in this part was shorter than when the information indicated that the second noun was the antecedent, as in Sentences 19 and 21, regardless of the consistency of the information provided in this part and the semantic preference implied in the verb. This result showed that when comprehending a sentence, the subject was not sensitive to semantic information, but benefited from the structural factor considerably. Chen and Hung maintained that probably this was because the task in this experiment reflected the automatic on-line processing course, which was different from having to rely on semantic information in completing a sentence. In fact, Chen and Hung's finding is compatible with the parallel function hypothesis but different from the results of Miao's (1996a, 1996b) studies. The discrepancy may be due to the different methodologies, namely the on-line approach and the off-line approach.

THE RELATIONSHIP BETWEEN SEMANTIC AND SYNTACTIC ANALYSIS IN SENTENCE COMPREHENSION

It is widely accepted that people carry out semantic as well as syntactic analysis in sentence comprehension, and the result of syntactic analysis is constrained by semantic factors. However, in psycholinguistics, there are two opposing positions on the relationship between semantic and syntactic processing. One of them argues that syntactic analysis follows its own principles independently and autonomously. The role of the semantic factor lies in inspecting the plausibility of the result of syntactic processing, rather than affecting the process of syntactic analysis (Ferreira & Clifton, 1986). This position is called the *modularity model*. The other position, called the *interactional model*, argues that semantic and syntactic analysis interact with each other, and semantic factors guide syntactic processing even directly (Rayner, Carlson, & Frazier, 1983). Both are supported by a number of experiments using materials of other languages. Peng and Liu (1993) investigated the relationship of syntactic and semantic analysis in Chinese sentence comprehension. In their first experiment, the subjects' self-paced reading paradigm was used. All the test sentences included temporary ambiguity that the noun following the first verb in the sentence could be interpreted as the direct object of the verb or the modifier of the object, and it was followed by an ambiguity-removing word. In some sentences, like Sentence 22, the subject of the sentence makes the noun following the verb plausible to be the object of the verb, and in other sentences, like Sentence 23, the subject of the sentence makes it implausible. The subjects were asked to control sentence presentation word by word and to judge the grammaticality of the sentence fragment that had been presented.

22. 董事长解雇了李经理十分信任的一个工人。
    (Dong shi zhang jiegu le Li jingli shifen xinren de yige gongren.) (The chairman of the board of directors fired manager Li very trusted worker. The meaning of the sentence should be, "The chairman of the board of directors fired a worker whom manager Li trusted very much.")
23. 车间主任解雇了李经理十分信任的一个工人。
    (Chejian zhuren jiegu le li jingli shifen xinren de yiye gongren.) (The head of the workshop fired manager Li very trusted worker. Its meaning is "The head of the workshop fired a worker whom manager Li trusted in very much.")

According to the modularity theory, the subjects should take the ambiguous word, manager Li in the examples, as the object of the verb "fired," which is not affected by the semantic information provided by the subject of the sentence (the chairman of the board of directors and the head of the workshop). Only when encountering the ambiguity-removing word *very* afterwards, did they find the original analysis inappropriate and set it right. As a result, the reading time and the percentages of judgment error of the ambiguous and the ambiguity-removing words in both types of sentences should be the same. In contrast, according to the interactive theory, when reading ambiguous words in different sentences, the subjects carried out syntactic analysis differently because of the influence of semantic information. For example, the second noun, manager Li, in the first sentence, was taken as the object of the verb, whereas it was taken as the modifier of the object in the second sentence. Consequently, when the ambiguity-removing word was presented afterwards, there should be different reading time and percentage of judgment error in different sentences. For the first sentence, the subject had to reanalyze the sentence, so the time spent on the ambiguity-removing word should be longer and the error percentage should be higher than those for the second sentence, where the subject did not need to correct the original analysis. As a matter of fact, the data obtained in the experiment contradicted the expectation of the modularity model, but basically coincided with the interactive model. In the second experiment, they asked their subjects to read and understand sentences by using the self-paced method, and then tested the subjects' comprehension. The result confirmed the findings of the first experiment. They argued that their study supported the weak interactive model of syntactic and semantic processing in sentence comprehension. The syntactic processing brought about all the possible structures. When the semantic bias existed, the structure that conformed with the lexical meaning concerned was chosen and the implausible structures refused. Semantic information exercised immediate influence on the process of syntactic analysis, and semantic constraints constantly affected syntactic analysis in sentence comprehension. I consider, however, the result can be also accounted for by a strong version of the interactive model, which maintains that syntactic processing does not give all the possible structures, but only produces one structure in accordance with lexical meaning.

GENERAL REVIEW

As I mentioned in the beginning of the chapter, the studies of Chinese sentence understanding are much fewer than those of character and word recognition, and most studies in this field explore the roles of factors influencing sentence understanding. Judging from the results of the studies, semantics seems to be an essentical factor. Native Chinese are sensitive to semantic information implied by words and sentences, and more dependent on lexical meaning in understanding. This may be related to the properties specific to the Chinese language: The parts of speech in Chinese are vague and unstable; a word can be used as a noun or a verb in different contexts without any morphological change; the boundary between words is indistinct in written Chinese; Chinese grammar is flexible, especially in speech. Consequently, Chinese speakers usually make use of semantic cues to understand sentences.

However, the evidence is not sufficient to make a final conclusion on this issue. Most of the test materials used in some studies are simple sentences; moreover, most of the sentences presented to the subjects are odd in syntax or violate Chinese grammatic rules and have never been encountered in natural conditions. In fact, these materials are word sequences rather than sentences. When the subjects are required to interprete the sequences, they have to resort to lexical meaning and event probability, because there are no syntactic cues available. Therefore, the results of these studies may be partially due to the artificiality of the materials and may not reflect the real picture of sentence comprehension. Second, some studies have shown that the syntactic factors are predominant in certain cases. For example, the role of the passive marker *bei* outweighs other cues including animacy; the pronoun assignment is not affected by semantic information, whereas the structural factor or word order is a determinant. Third, only a few factors have been investigated. Many other factors—pragmatic, phonetic, and suprasegmental cues, as well as those within the areas of syntax and semantics—may contribute to sentence understanding. The roles of a variety of factors and their interaction are complicated. Whether semantic information plays a key role in sentence understanding remains a problem that needs further studying, and its key role, if any, is worth inquiring into.

In addition to investigating the contribution of linguistic and nonlinguistic factors in Chinese sentence comprehension, more attention should be paid to how these factors exert their effects on and interact with each other, to the internal process and mechanism, and to the time course of comprehension. In doing so, the on-line approach is particularly essential. The off-line approach, frequently used in mainland China, is able to find the outcomes of comprehension and infer some peculiarities of comprehension from the outcomes. Nevertheless, it is not able to reveal the process and mechanism of comprehension. Even if the off-line aproach is used in investigating the problem of influential factors, the approach has many limitations. For example, a certain factor may have its influence on the process of comprehension, but the influence may not be displayed in the outcome. Thus it cannot be found by the off-line approach. On-line methodology—such as self-paced reading, the moving-window technique, and the gating technique—has been applied to the study of Chinese sentence comprehension by some researchers and has had a favorable beginning, but more on-line studies are needed.

One of the purposes of studying Chinese sentence comprehension, probably the most significant one, is to explore whether understanding Chinese sentences is different from understanding sentences in other languages. Therefore, the study of Chinese sentence comprehension should emphasize the effect of the peculiarities of the Chinese language on sentence comprehension. Unfortunately, our work in this respect is far from enough and lags behind the study of character and word recognition. Quite a few studies used Chinese sentences as the experiment materials, the purpose of which, however, did not focus on seeking for the common nature of sentence comprehension or the specific nature of Chinese sentence comprehension on the basis of analyzing the characteristics of Chinese sentences. (I did not introduce those studies in this chapter.) Many linguists agree that having no inflection is a fundamental feature of Chinese; Chinese is generally considered a topic-prominent language because the structure of many Chinese sentences is topic–comment, rather than subject–predicate; the modifier is positioned in front of the modified in Chinese, which is different from some other languages. Do these peculiarities affect Chinese sentence comprehension? And how do they affect it? We are unable to answer these questions at present because we have not yet investigated them systematically. I maintain that in order to grasp the nature of Chinese sentence comprehension and establish a perfect language comprehension theory, it is necessary to study Chinese sentence comprehension in accordance with the specific features of the Chinese language and reveal the universalities and particularities of sentence comprehension.

REFERENCES

Bates, E., & MacWhinney, B. (1987). Competition, variation, and language learning. In B. MacWhinney (Ed.), *Mechanisms of language acquisition* (pp. 157–193). Hillsdale, NJ: Lawrence Erlbaum Associates.

Bates, E., MacWhinney, B., Caselli, C., Devescovi, A., Natale, F., & Venza, V. (1984). A cross-linguistic study of development of sentence interpretation strategies. *Child Development, 55*, 341–354.

Bates, E., McNew, S., MacWhinney, B., Devescovi, A., & Smith, S. (1982). Functional constraints on sentence processing: A cross-linguistic study. *Cognition, 11*, 245–299.

Caramazza, A., Grober, E., & Garvey, C. (1977). Comprehension of anaphoric pronouns. *Journal of Verbal Learning and Verbal Behavior, 16*, 601–609.

Chang, H. W. (1992). The acquisition of Chinese syntax. In H.-C. Chen & O. J. L. Tzeng (Eds.), *Language processing in Chinese* (pp. 277–311). Amsterdam: Elsevier Science Publishers.

Chen, H.-C., & Hung, W. W. (1995). Syntactic processes in Chinese reading: Some preliminary studies. *Psychological Science, 6*, 321–325.

Chen, S., Tzeng, O., & Bates, E. (1990). Sentence interpretation in Chinese aphasia. In *Proceedings of the Tenth Meeting of the Second Language Research Forum*. Eugene: University of Oregon, Department of Linguistics,

Clark, H. H., & Sengul, C. J. (1971). In search of referents for nouns and pronouns. *Memory & Cognition, 7*, 35–41.

Ehrlich, K., & Rayner, K. (1983). Pronoun assignment and semantic integration. *Journal of Verbal Learning and Verbal Behavior, 22*, 75–87.

Ferreira, F., & Clifton, C. (1986). The independence of syntactic processing. *Journal of Memory and Language, 25*, 348–368.

Garvey, C., Caramazza, A., & Yates, J. (1976). Factors influencing assignment of pronoun antecedents. *Cognition, 3*, 227–243.

Grober, E., Beardsley, W., & Caramazza, A. (1978). Parallel function strategy in pronoun assignment. *Cognition, 6*, 117–133.

Hirst, W., & Brill, G. A. (1980). Contextual aspects of pronoun assignment. *Journal of Verbal Learning and Verbal Behavior, 19*, 168–175.

Li, P. (1994). Understanding the time course of sentence comprehension: A sentence gating study in Mandarin Chinese. In H.-W. Chang, J.-T. Huang, C.-W. Hue, & O. J. L. Tzeng (Eds.), *Advances in the study of Chinese language processing* (Vol. 1, pp. 285–320). Taipei: National Taiwan University, Department of Psychology.

Li, P., Bates, E., Liu, H., & MacWhinney, B. (1992). Cues as fundamental constraints on sentence processing in Chinese. In H.-C. Chen & O. J. L. Tzeng (Eds.), *Language processing in Chinese* (pp. 207–234). Amsterdam: Elsevier Science Publishers.

Li, P., Bates, E., & MacWhinney, B. (1993). Processing a language without inflections: A reaction time study of sentence interpretation in Chinese. *Journal of Memory and Language, 32*, 169–192.

MacWhinney, B. (1987). The competition model. In B. MacWhinney (Ed.), *Mechanisms of language acquisition* (pp. 249–308). Hillsdale, NJ: Lawrence Erlbaum Associates.

MacWhinney, B., Bates, E., & Kliegl, R. (1984). Cue validity and sentence interpretation in English, German and Italian. *Journal of Verbal Learning and Verbal Behavior, 23*, 127–150.

Miao, X. C. (1981). Word order and semantic strategies in Chinese sentence comprehension. *International Journal of Psycholinguistics, 8*, 109–122.

Miao, X. C. (1996a). The effect of sentence semantics and distance between pronoun and its antecedent on pronoun processing. *Psychological Science, 2*, 71–74.

Miao, X. C. (1996b). A study of semantic and grammatical factors influencing pronoun processing. *Acta Psychologica Sinica, 4*, 352–358.

Miao, X. C., Chen, G. P., & Ying, H. C. (1984). Reexamination of the roles of word order and lexical meaning in Chinese sentence comprehension. *Report of Psychological Sciences, 6*, 1–7.

Miao, X. C., & Song, Z. G. (1995). The influence of verb meaning and sentence grammar on pronoun processing. *Psychological Science, 4*, 197–200.

Peng, D.-L., & Liu, S. L. (1993). Syntactic and semantic analysis in Chinese sentence comprehension. *Acta Psychologica Sinica, 2*, 132–139.

Rayner, K., Carlson, M., & Frazier, L. (1983). The interaction of syntax and semantics during sentence processing. *Journal of Verbal Learning and Verbal Behavior, 22*, 358–374.

Sheldon, A. (1974). The role of parallel function in the acquisition of relative clause in English. *Journal of Verbal Learning and Verbal Behavior, 13*, 272–281.

Author Index

Subject Index

T

U

V

W